MANAGING RISK AND RELIABILITY OF PROCESS PLANTS

MANAGING RISK AND RELIABILITY OF PROCESS PLANTS

Mark Tweeddale

AMSTERDAM BOSTON HEIDELBERG LONDON NEW YORK OXFORD
PARIS SAN DIEGO SAN FRANCISCO SINGAPORE SYDNEY TOKYO

Gulf Professional Publishing is an imprint of Elsevier.

Recognizing the importance of preserving what has been written, Elsevier prints its books on acid-free paper whenever possible.

Library of Congress Cataloging-in-Publication Data
Tweeddale, Mark.
Managing risk and reliability of process plants / by Mark Tweeddale.
p. cm.
Includes bibliographical references.
ISBN-13: 978-0-7506-7734-9 ISBN-10: 0–7506–7734–1 (acid-free paper)
1. Chemical plants—Management. I. Title.

TP155.5.T788 2003
660′.2804—dc21

2003048309

ISBN-13: 978-0-7506-7734-9
ISBN-10: 0-7506-7734-1

British Library Cataloguing-in-Publication Data

A catalogue record for this book is available from the British Library.

The publisher offers special discounts on bulk orders of this book.
For information, please contact:

Manager of Special Sales
Elsevier
200 Wheeler Road
Burlington, MA 01803
Tel: 781-313-4700
Fax: 781-313-4882

For information on all Gulf Professional Publishing available, contact our World Wide Web home page at:
http://www.bgulfpph.com

Typeset by Integra Software Services Pvt. Ltd, Pondicherry, India
www.integra-india.com

Transferred to Digital Printing 2009.

To my wife Helen,
who first suggested I write this book,
and was patient with my preoccupation
while I was doing so.

Contents

Acknowledgments

The material in this text is what I have learned and assembled over the years from working with numerous colleagues and clients in industry, government, and academia. Their assistance is most gratefully acknowledged.

Particularly I wish to acknowledge the help, encouragement, and inspiration I have continually received from Dr. Trevor Kletz since I first became involved with process risk and reliability.

Finally, I am grateful to Professor Beverley Ronalds, of the Centre for Oil and Gas Engineering at the University of Western Australia, who provoked me into getting this task finished.

Mark Tweeddale
December 2002

Foreword

Many of us have a fascination with disasters. The immensity of the consequences stretches and excites the mind. Usually it is possible to say "But if (something) had been different, it would have been much worse!" or "If only (something else) had not been the case, it would not have happened at all!" For some nontechnical people there is the delightful intellectual challenge of using media reports to compose conspiracy theories of how the disaster came about.

But for process plant professionals, what is more intellectually fascinating is investigating and understanding the pattern of the causes of such disasters, and how to avoid them. It is said that "the price of freedom is eternal vigilance." Similarly it may be concluded that "the price of safety and reliability is eternal attention to detail." But it must be *structured* and *managed* attention to *specific* detail.

Choose any process plant where there has been a major incident involving loss of life, or major environmental damage, or major damage to property or loss of production. On the day before the disaster the manager would, almost certainly, have believed that the plant was operating safely, and could have produced evidence that led him or her to have that view. Yet an investigating team would almost certainly find, immediately after the disaster, that there had been many warning signs that the plant was seriously at risk, clearly evident for days or weeks beforehand.

A disaster results in management taking prompt and energetic action to "ensure that it does not happen again." But with time and staff movements, memories fade, and the warning signs reappear and are again ignored.

This book aims to provide a structured approach to recognition of those early warning signs, and for management to plant risks and of reliability, thus shutting the stable door *before* the horse bolts, rather than afterward.

OBJECTIVES OF RISK MANAGEMENT

The objectives of risk management are *to maximize the likelihood of favorable outcomes, and to minimize the likelihood of unfavorable outcomes.*

Although much of this text focuses on identification and minimization of the potential for undesirable outcomes or incidents, the processes used are identical to those used to improve and maximize performance.

The favorable outcomes include safe and reliable operation, high availability of plant, environmentally sound performance, and minimal equipment damage.

OBJECTIVES OF THIS TEXT

This text aims to provide:

- a philosophical foundation from which to approach risk management of process plant such as is used in the oil, gas, chemical, and similar industries;
- an understanding of the managerial and risk principles that determine the form of a soundly structured program of risk management;
- a knowledge of the specific approaches that may be taken to analyze, assess, reduce, and manage loss of safety, reliability, and other risks on a process plant;
- an awareness of the limitations of those approaches and methods; and
- a list of references enabling further study.

Chapter 1

Introduction

When you have studied this chapter, you will understand:

- why process plant risk management is important;
- the difference between classical safety management and process risk management;
- some of the introductory concepts;
- the generic process of management of risks of many types, including process plant risks;
- the relationship between risk and reliability; and
- the different approaches to management of (a) high risks and (b) those risks that are low but that could have serious consequences if realized.

BACKGROUND

The escape of toxic methyl isocyanate vapor from the Union Carbide India Limited plant at Bhopal in India on 3 December 1984 was the most serious chemical plant incident in history, causing thousands of deaths and many tens of thousands of severe injuries, many of them causing permanent incapacity.

This and the explosion at the Phillips Petroleum polyethylene plant at Pasadena on 23 October 1989, which killed 23 people and injured hundreds more, alerted management and governments to the need for much more than traditional occupational safety and health programs to provide safety for those working in, or living around, process plants.

However, this was not new knowledge. Major incidents have been occurring around the world for as long as hazardous materials have been processed. The American Institute of Chemical Engineers had for many years been conducting annual loss prevention symposiums, at which papers were presented by U.S. and international speakers addressing the potential for such incidents and exploring the means of preventing them. As a result of earlier incidents in the United Kingdom and Europe, governments there were drafting and implementing legislation to require organizations operating such facilities to demonstrate that they could, and would, operate them safely.

Loss prevention is not only concerned with incidents that cause injury to people. It covers all forms of loss, including damage to the environment and property, and interruption to production caused by major failures of a plant, even when there is no injury to people or damage to the surroundings. Avoidance or minimization of the risks of all these types of incident is embraced by the field of process-plant risk and reliability management.

1.1 THE SITUATION

1.1.1 Reliability

With ever-increasing competitive pressures worldwide, it is essential that plants operate with high reliability to maximize the return obtained for the capital investment.

Where an operation provides a continuous service, such as piped supply of gas to a consumer, any interruption to supply can have severe effects, in proportion not only to the lost service but to the profitability of downstream industries, community amenities, safety, environmental performance, etc.

1.1.2 Risk

There is a widespread belief in the community that any major breakdown or accident in an industrial or service organization (e.g., transport, power, water) is the result of negligence by someone. The community is very sensitive to perceived "negligence" by management of large organizations. The professional institutions commonly require members to put community responsibilities first, above those of the organization or the individual. Litigation to recover damages is now normal.

There is a trend toward seeking to prosecute the person seen as most responsible. This may not be the person closest to the incident, but is increasingly someone higher in the organization with the perceived responsibility of ensuring that more junior employees are trained and supervised in such a way that breakdowns and accidents will not occur.

No longer can an individual be confident of being protected by his or her organization; an organization may undertake to pay any fine, but where criminal negligence is proved in court, an individual may face a term in jail.

The old legal principle of "no liability without fault" has long been replaced with "the injured party must be compensated by the person or organization most involved in the accident, whether at fault or not."

However, there are approaches that can be taken to reduce to a very low level, the chance of being held liable. Similarly, there is also a growing awareness that to meet an organization's commercial objectives satisfactorily, it is important to identify and to quantify as well as practicable the exposure to commercial risk. It can then be decided whether to accept the financial risk of accidents or incidents, or to transfer the risk by insuring.

The approaches used to identify and assess risks of all kinds, and to reduce or transfer them, form the basis of *risk management.*

1.2 HANDLING THE SITUATION

Unreliability can result in:

- loss of profit for the organization operating the unreliable facility or operation;
- loss of profit for organizations relying on supply of goods or services from the unreliable operation;
- injury to people employed by the organization operating the unreliable facility or operation, or using the goods or services, or in the vicinity of the unreliable facility or operation or the goods or services provided by it.

In the increasingly litigious society of today, such unreliability can result in legal action, either to recover damages caused by the unreliability, or for criminal negligence which resulted in injury or environmental damage.

The best defense against a financial loss or a charge of negligence is to avoid mishaps, that is, to operate reliably.

When a mishap occurs, the best defense against a suggestion of negligence is to have taken all practicable care in the light of information available at the time.

Where there is the potential for a major mishap, it would be necessary to show that there had been proper attention to identifying the hazards, and that proper action had been taken to safeguard people, the environment, and property.

1.3 MANAGEMENT OF THE HAZARDS, OR THE POTENTIAL FOR MISHAP

The existence of hazards does not, in itself, necessitate action. A hazard is the *potential* to do harm. We are surrounded by hazards all our lives. Most of them we accept without concern.

The term "risk" implies probability, not certainty. Risk is defined by IChemE (1985) as "the likelihood of a specified undesirable event occurring within a specified period or in specified circumstances." It may be either a frequency (the number of specified events occurring in unit time, such as the frequency of explosions per year—possibly but not necessarily expressed as a small decimal quantity) or a probability (the probability of a specified event following a prior event, such as the probability of collapse of a nominated structure following an earthquake of nominated strength), depending on the circumstances.

If the probability of harm being done by a hazard is low enough, then the risk is low, and no action is needed. If the damage is potentially very serious financially, and if the probability is sufficient, then it may be decided that it is appropriate to insure. In principle, the assessed risk assists in determining the reasonable level of premium.

Risk management is the name given to a systematic approach to identifying hazards, assessing the risks from each, and deciding what (if anything) needs to be done.

1.4 WHY BOTHER WITH RISK MANAGEMENT, ANYWAY?

There are many reasons why organizations may be concerned with managing their risks. These range from avoidance of injury or the cost of replacing damaged equipment, to such matters as maintaining a good public image or avoiding legal claims or prosecution of senior managers for negligence.

These reasons can all be classified as one of three main types:

- legal
- commercial
- moral or ethical

1.4.1 Legal Requirements

The legal reasons for risk management will depend on the particular legal framework and legislation in the particular community, but are commonly of two types: statutory obligations, and the "duty of care." If the requirement of legislation is ignored (e.g., by not putting a guard on a machine), then prosecution is to be expected. On the other hand, the body of historical case law has established the principle that one must take care of others, even if there is no specific legislation covering the particular matter. An employer who does not take due care of his or her employees is liable. A manager found guilty of negligence may be sent to jail.

1.4.2 Commercial Requirements

A mishap on a process plant may have a variety of commercial implications:

- loss of profit from production lost due to plant downtime;
- the cost of damage to equipment, comprising replacement costs of spares, etc., and labor;
- the costs resulting from injury or loss of life, including damages claims and the effect on insurance premiums;
- the cost of environmental damage, including cleanup costs and the cost of additional equipment and procedures for environmental protection;
- the costs of legal action, including damages awarded for injury, failure to honor supply contracts, damage to property, etc., legal fees, cost of the time of staff defending the cases, and the opportunity cost of staff not being able to progress toward corporate goals;
- the costs of damaged public image, including public opposition to future developments.

1.4.3 Moral or Ethical Requirements

The cost of accidents is broader than simply the financial costs of compensation, etc. There is the human dimension of life: the value to relatives and friends of a fit and healthy person that cannot be replaced by money. No senior manager wants to look back on his career to see a trail of human wreckage, or of destroyed environment. However, sometimes a problem arises in the middle levels of management where it is (usually erroneously) believed that senior management is really only interested in short-term profit.

In addition, most professional engineering institutions have codes of ethics requiring their members to place their responsibility for the welfare, health, and safety of the wider community above sectional interests (such as those of their employer), private interests (such as their own), or the interests of other members. They may also specifically require members to take steps to inform themselves, clients, employers, and the community of the social and environmental consequences of what they do.

1.4.4 Three Variables: Cost, Risk, and Professional Skill

Sometimes a graph is drawn, conceptually displaying how expenditure to improve reliability and risk, and the costs of risked mishaps (or of unreliability) are inversely related. See Figure 1-1A. This is understood to imply that there is a simple relationship between cost and risk: if risk is to be reduced, then more must be spent.

There is, in fact, a third dimension: professional skill. There is, conceptually, not just one hyperbolic curve of risk vs cost, but a family of curves. The greater the professional skill, the lower the curve. After all, it is said that an engineer is someone who can do for $1 what anyone can do for $2. Use of systematic risk management methods facilitates not only finding the optimal region on the curve of risk vs cost, but also moving to a lower curve.

However, the above graphs are idealized concepts. The relationship between risk and cost in any particular situation may not be a hyperbola, and there is not always an identifiable way of moving to an improved (lower) curve.

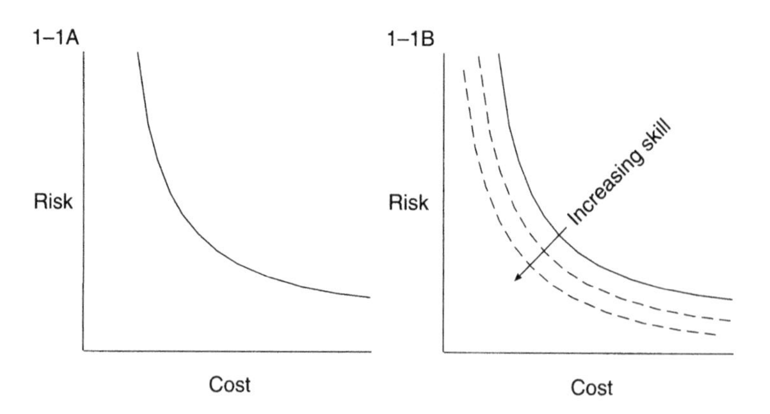

Figure 1-1. (A, B) Risk vs cost.

1.5 THE BENEFITS OF RISK MANAGEMENT

If undertaken systematically, by people from within the organization with a range of relevant experience and expertise and with commitment to act on the findings, the likelihood of serious mishap will be greatly reduced.

In particular:

- hazards to people, the environment, property, and production continuity will be comprehensively identified (although it is never possible to be certain of identifying all hazards);
- the risks resulting from the hazards will be ranked in order of magnitude, facilitating decisions about which to act on, and which to defer or accept;
- decisions are facilitated on the most cost-effective means of reducing the risks, and the nature and extent of the insurance appropriate to the residual risk;
- the procedures facilitate marshaling and organizing the efforts of the available people with relevant expertise (e.g., in operations, engineering, finance, insurance, management); and
- following the procedures, with an appropriate team of people, leads to dissemination of information about how to operate so as to minimize the risks, and facilitates consensus-building.

1.6 FIELDS OF RISK MANAGEMENT

The undesirable consequences of production or engineering activities for a process plant can take many forms. These depend on the nature of the activity. They can range from structural collapse of a building or a bridge (due to such causes as design error, abnormal and unexpected loading, or deterioration of condition with time); fire, explosion, or toxic escape (such as from a process plant or storage); machine failure or disintegration (such as turbine overspeed); failure to provide the required service (such as power blackout of a city, or failure of water supply or sewage treatment); financial collapse of a major contractor in an engineering project; major commercial or financial problems; and so on.

The undesirable consequences may result from:

- normal circumstances (such as continual and even licensed emissions from vents on a process plant having an effect on the surroundings that may be immediately visible, or that may not be recognized for a number of years);

- frequent abnormal circumstances (such as careless operation of plant or equipment resulting in injuries, releases of environmentally damaging material, lost production, or damaged equipment); or
- improbable abnormal circumstances that may never have happened so far, but that are at least theoretically possible.

In the case of adverse effects of normal circumstances, in one sense there is no risk involved, but certainty. In another sense, the exposed people or environment may be exposed to a threat that is probabilistic in nature. For example, a continuous but low-concentration emission of a toxic chemical may be known to injure the health of an exposed population (e.g., by initiating cancer), but with any particular individual having a low probability of being affected.

One can envisage a circumstance where a continuous emission of a chemical is believed at the time to be harmless, but where it is discovered at some later date that the health of the population or the environment has been affected. This is not strictly a matter for risk management, in the sense that the effect is certain, not probabilistic, but the nature of the certainty is not known. Nevertheless, threats of this type are amenable to many of the nonquantitative approaches in the risk management tool kit.

1.7 SCOPE OF PROCESS RISK AND RELIABILITY MANAGEMENT

In this text, the focus is directed toward injury, loss of production, and damage of various kinds resulting from discrete sudden accidental events, such as the following.

- A sudden leak of a toxic gas may expose people downwind to high concentrations for a period of minutes or hours, causing temporary or permanent incapacity or fatal injury.
- A sudden leak of gasoline may ignite and expose people in the vicinity to high levels of heat radiation, causing burns of varying severity, possibly fatal.
- A turbine blade may fail, causing a substantial loss of production while the machine is repaired.
- A pump on an offshore oil platform may cease to operate at a time when the standby pump is undergoing maintenance. Production is thus curtailed until one of the pumps can be restored to service.

- The supply of gas from a distribution network in a city may be discontinued because of a large leak in a gas main. As a result, a glass manufacturing plant loses fuel for its furnaces. The furnaces cool down before gas can be restored, causing major damage to the furnaces. Due care is not taken by the distribution company to ensure that all gas appliances are turned off before supply is restored, so gas escapes into some houses, and a number of explosions occur, resulting in fatalities and property damage. In this case, loss of profit from the failure to supply gas to the customers is the least of the costs to the distribution company resulting from the unreliability of its operations.

The following examples illustrate what is not covered here, being more in the field of "occupational health."

- A plant manufacturing a toxic gas may have continuous emissions from vents and minute leaks around valve stems, etc., which result in a continuous low level of exposure of people working in the area. This exposure may have a long-term effect on their health.
- A process operator may be required to work in a hot environment for long periods, with an adverse effect on his health.
- A maintenance tradesman may be required to work for long periods in a noisy environment, resulting in impairment of hearing through not wearing hearing protection.

1.8 THE RISK SPECTRUM

If one studies the history of hazardous incidents (covering "losses" in the broadest sense) in any class of engineering activity, and classifies the incidents according to the magnitude of the loss or the severity of the effect, one may note that the most serious incidents are relatively rare, and the most common types of incident are relatively minor. (This is, of course, most fortunate!) This can be illustrated in Figure 1-2, which is for a hypothetical organization, but is typical of what is found with numerous organizations.

The lower right-hand end of the line is the field of "occupational safety" or "workplace risk," where the severity of accidents is often limited to the energy available to a person in his or her workplace and may result from personal strength, hand tools, etc. The top left-hand end of the line is the field of major hazards, or "process safety," where the severity of accidents results from the energy or toxicity of the materials in process, or from the sheer scale of the operation. The position of the line at the very top of the graph is unclear, as there have been no accidents of that scale.

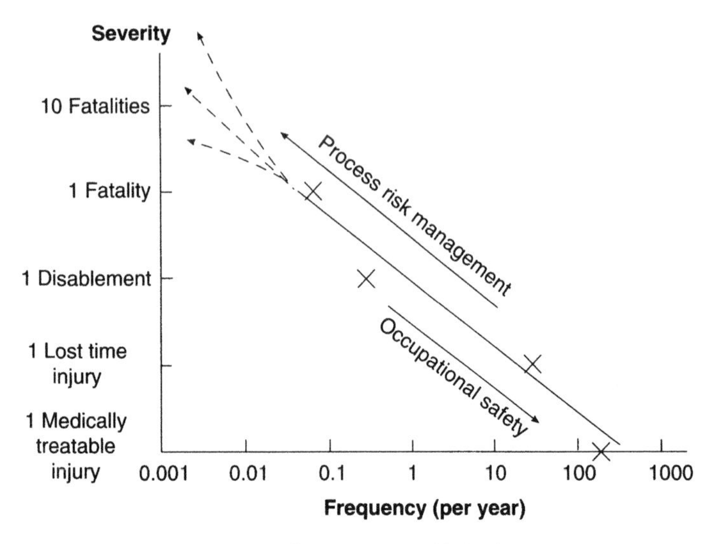

Figure 1-2. Spectrum of injuries.

This can be illustrated another way, as in Figure 1-3, where the risks may be of any type: injury, environmental damage, plant damage, production interruption, etc.

For an illustration of this, consider the field of industrial safety. Minor accidents are more common than accidents with multiple fatalities. In fact, any particular organization may never have had a fatal accident. Paradoxically this leads to a problem for management.

Management of industrial safety has, in the past, mainly focused on the "lost time injury" or the "medically treatable injury," and the numbers of such accidents have been recorded and trends studied. We can get a quick (if rough) feel for the success of our efforts in managing such accidents by

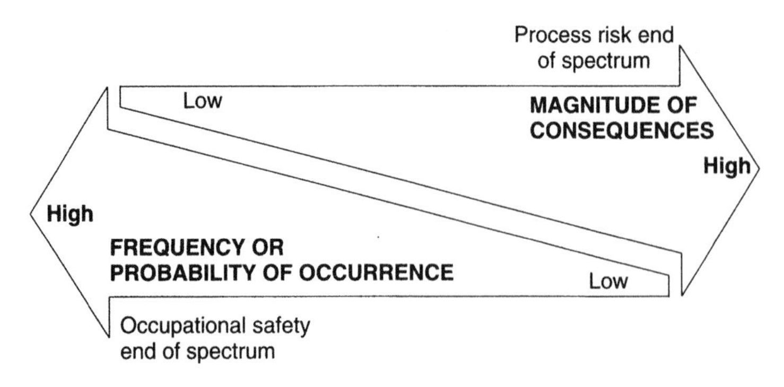

Figure 1-3. Generalized spectrum of risks.

watching whether the frequency rises or falls as time passes. But in the case of very serious accidents (e.g., "disasters"), we cannot do that, as we probably have never had such an accident. We cannot compare the number of explosions we have had this year with the number for the same period last year.

This absence of history is no indication whether we are likely to have such a disaster in the coming year. Further, the rate of lost-time accidents in an oil refinery (for example) is no indication of the risk of a major fire or explosion, as the two classes of incident arise from substantially different causes. The lost-time accidents tend to arise from the activities of people in their workplaces, such as incorrect use of tools. (Managing risks of this sort of accident calls for both managerial skills and detailed knowledge of safe working practice.) The major accidents tend to depend on large inherent hazards, such as the potential energy of large elevated structures, the combustion energy of fuels and chemicals, the toxicity of some materials, or the kinetic energy of large rotating machines. (Managing the risks of this sort of accident calls for both managerial skills and technical understanding of the technology involved.)

In the field of industrial safety, it is valuable to draw the distinction between the subfield of "occupational safety" (i.e., the relatively frequent but relatively minor workplace accidents) and "technical safety" (i.e., the improbable but very serious "disasters").

In the occupational safety field, a safety officer or manager may very well have a background in first-line supervision or in the trades, as that is excellent experience for seeing the nature of occupational hazards, and good and bad ways of working in their presence. In the technical safety field, it is necessary that safety officers or managers have a strong technical background, so as to understand the nature of the process hazards and the technical and managerial standards and controls needed. But, in some cases, organizations have made the mistake of believing that there is no place for first-line supervisors, operators, and tradespeople in maintaining or improving technical safety. This is discussed further in Chapter 11, "Introduction to 'Software' and the 'Human Factor.'"

Because it is not possible to assess the level of process safety by counting the number of major accidents as they happen (because they haven't happened yet!), another approach is used: risk assessment, in which the probability of a major accident is assessed and expressed typically as a probability per year, or a (small decimal) frequency.

It is important to note that an organization's performance in occupational safety (as measured perhaps by the Lost Time Injury Rate or the Medical

Injury Rate) is no indication of the level of process safety. There have been many organizations with an excellent occupational safety record that were found, after having a major incident, to have had high risks of such incidents. This is discussed more fully in Chapter 10.

1.9 STEPS IN RISK MANAGEMENT OF A PROCESS PLANT

The steps in risk management of a process plant are shown in Figure 1-4 and outlined below. The paragraph numbering corresponds with the numbering of the elements in the figure.

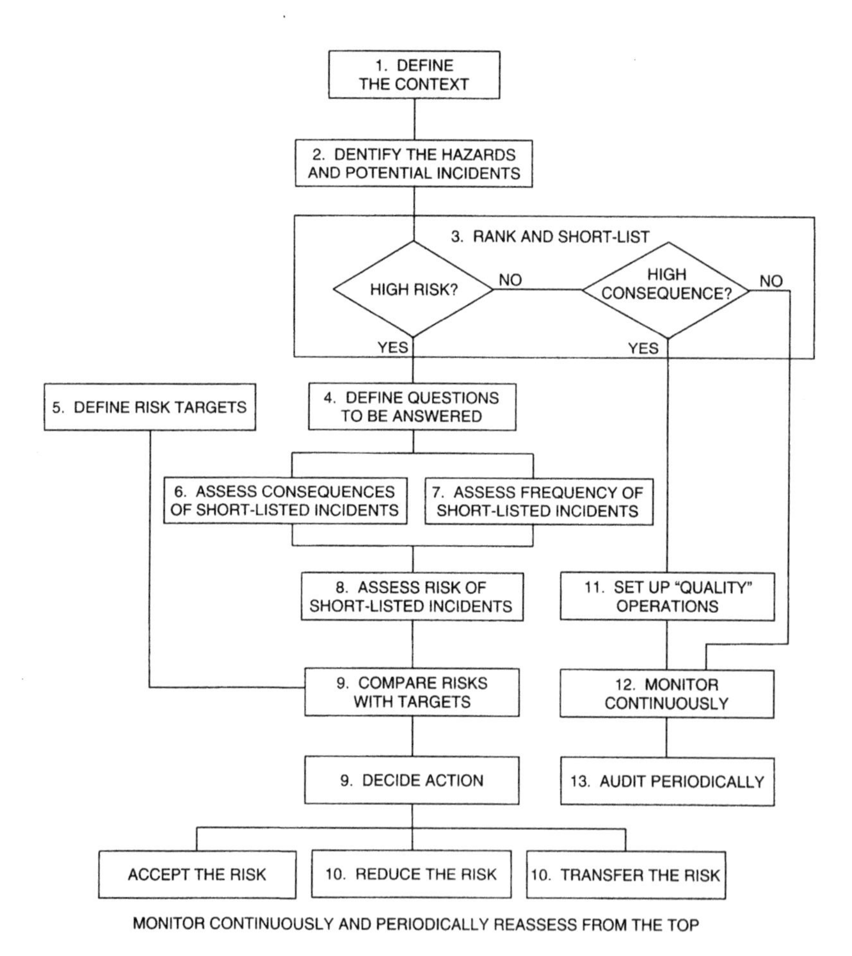

Figure 1-4. Flowchart of process risk management for those scenarios warranting full risk assessment.

1. Review the situation, including the statutory requirements and the expectations of the "stakeholders" (employees, public, customers, suppliers, interest groups, etc.) and the level of performance to date, currently, and probable future performance, as a way of starting to establish the extent of the need for risk management and the nature and scope of the program to be undertaken.

2. Identify and list the hazards of the materials handled, the processes used, and the operations performed. The result of this step is a list of the inherent hazards (e.g., fire, explosion, toxicity) of the materials, processes, and operations, and a list of the hazardous incidents or accidents which could involve those inherent hazards (e.g., collision with a fuel stock tank leading to a fire).

This can be done in a variety of ways, dependent to some extent on the operation being studied. For example, a list of potential hazardous incidents arising from a railway operation will be very different from the list of potential hazardous incidents in an oil refinery.

3. Short-list the identified hazardous incidents or scenarios in a rational and systematic manner so as to determine which:

(a) warrant full risk assessment (i.e., the scenarios with the highest apparent risks, and those with the highest potential for risk reduction per unit expenditure of money or staff resources, etc.); or

(b) warrant careful review of the "fitness for purpose" (i.e., quality) of their design and operation, followed by incorporation into management systems for routine monitoring and auditing (i.e., the scenarios with the highest potential consequences); or

(c) are of lower priority but nevertheless need ongoing monitoring and auditing.

4. Identify the types of questions that the risk assessment is expected to answer, or the decisions for which it is expected to provide information. Then identify the types of persons who will need that information, and hence the form in which that information needs to be presented. Typical people will include managers, other employees, statutory authorities, local residents, interest groups, customers, and suppliers. Note that some of the questions and decisions will need quantified information, some will need descriptive (nonquantified) information, and some will need both. The information probably will include both tangible, objective facts and intangible, subjective opinions.

5. Define quantified risk or safety targets. These are usually expressed as the probability of a defined undesired event in a defined time period, such as the risk of death per year. The nature of the risk targets used will depend on the nature of the short-listed hazardous scenarios. For example, criteria for public safety in the vicinity of flammable or explosive storages could be specified in terms of probability of fatality per year, such as 10^{-6} p.a. or 1 per million per year. A criterion for environmental damage resulting from a particular process material could be in terms of the frequency of discharge to the environment of a defined quantity of the material—for example, 200 liters—but depending on its ecotoxicity. A criterion for risk with contract work could be expressed in terms of the probability of a cost overrun exceeding some nominated level, such as 20%.

6. For each short-listed potential hazardous incident or scenario, assess the likely severity of the consequences, using methods that are appropriate to the type of incident. For example, a fire resulting from an oil leak can be assessed in terms of the heat radiated. From this can be calculated the heat radiation intensity at various distances, and the probability of someone being killed or seriously exposed. In the case of a leak of a toxic gas, the rate of escape can be calculated, and the concentration at various distances downwind determined for various types of weather. From these concentrations can be calculated the probability of someone being killed in the event of exposure for a defined time.

7. For each listed potential hazardous incident, assess the probability or frequency of occurrence. This can be done in a variety of ways, depending on the nature of the incident. For example, the frequency of a fire involving a leaking oil pipe could be estimated from the length, diameter, and duty of the pipe, and the probability of ignition of the spill in that particular location.

8. Assess the risk to whatever is under threat (the exposed person, environment, plant, production continuity, or property) by multiplying the severity and the frequency of each hazardous incident, then determining the total exposure from all the incidents. For example, if a person is potentially exposed to both a fire and a toxic gas escape, the cumulative fatality risk could be assessed broadly as follows: If the probability of fatal injury from heat exposure, in the event of the fire, was assessed to be 10%, and the frequency of such fires was assessed to be 1 in 1000 per year, then the risk from fires would 1 in 10,000 per year, or 0.0001 per year. If the probability of fatal exposure to the toxic gas were assessed as 1% per escape, and the frequency of such escapes were assessed to be 1 in 500 per year, then the risk would be 1 in 50,000 per year, or 0.00002

per year. The cumulative fatality risk to the person would be 0.00012 per year.

9. Compare the assessed cumulative risk against the criteria or targets, and determine the extent of the need for reduction of risk. For example, if the fatality risk criterion for the person had been set at around 1 in 1,000,000 per year, that is, 0.000001 per year, then the assessed risk would clearly be much greater (by a factor of around 120) and would need reduction.

10. By examination of the factors that have made most contribution to the assessed risk, determine the most fruitful areas for risk reduction, and arrange appropriate investigation and action.

For example, in the hypothetical case given above, the largest component of the assessed risk was due to fire. It would be wise to give attention first to means of reducing the severity or frequency (or both) of fire, before considering the component of risk from the potential gas escape.

For Those Scenarios Warranting Monitoring and Auditing

11. Establish methods of operation, maintenance, supervision, and management such that the likelihood of a mishap is essentially low (i.e., establish operations that are "fit for their purpose").

12. Define routine monitoring requirements (i.e., what is to be monitored, how often, by whom, against what standard, etc.).

13. Define periodical auditing arrangements (i.e., what is to be audited, how often, by whom, etc.).

Undertake this entire process in a continuing cycle, monitoring and reviewing the progress being made, and being alert for any signs that important risks may have been overlooked, or that otherwise low risks have been rising.

1.10 RISK MANAGEMENT WITHOUT NUMBERS

The steps of risk management do not need to be undertaken in a quantitative way. In many cases, it is not possible to quantify mathematically the consequences or the frequency of hazardous events. In such cases, judgments can be formed of whether the expected consequences could be "very severe," "moderate," or "minor"; the frequency could be "highly likely," "likely," "possible," "just credible," "not credible," etc. Criteria can be defined in terms of combinations of the above.

(Note, however, that there have been many misleading risk charts prepared where the intervals between the various levels of consequence and of likelihood have not been identical, resulting in a distorted tabulation of risks. This is discussed more fully in Chapter 3.)

Provided that the systematic approach outlined above is followed, and that careful thought is given to the magnitude and likelihood of the possible hazardous incident, the risks can be properly managed. Where more rigorous quantification is possible, it sharpens the focus and makes it easier to be cost-effective and time-effective in risk reduction and general risk management. But quantification is not an essential feature of risk management. This is discussed very fully by Grose (1987).

1.11 SOME ILLUSTRATIONS OF THE APPROACH

The steps outlined above may seem, at first sight, to be complicated and laborious. In fact, they are simply the steps that we take intuitively every day. The following examples illustrate this:

• When driving on a stretch of road known to be hazardous, you have a mental picture of ideal driving conditions. These may include good visibility, dry road, light traffic, and orderly driving. You notice in your rear-view mirror that a car some distance behind you is catching up fast. It is weaving all over the road, and the driver appears to be holding a can (beer?) in his right hand and to be drinking from it. You intuitively assess the possible consequences of this car getting close as being serious to you. You intuitively assess the likelihood of the consequences being realized as substantial. You decide that the risk is high compared with your mental image of safe driving. You decide your course of action: the risk is too high to be accepted; you don't take much comfort in your being insured; so you take evasive action such as turning off onto a side road at once.

• You may be managing an engineering contracting section that is preparing tenders for a variety of projects. You know that there is always a chance of a contract turning out to be more expensive than the estimate used for the quotation. You know that you will not stay in business if a significant proportion of your larger contracts overrun on cost. You are preparing the cost estimate for a very large contract. You judge that a 10% overrun would be unacceptable, and that a 25% overrun would be extremely damaging. You see that some parts of the work will be difficult to estimate with any real confidence. Thus the potential consequences of this contract could be very serious, and the likelihood is significant. Thus you decide that the risk cannot be accepted as it stands, and you consider your options, such as quoting some parts of the work with a specification of what that covers, and specifically providing for variations to cover the contingencies, or quoting high for those sections of the work.

- You may be responsible for designing a warehousing and distribution facility for a company handling an environmentally active chemical, such as a herbicide. The facility is located close to the shores of a lake. You are aware that it is possible that materials handling operations, or a warehouse fire, could result in material being released to the lake, either through the drainage system or in runoff of contaminated firefighting water. The consequences would be most serious, both environmentally and, as a result, commercially. The nature of the operations is such that small spills are very credible, and a fire is possible if not to be expected. You decide that some risk reduction is needed and install environment-protecting features such as curbing and bunding; fire safety and firefighting equipment including sprinkler systems; and a drainage collection pit sufficient to hold a large quantity of spent firefighting water. You also take out environmental impairment insurance.

Risk management, following the steps set out above, simply systematizes our intuitive approach. Some of the steps are undertaken mathematically where practicable, as a means of sharpening our focus, and reducing the area of uncertainty between "low risk" on one hand, and "unacceptably high risk" on the other.

1.12 DEFINE THE CONTEXT

The first step when starting the risk management process, it is necessary to determine the following:

- the extent of the need for risk management; and
- the nature and scope of the work to be undertaken.

To do this, it is necessary to review the situation in which the organization finds itself, including

- the statutory requirements, present and impending;
- the broad nature of risks faced by the organization (e.g., injury, environmental, inability to supply customers, property damage);
- the nature and the expectations of the "stakeholders" (employees, public, customers, suppliers, interest groups, etc.); and
- the level of performance to date and currently, as well as probable future performance.

For example, if there has not previously been a systematic program of risk management, the first approach may be fairly broad, to identify the most critical risks for treatment and the most important areas for monitoring and auditing. A subsequent approach may then explore it more thoroughly. Again, it may be decided that an initial study should focus on risks to people and the environment, with a later study focusing on plant reliability.

It is quite possible that the context, and the nature and scope of the risk management work to be undertaken at a particular time, may need to be revised after undertaking step 2 (Identify the Hazards and Potential Incidents) or step 3 (Rank and Short-List) to ensure that the work is most closely aligned with the insights gained during those steps.

REFERENCES

Grose, V. L., *Managing Risk.* Prentice-Hall, Upper Saddle River, NJ, 1987.

IChemE, "Nomenclature for Hazard and Risk Assessment in the Process Industries." Monograph, Institution of Chemical Engineers, Rugby, UK, 1985.

Chapter 2

Hazard Identification

When you have studied this chapter, you will understand:

- the main types and causes of hazardous incident in process plants; and
- how to identify hazards and risks.

This chapter addresses step 2, "Identify the Hazards and Potential Incidents," of risk management in Figure 1-4.

2.1 INTRODUCTION

2.1.1 Situation

A common question asked by senior managers is: "Have any of our operations any major hazards, that is, the potential to cause a major incident that would be damaging to us or to others?"

If the answer to that question is "Yes!" then the next question is: "How should we get started, and where?"

This section aims to answer those questions.

2.1.2 Sources of Major Hazard

The types of major hazard (to people, the environment, and property) inherent in operations of industrial and other activities are generally derived from the energy of the "systems" in operation. The types of energy include:

- kinetic energy (e.g., the momentum of moving vehicles, such as in a collision resulting in damage to the vehicle and to the object which is struck);

- "potential energy" (e.g., the energy of heavy weights that can fall, such as structures, materials being handled);
- chemical energy (e.g., the heat energy obtained from combustion of materials or goods being transported, which may result in fires, explosions, fireballs, etc.);
- "toxic energy" (this is a rather unfortunate term, but includes the toxic effect of chemicals, asbestos, etc., on people and the environment, such as may arise from toxic gas escapes or spillages of environmentally active materials to drainage systems).

There are other types of loss, such as:

- loss of business (e.g., due to loss of production, loss of a customer, or a customer having to cut back on business operations because of its own commercial problems or decisions, loss due to contractual difficulties);
- financial loss (e.g., due to physical damage to plant necessitating repair or replacement, changes in exchange rates, unwise investments);
- contractual loss (financial) as a supplier of goods or services (e.g., due to underestimation of the cost of supplying the service, or difficulties in supplying it);
- contractual loss (performance) as a supplier of goods or services (e.g., failure to meet the contractual requirements in relation to delivery time, quality, etc., where there is a contractual penalty, or some other form of loss such as due to loss of a customer);
- contractual loss (financial and performance), such as a client suffering loss as a result of failure of a contractor to deliver on time, or of adequate quality, or as a result of cost overrun, where that is to be borne by the client;
- legal liability (e.g., court action because of failure to comply with statutory requirements, or the duty of care, or because of some civil action for damages);
- many others, depending on the branch of engineering or the type of activity undertaken.

The first step in risk management is to identify the inherent hazards of the operation or activity being studied, whether the activity is physical, commercial, contractual, or legal.

It is helpful if the hazards (and other types of potential sources of loss) inherent in any type of operation are classified in terms of their potential types of impact as set out above, both in relation to the nature of the impact (fire, explosion, toxic escape, adverse publicity, etc.) and what it is impacting on (people, property, environment, commercial, etc.).

2.2 TYPES OF IMPACT

Hazards may have the potential for one or more types of impact. These include the following:

- Death or injury to people:

  employees and contractors (generally on site), or

  members of the public (generally off site).
- Damage to property.
- Damage to the environment.
- Loss of production and hence loss of profit from sales.
- Inability to honor contracts to provide service.
- Charges or legal claims arising from any of the above:

  criminal negligence,

  damages, or

  third-party suits. For example, suppose an organization has a contract with customer A. Difficulty in meeting the contractual requirements of the customer prevents customer A from meeting his contractual responsibilities to third party B. Third party B can take action against A and against the first organization for damages.
- Loss of reputation as a reliable supplier of service. For example, performance leading to a perceived inability to provide a consistently high standard of service in terms of quantity, quality, and cost.
- Failure to obtain contracted service from contractors. For example, if a selected contractor or subcontractor is financially unsound, becomes insolvent, and ceases business, an organization may be unable to complete a project on time and may suffer business loss as a result.
- Financial losses related to contractual obligations (exchange rates, negotiation, etc.). For example, if an organization enters into a contract to purchase large capital equipment from overseas, and the local currency depreciates substantially before delivery, the price to be paid by the organization could increase dramatically. Another example would be difficulties encountered with new technology, which may lead to delays in implementation, extended commissioning, and continuing "teething troubles."

- Loss of revenue from customers. For example, if a good and satisfied customer experiences business difficulties, the organization supplying the customer will suffer commercially.

2.3 TYPICAL TYPES OF INCIDENTS LEADING TO THE IMPACT

For any particular field of engineering or management, it is necessary to consider the possible types of incidents specific to that field. However, it is important not to overlook a variety of incident types that can occur on a wide range of types of facility, operation, or activity. These include the following:

- Fire
- Explosion
- Toxic escape (and spillage of hazardous materials generally):

  spill of toxic material into watercourse, groundwater, etc.;

  runoff of contaminated firefighting water from storage of environmentally damaging materials (e.g., oils, chemicals);

  escape of materials to the atmosphere;

  toxic or irritant smoke from fire at operating facilities, stores, workshops, etc.;

  improper disposal of waste, etc.
- Collision and other forms of mechanical impact
- Public misadventure
- Electrical accident:

  employee electrocution, or

  electrocution of a member of the public
- Loss of utilities:

  electric power,

  water,

  drainage and waste water treatment facilities, or

  manufacturing facilities (steam, cooling, computer service, etc.)
- Structural collapse or damage:

  bridges,

  earthworks,

buildings, or

marine structures

- Mechanical breakdown
- Natural disaster:

flood,

wildfire,

earth instability (earth fall, earth slip, "washaway," etc.),

earthquake,

windstorm, cyclone, etc.

- Shrinkage or disappearance of markets
- Loss of market share to competitors (new competitors, new technology, etc.)
- Overcommitment of resources (equipment, people, etc.)
- Industrial action
- Sabotage, vandalism, terrorism, etc.
- Pilfering, theft, etc.
- Numerous other possibilities, some outside and some inside the control of the organization

2.4 TYPES OF PROCESS PLANT INCIDENTS

2.4.1 Introduction

The main types of hazardous incident that may occur in a process plant and associated storage, warehousing and transport operations, are:

- major fire;
- BLEVE or fireball;
- flash fire;
- vapor cloud explosion (confined or unconfined);
- dust explosion;
- other types of explosion;
- toxic gas escape;
- toxic fumes from fires;

- chronic toxic exposure in the workplace (not strictly covered by this course, but mentioned for completeness);
- damage to the environment from abnormal gaseous or liquid escape;
- "domino" incidents; and
- major equipment breakdown.

These are described in turn below.

2.4.2 Major Fires

Fire is the most likely type of incident in most forms of "hazardous" industry, because of the range of flammable gases, liquids, and solids handled, processed, or stored.

The characteristics of a fire depend on both the nature of the plant or facility, and the nature of the material being burned. Several types of fire are discussed below.

2.4.2.1 Flammable Liquid Fires

A leak of flammable liquid may be ignited. In the case of a large leak, the liquid will tend to accumulate on the ground until the area of the burning pool is sufficient for the rate of combustion to balance the rate of escape (less any material flowing to drain).

Ignition depends on the particular design and layout of the plant, and in practice only a small proportion of leaks of flammable liquid ignite, as an ignition source needs to be sufficiently close to the leaked liquid for the vapor around the liquid to be at flammable concentration at the ignition source.

Ignition is aided by a high vapor pressure (volatile liquid, elevated temperature) and by conditions which impede dispersion of the vapor, allowing a flammable concentration to build up.

Pool fires tend to be smoky, with the smokiness increasing with the pool diameter, because of the difficulty of supplying a large fire with sufficient oxygen for full combustion.

As a general rule, the greater the diameter of the pool of burning liquid, the higher the flame, and the greater the consequent reach of the flame, the amount of heat radiated, and the area and extent of damage. Therefore it is desirable, from a fire safety viewpoint, to design the layout and drainage to minimize the area of any pool in the event of a spill.

In the process industry, typical sources of flammable liquid fires include:

- storage tank farms;
- drum stores;
- processing plants;
- road and rail tanker bays;
- road and rail transport;
- shipping and port incidents; and
- cross-country pipelines.

In tank farms, typical locations of fires include:

- at the circumferential seals and cracks in the roofs of floating roof tanks,
- at spillages in bunds, or
- at pumps, for example, due to seal leaks.

Tank farms should be relatively safe, as operating conditions are usually not onerous; they are usually operated without constant human activity such as maintenance or transport; and they are often relatively isolated from the public. (Unfortunately this relative safety sometimes leads to relaxed management, and hence to increased risks.) However, in the event of failure of tank roofs, such as in floating-roof tanks, or in the case of fixed-roof tanks suffering an internal explosion, severe and long-lasting fires have resulted.

Drum stores are commonplace throughout industry, both process and manufacturing, and in warehousing. Occasionally they catch fire. In September 1981, a large fire at a drum store at Stalybridge, UK, demonstrated the dangers of drums exposed to fire. Such drums become heated and pressurized, ultimately bursting and often rocketing, producing fireballs and trailing burning liquid. In the Stalybridge incident, the surrounding area was able to be evacuated before the fire reached its full severity, and no member of the public was reported to be injured. However, a truck driver was killed.

A large leak of flammable liquid at an oil refinery or other processing plant can result in a very serious fire, causing damage to the plant and overheating other pipes and vessels, resulting in further leaks and escalation of the fire. However, unless the public is unusually close to the plant, they would not normally be at significant risk from heat radiation or flame impingement from the fire. (Note, however, that secondary effects, such as toxic combustion products or BLEVEs caused by the fire impinging on other vessels, etc., could cause risk to the public.)

Although bays for loading or unloading road or rail tankers handling flammable liquids are of apparently simple design, because of the constant coupling and uncoupling, the batch nature of the operations, the movement of large vehicles, and the extensive human involvement, there are numerous potential causes of large leaks and fires. For this reason, such bays are usually equipped with a range of control, protective, and firefighting equipment and resources.

Road and rail transport of flammable liquids, because of the chance of collision and loss of containment, have the potential for large fires. In practice, the frequency of such fires has been relatively low in relation to the tonnage of materials handled and the number of vehicles involved. Although there is often public concern about the risks, the actual risks appear to have been very low, as evidenced by the very low number of fatalities to members of the public caused by transport accidents in which the hazardous nature of the cargo was a factor. These historically low risks have resulted from the care that is taken with steps 11, 12, and 13 in Figure 1-4.

The degree of risk from a leak of flammable liquid from a cross-country pipeline depends on a variety of engineering, process, and management factors (such as flow rate, pressure, material properties, and quality of maintenance) and on other factors that are very difficult to estimate (such as the probability of ignition of the pool formed by the leaked liquid).

2.4.2.2 Liquefied Flammable Gas Fires

When a pressurized liquefied flammable gas escapes, such as LPG from a camping cylinder, a proportion of it "flashes" into vapor instantly, using the sensible heat (due to its temperature being above its atmospheric-pressure boiling point) to supply the latent heat of vaporization. Thus the leak takes the form of a spray of droplets chilled to the atmospheric-pressure boiling temperature and chilled vapor. If this ignites, it forms a turbulent, luminous, and very intense flame.

This is the type of fire that can occur at liquefied flammable gas facilities such as the following:

- pressurized storage such as LPG stock tanks at bulk depots;
- processing plants, such as refineries and liquid separation units;
- tanker loading or unloading facilities;

- transport; and
- cross-country pipelines.

Note that liquefied flammable gases are often stored and handled under refrigeration. If the temperature is less than the atmospheric-pressure boiling point, a leak will not flash into vapor and spray immediately on emergence. However, on contact with the ground or surrounding structures, such a leak will be rapidly evaporated by absorption of sensible heat. If the leak is ignited, an intense fire will result, though not as turbulent and luminous as in the case of a flashing liquid leak.

2.4.2.3 Flammable Gas Fires

A leak of flammable gas, if ignited, will burn with an intensity depending on the nature of the gas and the nature of the leak (e.g., a turbulent leak will mix better with the air and produce a more intense and luminous flame than a nonturbulent leak).

It has been observed that a leak of flammable gas into the open air is highly unlikely to accumulate sufficiently to explode if ignited. A very high escape rate would be needed. But if the escape is into a confined space, even a small mass can result in a very damaging explosion. A leak of around 350 kg of ethylene from a pipeline of 3 mm dia into a compressor house resulted in an explosion that destroyed the building and the structure supporting the compressors, costing around $20M in material damage and the same amount in lost business.

2.4.2.4 Flammable Powder Fires

Fires involving flammable or combustible powder are not believed to place the public at risk, but could be a threat to employees. However, if the fire is tackled with a jet of water, the powder can be stirred up and lead to a dust explosion. Such fires must be fought with a fine spray, so as not to stir up the powder.

2.4.2.5 Highly Reactive Material Fires

A large range of highly reactive process materials are used in processes, stored, and transported by the process industry.

Organic peroxides are commonly used as catalysts. The amount in the process at any time is often very small, but larger amounts may be stored. It is important that the correct storage conditions be maintained, or the peroxides can start to decompose and catch fire. Fighting such a fire calls for special approaches.

Some catalysts are "pyrophoric," that is, they ignite spontaneously if exposed to air. They are stored in special conditions, such as in a nitrogen atmosphere.

Ethylene oxide, which is a low-boiling-point liquid, needs special care. It is usually stored under refrigeration with close control and protective systems. If it is heated, or in the presence of some impurities, an accelerating decomposition can start, resulting in a serious fire or explosion.

In considering the nature of hazards due to a specific highly reactive material, it is necessary first to study the properties and the nature of the reactivity of that material, and the manner in which it is stored and handled.

2.4.3 BLEVEs or Fireballs

BLEVE stands for boiling liquid expanding vapor explosion. In some ways this is a misnomer.

Strictly, the definition of a BLEVE, as suggested by the IChemE (1985), is:

> ...the sudden rupture of a vessel/system containing liquefied flammable gas under pressure due to flame impingement. The pressure burst and the flashing of the liquid to vapor creates a blast wave and potential missile damage, and immediate ignition of the expanding fuel–air mixture leads to intense combustion creating a fireball.

The blast wave has generally been much less damaging than the fireball. Large fragments of the ruptured vessel have, however, been projected for hundreds of meters, trailing burning fuel, and have caused severe localized damage.

However, because the main damage is due to the fireball, the practical definition of a BLEVE could be: "the fireball that results when a pressurized container of a liquefied flammable gas, or of a flammable liquid held at a temperature above its normal atmospheric-pressure boiling point, is ruptured due to weakening of the container by exposure to fire."

If an LPG camping cylinder ruptures due to fire exposure, a very damaging BLEVE or fireball is the result.

Pressure storages for LPG and similar material need to incorporate a number of specific design measures to minimize the risk of BLEVE.

The effect of a BLEVE is very high heat radiation of a short (e.g., 5–20 seconds) duration. The intensity of the heat radiation can burn people and start property fires within a substantial distance of the vessel that caused the BLEVE.

2.4.4 Flash Fires

A cloud of flammable vapor can result in many ways, such as:

- evaporation of a leak of liquefied flammable gas;
- evaporation of a leak of flammable liquid heated to a temperature close to or above its atmospheric-pressure boiling point;
- a leak of flammable gas; or
- a large spill of volatile flammable liquid.

If such a cloud is ignited in the open air, the most likely result is a "flash fire" as the gas or vapor burns quickly with a soft *whoomph*. It is unlikely to burn fast enough to produce a damaging pressure wave or shock wave—a *bang*.

Such a fire, though brief, can kill or seriously injure anyone within the burning cloud; note that the cloud expands while burning. The effects are confined almost entirely to the area covered by the burning cloud.

2.4.5 Vapor Cloud Explosion

If a cloud of flammable vapor is ignited, it may burn as a flash fire without a blast wave, or it may explode with a damaging blast wave.

The likelihood of explosion is increased if one or more of the following conditions apply:

- confinement, partial confinement, or obstructions in the vicinity;
- large amount of vapor;
- mixture close to the *stoichiometric* composition, where the amount of oxygen exactly matches the amount of fuel to be burned;
- turbulence in the burning cloud; or
- the vapor being of a material that normally burns with a high flame speed.

In recent years, laboratory-scale and large-scale tests have shown that a cloud of hydrocarbon vapor in the open air, and unobstructed by buildings, pipework, vessels, etc., will not explode. However, if a vapor cloud is obstructed, turbulence is created by the expansion of the burning cloud. This causes a highly irregular flame front, with a larger surface area, leading to a greatly increased mass rate of combustion and then to generation of a pressure wave. Once the flame front moves beyond the obstructed region, the flame front becomes more regular, and the mass rate of combustion decays. It is now evident that, except in very unusual circumstances, it is only that part of the cloud that is within the obstructed region that contributes to the pressure or blast wave.

There have been very few incidents in which an unconfined cloud of less than 5 tonnes of vapor has caused significant blast damage, though, in a confined or obstructed space, a cloud of a few kilograms is likely to explode violently.

The effect of a vapor cloud explosion is physical damage due primarily to the blast wave. In a chemical plant, this damage usually starts leaks that result in very damaging fires. In adjacent residential areas, the damage may range from broken windows to more serious structural damage.

A vapor cloud explosion may fatally injure people by:

1. projecting missiles at them;
2. throwing people violently against solid objects;
3. collapsing buildings around them; or
4. enveloping them in the burning or exploding cloud.

Unlike conventional explosives, an unconfined vapor cloud explosion does not generate sufficient pressure to be of itself seriously injurious to the human body.

In the open air, it is very unlikely that a spill of flammable liquid could generate vapor fast enough for a flammable cloud to become large enough to explode if ignited. However, such a spill indoors can be very hazardous, as the vapor can accumulate.

To reduce the risks from explosion, there are two main approaches:

(a) reduce the amount of material that can leak; or

(b) reduce the confinement or the obstructions in the vicinity; or both.

The larger the mass of fuel in the cloud:

- the more likely the cloud is to find an ignition source;
- the more likely the ignited cloud is to explode; and
- the more damage the exploding cloud will do.

The greater the confinement:

- the more likely is ignition, as the cloud cannot disperse;
- the more likely is the ignited cloud to explode; and
- the greater the explosive pressures, as the pressure is contained.

2.4.6 Dust Explosions

A mixture of flammable or combustible dust in air can explode if ignited. Many historical dust explosions have been in the form of repeated explosions following an initial small explosion. The cause of this is that the first explosion shakes loose dust that has accumulated on beams, etc., allowing a second or third explosion to occur in quick succession. These later explosions can be more damaging than the first one.

More than 30 people were killed in an explosion of wheat dust in a silo building.

The damage radius of a dust explosion is usually limited to the building in which it occurs, and to a very short range outside.

2.4.7 Other Explosions

Some highly reactive materials, such as ethylene oxide, organic peroxides, or calcium hypochlorite can decompose explosively if heated, for example, in poorly controlled storage conditions, or if they become contaminated.

2.4.8 Toxic Gas Escapes

In processing oil and gas, and in the manufacture of products widely used in the community, the process industry commonly handles toxic gases, including:

- hydrogen sulfide;
- hydrogen fluoride;
- chlorine;
- hydrogen chloride;
- vinyl chloride;
- ammonia; and
- carbon monoxide.

Hydrogen sulfide, hydrogen fluoride, chlorine, hydrogen chloride, and ammonia have an acute effect, causing coughing and difficulty with breathing at fairly low concentration. Although distressing, this has the effect of giving a warning. Carbon monoxide, such as is produced in the exhausts of cars, also has an acute effect, but is not detected by such symptoms. Hydrogen sulfide has a particularly dangerous property: it may desensitize one's sense of smell, so that one is unaware of being exposed to it.

Vinyl chloride only causes acute effects in relatively high concentrations. (Long-term exposure to low levels is potentially very serious. This is a hazard for employees, but is not usually faced by the public.)

Conditions favoring rapid dispersion to a harmless concentration are:

- strong hot sunshine, which assists vertical dispersion; and
- strong wind, which assists horizontal dispersion;

The worst conditions for rapid dispersion are:

- smoggy or foggy weather and
- crisp clear nights or early mornings.

2.4.9 Toxic Fumes from Fires

Some commonplace and normally safe materials generate toxic fumes when burned. In the event of a fire at a warehouse that contains such materials (e.g., PVC, polyurethane) the smoke from the fire could perhaps cause public nuisance and concern for health. This problem is not limited to materials used in the process industry; the smoke from house fires also usually contains carbon monoxide, hydrogen chloride from burning plastics, etc.

Smoke from a large warehouse fire may also contain some unburned toxic material (e.g., pesticides, herbicides) vaporized by the heat.

The smoke normally initially rises due to its temperature, taking the highest concentration of toxic or irritant materials away from people closest to the burning installation. When the smoke has dispersed to a lower concentration, and the temperature is thereby lowered, the smoke plume may spread downward, too, resulting in exposure of people on the ground at a distance. At such a distance the concentration is normally greatly reduced.

2.4.10 Chronic Toxic Exposure

This type of problem is characterized by exposure over a long period to toxic materials at concentrations that do not produce any apparent health effect or inconvenience in the short term. Examples are asbestos, vinyl chloride, and benzene.

This type of problem is typical of the occupational health field and commonly results from the usual or "normal" conditions on the plant presenting excessive exposure. There is not a "risk" of excessive exposure, but a certainty, if the person stays in the area sufficiently long.

It is important to be aware of the need to consult occupational health specialists when designing plants. This text does not aim to teach occupational health practice.

2.4.11 Damage to the Environment due to Toxic Liquid or Gas Release

There is rarely a significant hazard to people from a sudden escape of a toxic liquid, unless it has the potential to find its way into the drinking water or the food chain. An example, however, could be a large escape of a corrosive material, such as a strong acid or a strongly alkaline material, where it could flow onto a public roadway or other area where people could walk into it without recognizing its hazard, and then be unable to get away.

However, a leak of biologically active material into the environment can have very serious effects—for example, a pesticide escaping from a processing plant into the factory's effluent system or storm water drains and thus into the external environment. Serious escapes have occurred as a result of warehouse fires, when the surface runoff of firefighting water has been contaminated by biologically active materials involved in the fires. In both such types of incident, forethought in the design stage and good operating practice and management can minimize both the potential (i.e., the hazard) and the risks.

When considering whether this type of hazard exists in a proposed or existing facility, it is necessary to consider the biological activity of each material, paying special attention to pharmaceutical, veterinary, pesticide, and herbicide materials.

2.4.12 "Domino" Incidents

There has been much public concern on occasions about the potential for "domino" incidents, in which a fire or explosion at one plant initiates another hazardous incident at another plant, and so on, leading to a "leapfrog" effect resulting in damage far more widespread than was caused by the initial incident.

There does not appear to be a generally accepted definition of a domino incident. In one sense, a BLEVE is a domino incident, in that a small fire leads to a major fireball. In another sense, the fires that result from a vapor cloud explosion could be considered domino incidents.

However, there is a good case for distinguishing between incidents which escalate at effectively the same location, or which occur in the damage zone of the first incident without extending the area of damage, and the type of succeeding incident that occurs at a markedly different location from the initiating incident and results in substantial damage beyond the initial damage, or a marked increase in the initial damage.

A domino incident is regarded here as one that results from an initial damaging incident; that occurs at a markedly different location from that initial incident; and that results in a marked increase in impact or damage over that of the initial incident.

Examples of a domino incident would thus be:

- an explosion at one plant, projecting a missile that damages a plant beyond the initial severe blast zone, causing a leak and a fire that greatly extends the damage zone of the initial explosion; or
- a leak of toxic gas, which injures people in a zone, and that incapacitates the operators of a plant that then catches fire. Although the fire is within the zone affected by the gas, the damage is markedly increased and is at a location markedly different from the source of the escape of gas.

The following are not regarded here as domino incidents, but as expected consequences of the first incident, or as escalation of the initial incident and part of it.

- An explosion that damages a plant in the blast zone, which then catches fire and burns, causing further damage in the heavily damaged zone. (This is the usual outcome of a vapor cloud explosion.)
- A BLEVE, in which an initial fire leads to a fireball, causing extensive damage centered on the same point.
- A small fire on a plant that damages pipework, leading to leaks and a larger and more damaging fire.

Domino incidents, as defined above, are much less likely than is commonly believed by the public. (In preparation of one database of around 1300 incidents related to the process industry, only about 5 incidents appeared to be domino incidents. An example was a marine tanker carrying crude oil that exploded while unloading, and from which a projectile (steel sheet) ruptured a storage tank on shore at a distance, resulting in a large fire remote from the zone of damage of the ship explosion.)

2.4.13 Major Equipment Breakdown

Breakdowns of equipment form a spectrum, just like accidents to people. There are relatively frequent minor breakdowns at one end of the spectrum, with rare or unlikely major failures at the other. The combined effect of numerous minor breakdowns on production and on maintenance cost can be very serious, just like the rare major breakdown.

The relatively frequent, minor breakdowns are managed using a structured approach very similar to that used for occupational safety; this approach is beyond the scope of this text.

The risks of major breakdown can be examined and managed using the same structured approach as the risks to the safety of people and of major damage to the environment, plant, and property.

One approach, used in Reliability Centered Maintenance (RCM), calls for a systematic listing, for each item of equipment, of its required capabilities or functions. These required capabilities or functions are described in "verb–noun" pairs. Some examples are listed below as illustrations.

Pipeline: *conveys* liquid (or small solids, or gas, etc.); *contains* liquid (or small solids, or gas, etc.)

Pump: *increases* pressure; *contains* material

Then, for each function, the required standard of performance is defined, and the potential causes of loss of that function identified. These may be of a variety of types, for example, technical, operational, or managerial.

One particular value of defining the required capabilities or functions is that this aids communication with nontechnical people, such as some senior managers or accountants. Engineers are sometimes believed by nonengineers to be people who spend money keeping their plant and equipment in good condition for its own sake, rather than people who aim to keep the equipment capable of performing whatever functions are essential for the continuance of the business.

2.4.14 General Comment

In order to keep these hazards in proportion, there are several important points to note.

- These hazards are inherent in many of the materials or equipment without which our community could not operate effectively: for example, domestic gas appliances, gasoline for cars, LPG for camping and cigarette lighters, plastics, chlorine for bacteria-free drinking water, pharmaceutical supplies, and pesticides and herbicides used in growing crops.

- Recognition of the hazards at the outset of design enables incorporation of engineering features, and adoption of operational, maintenance, and management procedures and standards, which can reduce the risk of the hazard being realized in practice to a level where it is less than numerous risks which we all accept every day. For example, a typical worker in a chemical factory is broadly as safe (or safer) at work, per hour exposed, as he or she is at home.
- There are numerous other risks in the community that we all accept each day without concern. Aircraft fly over cities. Towns are built downstream from dams. There is a history of aircraft crashing and of dams failing, so we insist on high safety standards. We do not prohibit use of aircraft, or construction of dams, because we recognize the value we get from them. In the same way, we need to insist on high safety standards for our process industries, but recognize the value we all get from them, and not place irrational obstacles in the way of their further development.

Figure 2-1 demonstrates the types of potential hazardous incidents.

2.5 APPROACHES TO SYSTEMATIC IDENTIFICATION OF HAZARDS AND RISKS

2.5.1 Introduction

It is not possible to be certain of identifying all hazards and risks. However, it is possible to be reasonably confident that no major hazards or risks are being overlooked if a systematic approach is taken.

There are many possible approaches to identification of hazards and risks. Some are described by Wells (1996). In general, the more detail that is available about the process and the plant, the more detailed the identification process can be, and the greater the confidence that significant hazards or risks have not been overlooked.

All these possible approaches, if they are to be successful, must employ a number of fundamental principles.

1. The approach must make use of a team of people with a wide variety of relevant background, so as to minimize the problem of the limited view

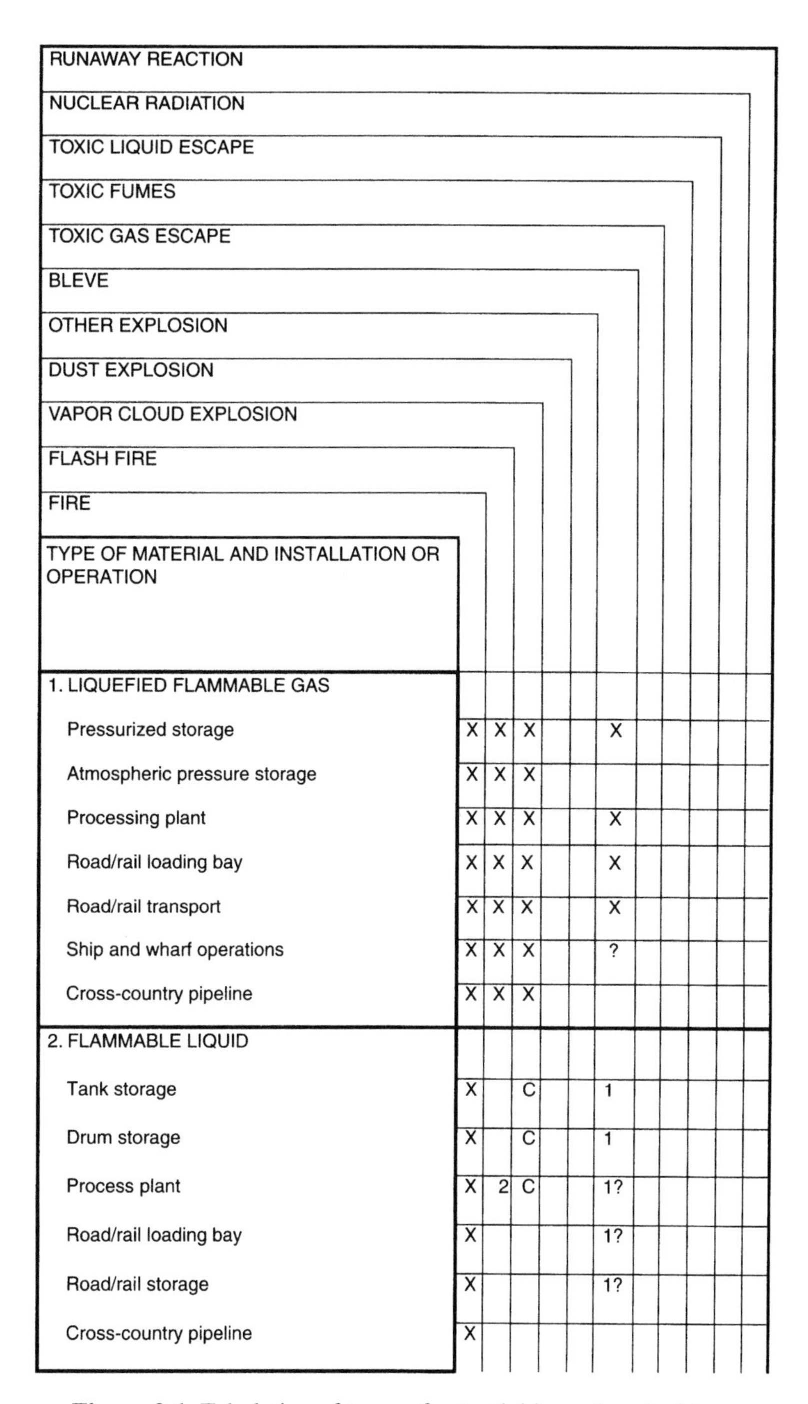

TYPE OF MATERIAL AND INSTALLATION OR OPERATION	FIRE	FLASH FIRE	VAPOR CLOUD EXPLOSION	DUST EXPLOSION	OTHER EXPLOSION	BLEVE	TOXIC GAS ESCAPE	TOXIC FUMES	TOXIC LIQUID ESCAPE	NUCLEAR RADIATION	RUNAWAY REACTION
1. LIQUEFIED FLAMMABLE GAS											
Pressurized storage	X	X	X			X					
Atmospheric pressure storage	X	X	X								
Processing plant	X	X	X			X					
Road/rail loading bay	X	X	X			X					
Road/rail transport	X	X	X			X					
Ship and wharf operations	X	X	X			?					
Cross-country pipeline	X	X	X								
2. FLAMMABLE LIQUID											
Tank storage	X		C			1					
Drum storage	X		C			1					
Process plant	X	2	C			1?					
Road/rail loading bay	X					1?					
Road/rail storage	X					1?					
Cross-country pipeline	X										

Figure 2-1. Tabulation of types of potential hazardous incident.

(figure continued on next page)

3. FLAMMABLE GAS: STORAGE AND PROCESSING		X	X	X							
4. COMBUSTIBLE POWDER (or dust-producing solids): STORAGE AND PROCESSING		X	?		X						
5. HIGHLY REACTIVE MATERIALS: STORAGE AND PROCESSING		X				X					
7. EXOTHERMIC REACTION		X				X					X
6. TOXIC GAS: STORAGE AND PROCESSING								X			
7. MATERIALS WITH TOXIC COMBUSTION PRODUCTS: STORAGE AND PROCESSING								X			
8. TOXIC OR BIOLOGICALLY ACTIVE MATERIALS: STORAGE AND PROCESSING									X		
9. RADIOACTIVE MATERIALS (e.g., in instrumentation or used in testing)										X	

KEY: X Hazard exists ? Possible C In confined space
1 If not vented 2 If heated and volatile

Figure 2-1. Continued.

each person has because of limited experience. (We all have "blind spots" in our knowledge and are unable to recognize that we are overlooking things outside our field of view, as determined by our background experience, which in turn determines our direction of view. But by having several people with different viewpoints, we aim to fill in the blind spots.)

2. Start by looking for the potential for mishap, without regard for likelihood. For example, if we are looking for hazards to people, look for sources of energy, whether kinetic, potential, chemical, or toxic. If we are looking for financial hazards, look for the activities with most financial importance, or with the greatest cash flows.

3. Having listed the inherent hazards, use a systematic approach that is as detailed as is practicable with the information available at the time about the process and plant.

4. Consciously look for deviations from normal, preferably aided by a checklist of important properties which could deviate, or of typical problems that have occurred in some related area at some time. For example, properties relating to transport include speed, direction, and load, and properties related to contracts include time to completion, sequence of completion of different sections of the work, cost, and quality of design and execution.

Some examples of suggested checklists or worksheets for process plants are given in this chapter. By understanding the principles adopted in the preparation of checklists or worksheets, it is possible to prepare these for any type of activity, whether in engineering, management, or some other field.

In some cases, it is very helpful to prepare these checklists as part of a spreadsheet used for ranking and short-listing of risks (see Chapter 3).

2.5.2 Identification of Major Hazard Inventories and Activities

A major incident cannot occur without a hazardous inventory of material, or a hazardous energy source of significant size. An incident can occur if such an inventory or energy source suffers:

- a loss of containment, or
- a loss of control

The question is: "How big must an inventory be to present a major hazard?"

This has been addressed in much the same way in a number of countries. An early example was in the United Kingdom, where the Control of Industrial Major Accident Hazards (CIMAH) Regulations (see HMSO, 1985) listed many materials and types of material, and quantities were specified at which inventories of those materials would qualify as "major hazards" for the purpose of those regulations. This same type of approach has been used in the United States by both OSHA and EPA (see Appendix E).

2.5.3 Block Diagram Stage

A very good method for initial identification of hazards in a process (e.g., at the block diagram stage) is for a group of people with relevant

Table 2-1
Technological Risk Identification Worksheet: For Process Materials

Type of Technological Failure \ **Process Material**										
Operational incident • Fire										
• Vapor "flash" fire										
• Explosion: vapor cloud										
• Explosion: combustible dust										
• Explosion: unstable material										
• Explosion: other										
• Fireball (e.g., from LPG)										
• Self-ignition										
• Toxic liquid escape										
• Toxic gas escape										
• Toxic smoke										
• Chronic toxic exposure of people										
• Major corrosion										
• Nuclear radiation										
• Public nuisance: dust, smell, noise, etc.										
• Other										

experience to complete the Technological Risk Identification worksheets (Tables 2-1, 2-2, and 2-3) given in this chapter, putting a mark in each box to indicate the possible existence of a hazard for further consideration, short-listing, and possible assessment.

2.5.4 Identification of Hazards from a Process Flowsheet

The method of hazard identification described above has the advantages of simplicity and speed, and is a good initial and "broad-brush" approach. However, for a more detailed examination of a particular process, selected because of its identified potential for large incidents (e.g., because of its large inventories of hazardous material, or large rotating machinery with high energy), it is prudent to undertake a "Hazard and Operability Study" (Hazop), or to use some of the principles of such a study. See Appendix A.

If the complete Process and Instrumentation Diagrams (P&IDs) are available, then a full Hazop study is the best method of identifying hazards. Such a study is normal in the design stages of a new project.

Table 2-2
Technological Risk Identification Risk Identification Worksheet (Process Conditions): For Plant Sections

Plant Section / Type of Technological Hazard										
Available energy • High speed of travel										
• High speed of rotation										
• High kinetic energy										
• High potential energy										
• High chemical energy										
• Exothermic chemical reaction										
• Other										
Human and environmental toxicity • Toxic inventory										
• Environmentally hazardous material inventory										
• High corrosion potential										
• Other										
Rigorous requirements • High pressure										
• High vacuum										
• High temperature										
• Low temperature										
• Complex control system										
• Complex protective system										
• Tight tolerance requirement										
• Other										
Innovation hazards • Innovative process										
• Innovative equipment										
• Innovative configuration										
• Complex configuration										
• Other										

In the earlier stages of a new project, before the P&IDs have been finalized, it is often necessary to undertake a preliminary risk assessment using the information available at the time, including the broad process flowsheet. This stage of study is called the "Safety, Reliability, and Environmental Review" (see Chapter 9). At this stage, some of the principles of a Hazop study can be used to good effect. By having a small team of three or four people of varied experience (e.g., production, maintenance, design) review

Table 2-3
Technological Risk Identification Worksheet (Incident Type): For Process Operations

Type of Technological Failure \ Plant Section										
Production capacity • Inadequate peak rate										
• Inadequate availability; poor reliability										
• Turndown ratio										
• Other										
Quality • Inability to meet specification										
• Inconsistency of quality										
• Other										
Cost • Variable production costs excessive										
• Fixed costs excessive										
• Other										
Site-related mishap • Earthquake, landslip										
• Flood										
• Brushfire										
• Discovery of past contamination										
• Other										
Structural failure • Natural forces										
• Excessive production loads										
• Deterioration of structure										
• Other										
Operational incident • Fire										
• Explosion: solid, vapor, dust, etc.										
• Environmentally damaging leakage										
• Chronic toxic exposure of people										
• Nuclear radiation exposure										
• Public nuisance: dust, smell, noise, etc.										
• Other										
Equipment/machinery failure • Physical malfunction (component failure)										
• Control failure										
• Breakage, disintegration										
• Other										

the main inventories, pipelines, and operations on the flowsheet, asking the main questions (about deviations from normal operation) from the Hazop checklist, focusing on causes and consequences of loss of containment or loss of control, a range of hazardous incidents can be postulated as a basis for further analysis and assessment. Note that such a study cannot be called a Hazop study, as a Hazop study is, by definition, undertaken as a final audit on a detailed design that would otherwise be regarded as final and firm.

2.5.5 Detailed Identification of Hazards on an Existing Process Plant

Where it is necessary, because of significant inventories of hazardous material, to undertake a detailed identification of hazards on existing plant, in an existing hazardous operation (e.g., transport of hazardous materials), then the best method is a full Hazop study (see Appendix A) or FMEA (see Appendix B).

This can be a very valuable undertaking if involvement of process operators, maintenance tradespeople, and supervisors is sought, along with the usual cross section of professional skills in production, engineering, instrumentation, etc. Involving those who are closest to the day-by-day and hour-by-hour operation of the plant will provide better identification of hazards, and better communication of process principles and safety requirements.

Many organizations have progressively worked through Hazop studies of existing plants and operations, as the starting point of a program of hazard identification, analysis, assessment, and risk reduction.

REFERENCES

HMSO, "The Control of Major Industrial Accident Hazards Regulations 1982." No. 1902. HMSO, London, 1985.

IChemE, "Nomenclature for Hazard and Risk Assessment in the Process Industries." Monograph, Institution of Chemical Engineers, Rugby, UK, 1985.

Wells, G., "Hazard Identification and Risk Assessment." The Institution of Chemical Engineers, Rugby, UK, 1996.

Wood, S., and Tweeddale, H. M., "Rosebank Peninsula Risk Assessment Study." GCNZ Consultants Auckland, HMTweeddale Consulting Services, Sydney, March 1989.

Chapter 3

Ranking and Short-Listing of Risks

When you have studied this chapter, you will understand:

- how to scan the operations of a process plant to identify the main types of inherent hazard it presents;
- how to estimate very roughly how great the resulting risks are;
- how to rank and short-list those risks into two classes;
- what the significance of those two classes of risk is; and
- how to involve plant staff at all levels in that task so as to build their understanding and commitment to effective risk management.

This chapter addresses step 3, "Rank and Short-List," and step 4, "Define Questions," of risk management in Figure 1-4. (It is also an effective means of meeting the OSHA requirement that "Employers shall determine and document the priority order for conducting process hazard analysis based on a rationale, which includes such considerations as extent of the process hazards, number of potentially affected employees...." See 29 CFR 1910.119(e).)

(The approach described here for step 3 is also a suitable method for coarse-scale identification of hazards and hazardous scenarios, i.e., step 2.)

3.1 INTRODUCTION

In any large organization or undertaking, there are usually far too many identifiable hazards and risks for them all to be investigated at the outset. Some will be pressing and will demand immediate allocation of resources (skilled people, money, etc.), and others will have to wait until the necessary resources are available. In the real world, it is never possible to do everything that needs doing at once.

A very powerful technique used to prioritize and short-list in many technical fields is the "Pareto Principle."

3.2 THE PARETO PRINCIPLE

The Pareto Principle was named by Dr. J. M. Juran (noted for his work in quality management) in honor of an Italian economist who noted that the majority of the wealth of the country was concentrated in the hands of a few of the families.

The principle applies very widely to many forms of activity. For example:

- the majority of the sales revenue of a company will come from a small proportion of the customers;
- the majority of the quality complaints about a product will result from a small proportion of the causes;
- the majority of the lost production time of a factory will be caused by a small proportion of the causes;
- most of the risk faced by an organization will arise from a few of the causes.

The Pareto Principle can be stated as: "Most of the effects are due to a few of the causes."

(This is sometimes known as the 80–20 rule, i.e., "80% of the problems are due to 20% of the causes." This is a little unfortunate, as the relationship is usually different from 80–20. It is often 70–30, or even 90–10.)

The Pareto Principle is used in management to focus attention on the most important tasks: those with the potential for most benefit. In risk management, the Pareto Principle is used as follows:

1. For the plant or activity concerned (e.g., the factory), list the sources of risk, for example, using "Rapid Ranking":

   Source A

   Source B

   ....

   Source Z

2. Assess the magnitude of the risk due to each source (see Table 3-1 and the histogram Figure 3-1).
3. Rank the sources in descending order of risk (see Table 3-2 and Figure 3-2).

Table 3-1

Activity	Risk Scores
A	1
B	78
C	4
D	2
E	6
F	0.5
G	9
H	0.4
I	6
J	2
K	7
L	0.2
M	32
N	3
O	0.2
P	1
Q	54
R	3
S	5
T	17
U	0.5
V	2
W	12
X	4
Y	0.1
Z	1

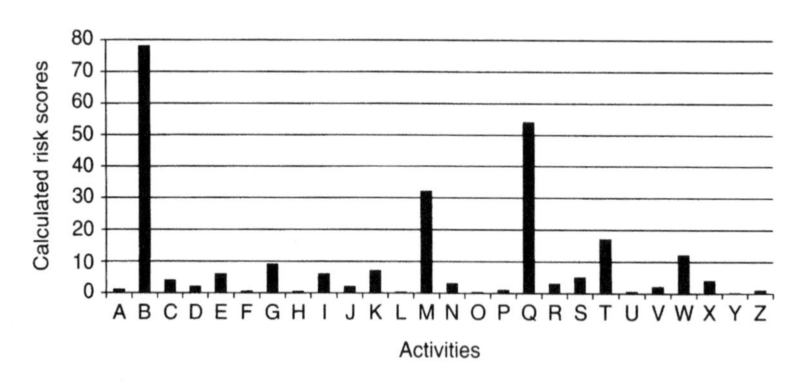

Figure 3-1. Calculated risk scores for activities.

Table 3-2

Activity	Risk Scores	Cumulative % Scores
B	78	31.4
Q	54	53.2
M	32	66.1
T	17	73.0
W	9	76.6
G	9	80.2
K	7	83.0
I	6	85.4
E	6	87.9
S	5	89.9
X	4	91.5
C	4	93.1
N	3	94.3
R	3	95.5
V	2	96.3
D	2	97.1
J	2	97.9
P	1	98.3
Z	1	98.8
A	1	99.2
F	0.5	99.4
U	0.5	99.6
H	0.5	99.8
L	0.3	99.9
O	0.2	100.0
Y	0.1	100.0

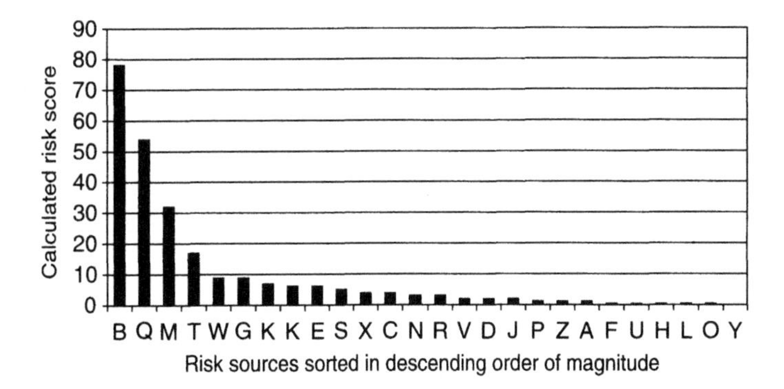

Figure 3-2. Sorted risk scores: individual.

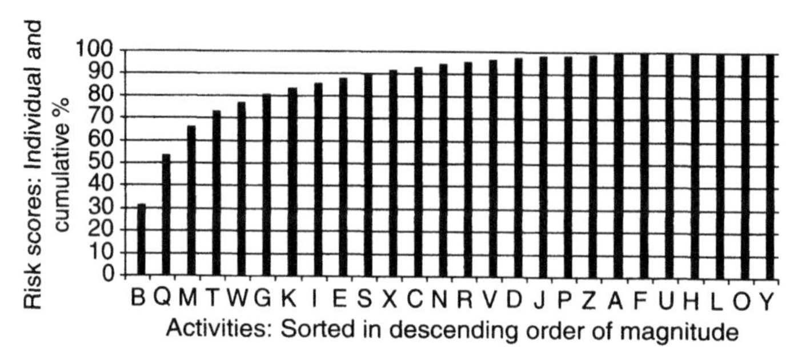

Figure 3-3. Sorted risk scores: individual and cumulative.

4. Starting from the top, calculate the cumulative total, and the cumulative percentage total (see Table 3-2 and Figure 3-3).
5. Note which sources have contributed to 80% (or some other high proportion) of the total risk. In the example, almost 80% of the total risk has been contributed by just six of the 26 sources, that is, 80% has been contributed by 25% of the sources.
6. Focus initial attention on those sources, as between them they contribute most to the risks. If attention were directed toward those lower on the list, the maximum risk reduction would be 25% even if they were totally eliminated (a highly unlikely outcome).

The managerial use of the Pareto Principle is to identify and concentrate on the "vital few" causes, and leave the "trivial many" until later, if ever.

3.3 TWO CLASSES OF RISKS FOR ATTENTION

It is important to recognize that there are two classes of risk that needs detailed attention, and each needs a different type of attention:

(a) High risks need investigation to determine why the risk is high (high consequences, high probability) and action to reduce the risks;

(b) Low risks, comprising a high potential severity combined with a low probability, need managerial action to ensure that the probability is indeed as low as believed, and that it remains low. This action may include establishment of "risk-based inspection," or "reliability centered maintenance."

This classification, and its importance, are discussed in more detail later.

3.4 RANKING THE HAZARDS AND THE ASSOCIATED RISK SCENARIOS

The approach outlined here can be undertaken manually on paper. However, there are real advantages in using a computer spreadsheet, with a format like that shown as Figure 3-4.

If undertaken appropriately, the approach has several benefits:

- increasing awareness of the hazards, and of engineering and managerial requirements for control of the risk;
- defining a cost-effective sequence of undertaking risk reduction work;
- defining priorities for development of systems for maximizing, monitoring and auditing ongoing risk standards;
- developing consensus about the appropriate priorities for risk reduction, and for ongoing risk management, thus facilitating prompt action, rather than ongoing debate and inaction.

To receive all these benefits, it is important to plan the way in which the steps are carried out, as well as what will be done in each step. The key principle is to involve from the outset all those who will have a key role in implementing the findings, and in maintaining the ongoing performance at a high level.

An effective way to undertake a more detailed review and ranking of the hazards on a site or in an operation is as follows:

1. With a small team of two or three people, for each plant or operation, list the hazardous materials and their inventories, or their energy sources (e.g., large rotating machines), or potential losses of any kind (e.g., the hazards and losses discussed in Chapter 2).
2. For each of these sources of hazard or potential loss, prepare a list of the types of hazardous incident or loss which could be envisaged.
3. For each such hazardous incident or potential loss, postulate and list a range of hazardous incidents, ranging from small and relatively likely to large and unlikely.
4. For each postulated hazardous incident, indicatively estimate the severity of the consequences of the incident (e.g., one chance in 10 of a person being killed, 1 person killed, 10 people killed). Note that the consequences may be estimated by the team directly, or using the "Delphi" method described later.

ANT: REVIEW TEAM: DATE: / /

Row number	Plant or product subsection or component	Class of fault or problem	Effect type	Conseq severity score	Specific scenario (i.e.how the problem type could occur with this component in this equipment	Initiation frequency score	Mitigating factors that minimize the chance of the estimated severity of consequences being realized	Conseq'ce realization probability score	Risk score
	B	C	E	G	D	H		J	K (=G+H+J)*

*The numbers are added if the logarithmic scoring systems shown in Section 3.5 are used, as this is equivalent to multiplying natural numbers. If a scoring system with natural numbers is used, the numbers are multiplied.

Figure 3-4. Example of general form of spreadsheet used in rapid ranking.

5. For each postulated hazardous incident, identify the initiating event or "cause" and indicatively estimate its frequency or likelihood (e.g., 1 per year, 1 per 10 years, 1 per 100 years, 1 per 1000 years). These estimates may be made directly, or using the Delphi method.
6. For each postulated hazardous incident, identify the various protective responses (both human and equipment) which would be expected to prevent the incident, if initiated, from resulting in consequences as severe as estimated in (4) above, then estimate indicatively the probability that the protective response would fail to prevent those consequences (e.g., 1 = no protective response, 0.1 = 1 chance in 10 that the protective response would fail, 0.01 = 1 chance in 100 that the protective response would fail—such as a simple trip system on its own). These estimates may be made directly, or using the Delphi method.
7. For each postulated hazardous incident, multiply the following to obtain a measure of the risk:
   - severity of consequences,
   - frequency of initiation, and
   - probability of failure of protective response.

   The higher the number, the higher the nominal risk.
8. Rank the incidents according to their indicatively estimated total risks. Then carefully review the ranking for consistency, remembering that the data used are very rough and indicative only and are bound to have inconsistencies at first. Make any adjustments to the data (and hence the risks) to obtain consensus that the ranking is a reasonable estimate.
9. Starting from the top, calculate the cumulative total risk from the list. Identify those incidents which amount to 80% (say) of the total risk, and regard these as the short-listed incidents for detailed hazard analysis, risk assessment, risk reduction, and continuing review by operating and managerial staff. (The Pareto Principle requires one to identify and focus attention on the "vital few" causes that contribute most to the total risks that the organization faces, and to leave until later the "trivial many" risks which contribute little in total.) Once one has reviewed the results so far, the spreadsheet can be extended with additional columns with estimates of the potential for risk reduction, the resources needed to achieve that reduction, the elapsed time required, etc., and new rankings can be decided by sorting the short-listed risks according to such indicators as the risk

reduction per unit of resources used, or the risk reduction per unit of elapsed time.

10. Rank the incidents according to their indicatively estimated consequence severity.
11. Where the incidents with large estimated consequences are not included in the short-list derived in step 9 (selection by total risk), they cannot be ignored, as they are "insignificant" risks only if the estimated (low) frequency of occurrence or the probability of failure of the protective response is correct, and if they are maintained at those low levels in the future.
12. These high-consequence low-frequency incidents are prime targets for an immediate audit, to ensure that the estimates of initiation frequency and response failure probability are indeed low, and for incorporation in a routine monitoring system to ensure that they remain low.

This may seem laborious, but for any individual plant it can be done quickly by the plant staff after very little assistance at the outset.

This general approach has been used successfully to focus attention on and to rank the hazards in many fields, from oil and gas, to steelworks, paper manufacture, electricity generation, road and rail transport, and sewage treatment. An outline of applications in oil/gas; petrochemical/chemical; an industrial estate with many small chemical factories; and paper manufacture is presented in Section 3.7. Details of a method suitable for application to reliability of major plant (e.g., of an oil-gas separation facility) are presented in Appendix C.

3.5 EXAMPLES OF SCORING SYSTEMS FOR USE IN RAPID RANKING

3.5.1 Introduction

The following tables for consequence, initiation frequency, and mitigation effectiveness use a logarithmic scale, that is, a score of 3 represents a quantity that is 10 times greater than one with a score of 2. To derive the risk score, the three scores are added.

Similar scoring systems can be developed using linear scales, in which case the scores must be multiplied.

3.5.2 Consequence Scoring Systems

See Tables 3-3 through 3-6.

Table 3-3
Consequence Scores for Injury (Acute Onset) or Public Health (Acute or Chronic Onset or Effects)

Description of Effect (of each incident)	Consequence Score
10 fatalities OR 100 serious permanent disabilities	6
One fatality, OR 10 serious permanent disabilities, OR 100 hospitalized	5
One serious permanent disability, OR 10 hospitalized, OR 100 visits to medical practitioner, OR large coverage in an early page of a national or major state newspaper, or equivalent coverage in other media	4
One hospitalization, OR 10 visits to medical practitioner, OR 100 mildly injured or feeling unwell, OR small coverage in national or major state newspaper, or leading story in a local newspaper	3
One person visits a medical practitioner, or 10 people mildly injured or feeling unwell, OR 100 complaints related to injury or health	2
One person mildly injured or feeling unwell, OR 10 complaints related to injury or health	1
One complaint related to injury or health	0

Table 3-4
Consequence Scores for Environmental Impact or Public Outcry

Description of Effect (of each incident)	Consequence Score
Main front page headline in national or major state newspaper, OR equivalent coverage in other media	5
Large coverage in an early page of a national or major state newspaper, or equivalent coverage in other media	4
Small coverage in national or major state newspaper, or leading story in a local newspaper	3
Small article on front page in local newspaper OR 100 complaints about environmental performance	2
Small article in inside page in local newspaper OR 10 complaints about environmental impact	1
One complaint about environmental impact	0

Table 3-5
Consequence Scores for the Costs of Consequential Damage; Loss of Materials, Loss of Production (e.g., through Unreliability) or Profit

Description of Effect (of each incident)	Consequence Score
\$10,000,000 repair cost of consequential damage OR damage to assets OR value of lost production	5
\$1,000,000 repair cost of consequential damage OR damage to assets OR value of lost production	4
\$100,000 repair cost of consequential damage OR damage to assets OR value of lost production	3
\$10,000 repair cost of consequential damage OR damage to assets OR value of lost production	2
\$1000 repair cost of consequential damage OR damage to assets OR value of lost production	1
\$100 repair cost of consequential damage OR damage to assets OR value of lost production	0

Table 3-6
Consequence Scores for Loss of Supply to the Market (Used Where the Costs of Interruption to Supply Are Severe and Not Easily Quantified)

Duration of Effect	Total (100%) Loss of Supply to Markets	Consequence Score Log_{10} (100% days)
1 day	1	0
1 week	7	1 (approx.—strictly 0.85)
1 month	30	1.5
3 months	90	2

3.5.3 Frequency Scoring Systems

See Tables 3-7 and 3-8.

Table 3-7
Initiation Frequency Scores

Initiation Frequency Description	Indicative Frequency per Year	Raw Score Log_{10} (frequency)	Adjusted Score (to eliminate negative quantities for frequency)
Roughly annually	1	0	5
Several times in a working career	0.1	–1	4
Unlikely, but not highly surprising, in a career	0.01	–2	3
Very unlikely in a career, but may happen somewhere in the world	0.001	–3	2
At the limit of credibility	0.0001	–4	1
Not really credible	0.00001	–5	0

Table 3-8
Mitigation Effectiveness Scores

Type of Mitigation in the Event of an Event Being Initiated	Mitigation Score
No mitigation: if initiated, the event will result in the estimated severity of consequences	0
10% probability that the event, if initiated, will result in the estimated severity of consequences. Normal manual response, if no special urgency	–1
1% probability that the event, if initiated, will result in the estimated severity of consequences. Normal instrumented protection system if tested regularly, or several independent manual responses	–2
0.1% probability that the event, if initiated, will result in the estimated severity of consequences. Several independent and diverse protection systems in parallel with very limited potential for "common mode" failures	–3
0.01% probability (virtually impossible) that the event, if initiated, will result in the estimated severity of consequences	–4

3.6 ESTIMATION OF THE MAGNITUDE OF THE CONSEQUENCES, OR THE FREQUENCY, OF OPERATIONAL LOSSES

3.6.1 Introduction

In principle, where the incident being estimated is not of a type for which mathematical methods exist for estimation of the consequence, then estimation or assessment of the magnitude of the consequences requires:

- reference to the history of similar incidents elsewhere;
- creative thought about what could occur, even if it has not happened elsewhere previously; and
- use of judgment.

For example, if a batch of molten metal escapes from its containment and flows over the ground, estimation of the injuries or fatalities which could result would be undertaken by envisaging the possible places where such an escape could occur, the probable number of people within the area of the spill, and their chances of escaping unhurt. The number of injuries can be estimated by judgment well enough for effective risk management, although there is no formal mathematical method available.

In coming to the estimate, account would be taken of the history of such escapes elsewhere, creative thought about the particular situation being considered, and application of judgment to that information.

In many cases, it will be found that there are a variety of outcomes to be considered, ranging from "worst possible" to "most likely." For example, a leak of LPG could result in a fire, which could be large or small, or a BLEVE of the stock tank, or a flash fire or vapor cloud explosion.

For each of these possible outcomes, it would be necessary to visualize the physical outcome and estimate the consequences in terms of any relevant units, such as fatalities, damage to plant and equipment, and public outcry.

In many of the scenarios that an organization will need to assess, there will be no better estimate possible than a judgment reached by a small group of suitably chosen staff, based on consideration of history in the organization and elsewhere, and creative thought. There will sometimes be no mathematical method available, especially for incidents of a type that has not been studied as intensely as have oil and chemical industry hazards.

3.6.2 Methods of Estimating for Short-Listing Purposes

Methods that can be used to aid formation of a sound judgment include:

- authorizing one person to prepare the estimates, and then having them reviewed;
- arranging a led group discussion, with the group arriving at a consensus; or
- using the "Delphi method" or some similar approach:

1. Each member is asked for an independent estimate without any discussion or reference to other opinions.
2. The estimates are collated.
3. The people with the high estimates and the people with the low estimates are each asked to explain the basis for their estimates.
4. After further discussion, the members usually reach consensus. If they do not, the process can be repeated, or it may be agreed that the average will be used as an acceptable compromise.

The Delphi method can be used for estimates of all types, where experience and opinion are the main available basis for the estimates.

The main weakness of the Delphi method is that the members of the group may have made very different assumptions in arriving at their estimates. It may be difficult to reconcile these assumptions, as all may be valid. The best method of reconciling these different assumptions is to incorporate them all by means of an "event tree" (see Chapter 6).

Group discussion (either in a led discussion or as part of the Delphi approach) is very helpful in determining the factors that can influence the outcome.

Approaches similar to those above can be used for estimating the cost of the damage arising from such incidents, first visualizing the extent of the damage (e.g., as a percentage of the new cost of plant, then multiplying that by the cost of that equipment, or as a repair cost directly).

3.6.3 Incidents Arising from Hazardous Materials

In the case of incidents arising from processing, storage or transport of hazardous materials, it is possible to estimate the potential consequences using methods developed over the past 15 years by the chemical and oil industries.

Formal assessment of the consequences of fires, explosions, toxic gas escapes, etc., requires use of mathematics and methods beyond the scope of short-listing.

3.6.4 Environmental Damage

In the case of many types of environmental risk, the extent and duration of any environmental impact is unclear, as the ecological effect of the material that could be accidentally released to the environment may not have been studied. Although such a study may be initiated as a result of hazard analysis identifying the risk, the study may require a long time and be very costly. It is often not possible to delay risk management action until such studies have been completed. Therefore a judgment will need to be made about the approximate scale of the effect ("large," "medium," or "small"; "long-term," "short-term," etc.), using benchmarks such as those outlined in Section 3.5. In making such judgments, it may be necessary to use information about similar materials, and to consult with statutory authorities for their views about the relative ecotoxicity.

3.6.5 Interruption to Supply of Goods or Services

The impact on a community, and hence on the organization, can be very large, especially in the case of national or state utilities such as generation and distribution of electric power, natural gas, water, and provision of drainage. Similarly, for commercial organizations, an interruption to the ability to supply customers with goods or services can lead to long-term loss of customers, in addition to the loss of income for the duration of the interruption. Estimation of the commercial effect of such losses can be done indicatively by judgment, considering the likely duration of the interruption, and the expected subsequent behavior of the "vital few" major customers. This will entail estimation of the amount of sales lost, the loss of revenue resulting from that, and the net change in production cost. (Note the application of the Pareto Principle in estimation of the consequences; don't be sidetracked by the difficulty of estimating the behavior of the "trivial many" customers.)

3.7 CASE STUDIES

3.7.1 Introduction

The following case studies, and examples of how the following approach can be extended, are discussed more fully by Tweeddale, Cameron, and Sylvester (1992).

3.7.2 Large Petrochemical and Chemical Factory

In the case of the single large petrochemical site mentioned previously, the work was undertaken entirely manually by small teams (e.g., the production superintendent and the plant engineer) on each plant, with some initial guidance in each case by a coordinator to ensure consistency of assessment.

Following completion of the assessment of each plant, the production manager of each area was asked to review the results and make the adjustments to ensure consistency. By this means, the people responsible had their attention focused on the hazards and found the undertaking very instructive. The work was not onerous; the largest plant took about 3 hours for the two people involved.

Then the ranked risks for the individual plants were discussed by the plant staff with senior management to have them ratified, and then reviewed across the entire site, and priorities were determined for more detailed inspection, auditing, hazard analysis and risk assessment, and risk reduction.

3.7.3 Oil-Gas Separation Facility

An oil-gas separation facility in an isolated location was the principal source of natural gas for a major city, to both industrial and domestic users. The design of the plant had been carefully reviewed from the safety viewpoint, and a program of work was in progress implementing the recommendations of those reviews.

In view of the reliance of the city on gas supplied by the facility, it was critical that the plant continue to operate without interruption to that supply. Some critical units had been duplicated in the recent past, thus providing improved security of supply in the event of breakdown of any single unit.

It was recognized that some parts of the plant were not provided with backup. Further, there were other sections of the plant where a single serious incident (e.g., fire) could conceivably damage critical equipment such that production would be interrupted. It was decided to examine the vulnerability of gas supply to the city to single incidents at the plant.

Such an examination, to be rigorous, would have required a large amount of highly detailed investigation, much of which would, in the event, be found to have been of plant sections where the potential for interruption of production was small. It was therefore decided to undertake a "screening" review, to identify those units or sections of the plant which presented the main risk of such interruption of production, and to define the options for each such unit or section, the options possibly including a more detailed and narrowly focused study.

The objectives of the assessment were:

- to identify which units at the facility appeared to present most risk of causing interruption to full supply of gas to the market;
- to identify the main options for eliminating or reducing or for managing those risks; and
- to provide adequate documented support for those findings to satisfy senior management.

The assessment was to be completed within around 2 months. Thus it would not be possible to investigate in detail. This meant that it would not be possible to prepare a comprehensive list of the possible scenarios by which continuity of supply could be lost; to undertake a statistically based quantitative assessment of either the extent or the probability of any interruption to supply; or to determine the absolute values of the risks of such interruptions. Thus the assessment was to develop a relative ranking, based on subjective, but experienced, judgment.

The end result of the assessment was:

- agreement about which of the plant units presented most risk of incidents causing loss of production of a scale that could have a serious impact on the market;
- a list of the options for reduction of the risks, awaiting decision and approval of expenditure on detailed evaluation of the options or more detailed examination of specific risks;
- a list of the risks that most needed careful ongoing management, to keep their likelihood low, and suggested initiatives to provide that level of management; and
- assembled documentation prepared in the course of the assessment, which enabled the bases and limitations of the assessment to be understood and penetrated.

Details of the method are set out in Appendix C.

3.7.4 Industrial Estate, Including Chemical Processing Factories

In this case, the work was aided by construction of a computer spreadsheet, broadly similar to that in Figure 3-4. First, the different installations were inspected.

1. For each installation, groups of rows were set out for the following types of activities:
   - transport of materials to and from the site
   - unloading or loading of vehicles
   - raw material storage
   - classified "dangerous goods" storage
   - production (several different areas studied separately)
   - finished product storage
2. For each installation and each type of activity, rows were set out for the following types of incident:
   - fire, with risks to people from heat
   - fire, with risks of nuisance to people from smoke
   - fire, with risks to the environment from firewater runoff
   - BLEVE, with risks to people from heat
   - BLEVE, with risks to property from heat

- explosion, with risks to people from blast
- explosion, with risks to property from blast
- toxic spillage, with risks to people
- toxic spillage, with risks to the environment

3. The rows were then completed, one by one, using the types of incident listed as an aid to the imagination in identifying possible incidents. Where no such incident could be postulated, the row was deleted.
4. After completion of the spreadsheet, the spreadsheet was sorted by type of effect (people, environment, property), by type of incident (fire, BLEVE, explosion, toxic spillage), and by the magnitude of the estimated risk, and checked for consistency. Because of the size of the spreadsheet, a significant number of adjustments were needed.
5. The sorted spreadsheet was then used as a basis for deciding which risks and types of risk warranted more detailed review, hazard analysis and risk assessment, and risk reduction.

In this case, three different types of effect were identified in the one review: risks to people, to the environment, and to property. After completion of the spreadsheet, the different types of incident were sorted together by the computer.

Different scales were used for the severity of the consequences of the three types of incident, as it was not possible to compare risks of fatality with risks to the environment or risks to property. For any particular task, such scales should be developed afresh to suit the situation of each organization, and not blindly copied from those used by others.

For more detail, see Wood and Tweeddale (1989).

3.7.5 Steelworks

For each plant in the steelworks, a small team was assembled, comprising two operational staff (e.g., operations manager, plant engineer), an engineering department representative, and a risk analyst to guide the meeting.

The plant staff outlined the flowsheet of the plant. Then, a photocopying whiteboard was set into rows and columns. The rows were labeled with the types of hazardous incident theoretically possible in any part of the plant. The columns were labeled broadly as shown in the sample spreadsheet of Figure 3-4.

The team considered each section of the plant in turn, with a complete whiteboard being completed per section.

The approach for any one section was to review each of the postulated hazardous incidents in turn, to determine whether it could occur in that section, and if so, how it could happen, how severe it would be, how frequently it would be initiated, and what the probability of failure of the protective response would be.

Thus, the whiteboard was completed for each section in turn. The information was transferred to a computer spreadsheet, and analysis continued as set out earlier. The analysis was used to determine the content of risk assessment training and the scope of detailed studies.

3.7.6 Gas/Liquid Separation Plant

A variation on the above approaches was used to rank and short-list possible causes of loss of supply of natural gas from a gas/liquid separation plant to a major city. For details, see Appendix C.

3.8 RISK MANAGEMENT WITHOUT NUMBERS

3.8.1 Introduction

Risk management can be undertaken very effectively even where it is not possible to quantify the severity and the frequency of occurrence of hazardous incidents.

As illustrated above, particularly in the case of the industrial estate, it is possible to ascribe a number to a judgment (or even a "gut feeling") about the magnitude or likelihood of a hazardous incident, by preparing lists of descriptions of effects and of likelihoods, and putting numbers to them. Of course, those numbers are no more valid than the judgment which derived them, but we are accustomed to making decisions every day about matters that we cannot quantify, relying solely on judgment.

By putting indicative numbers to the components of risk, it is possible to adopt a more systematic approach to ranking and short-listing risks for attention. For that reason it is strongly recommended, if you prefer to use purely descriptive approaches to risk management, that you seriously consider taking the step toward quantification described above.

3.8.2 Risk Matrix

One method of ranking risks without actually using numbers is to prepare a matrix of consequence (rated "High," "Medium," and "Low") against likelihood (rated "Very Likely," "Possible," "Very Unlikely"). Then a dividing line can be drawn to separate the high-risk elements from the others.

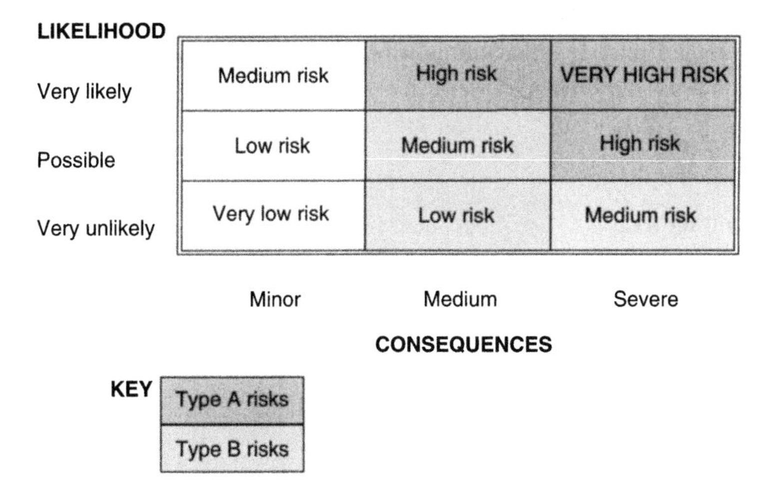

Figure 3-5. Risk matrix.

The apparent simplicity of the risk matrix in Figure 3-5 has led to development of many unsound copies. The diagonals in the matrix of Figure 3-5 are shown as having equal risk. This only applies if the intervals between the levels of likelihood (i.e., between "very likely" and "possible," and between "possible" and "very unlikely") are all equal and the same as the intervals between the various levels of consequence. This can be understood at once if notional values are inserted for the levels of likelihood and of consequence, and the resulting values of risk inserted into the cells in the matrix by multiplication.

The use of a nonquantified matrix such as that in Figure 3-5 is made much more reliable if there is an attempt notionally to quantify the approximate values of the levels, as a guide to those allocating those levels to identified risk scenarios. The result of not attempting to do so is often a very inconsistent risk ranking, without that inconsistency being recognized.

3.9 IDENTIFYING THE QUESTIONS TO BE ANSWERED IN THE RISK ASSESSMENT

This addresses step 4 in Figure 1-4.

When the important risks have been identified in the course of rapid ranking, it is wise to consider the objectives of further study, rather than proceeding ahead in a mechanical way with a "standard" risk assessment. This is equivalent to the first and most important step in quality assurance: identifying and understanding the client's requirements.

The purpose of risk assessment is to assist risk management. Risks are not managed only by management on behalf of management. There are many stakeholders involved with the process industry:

- management
- other employees
- contractors
- shareholders
- nearby members of the public
- public interest groups
- local, regional, and national government

For any particular process plant, many of these may legitimately have questions that need answering and decisions that need to be made. The logical approach to considering these is as follows:

- Identify the stakeholders, or anyone with a legitimate case for seeking answers to questions.
- Identify the types of questions they may wish to ask and the decisions which may need to be made (where practicable, do this by consultation with them).
- Determine the type of information needed for their questions to be answered and the decisions clarified.
- Determine the way in which that information should be assembled and presented for those questions to be clearly answered and those decisions clarified.

Examples of questions that may be asked for particular facilities include:

- Are the products to be handled at this facility able to be handled responsibly, or are they too hazardous for this situation?
- Should the proposed type of facility be permitted at all? Is it able to be constructed and operated consistent with industry codes, etc.?
- Is the proposed facility within the capability of this particular organization to operate without mishap?
- Should the proposed facility be permitted in the proposed location?
- If the facility is built there, what restrictions should be placed on surrounding land uses?
- Should the proposed facility be built to the planned design?
- How can the proposed design be made safer?

- What are the most cost-effective ways of reducing risks from the facility?
- What special operational requirements are there to avoid mishap?
- What should be routinely monitored and periodically audited to control the inherent hazards of the facility?
- How often should the facility be audited for safety? What special expertise should the auditors have?
- What external controls should be imposed on the operation of the facility?
- What does the public need to know about the facility?
- What are the most likely types of mishap needing emergency response?
- What is the scale of the worst credible incidents?
- What on-site and off-site arrangements need to be made to handle possible emergencies?

Then the further study is structured to provide the required information in a form that enables it to be used easily. This may be tabulated for managers, and in diagram form for nontechnical people. (But beware of being seen as "talking down" to people.) The information will be partly quantitative and partly qualitative. The two types of information are equally important.

One example of the application of these questions is described in Section 14.3 of Chapter 14. It is discussed more fully by Tweeddale (1993a,b).

REFERENCES

Tweeddale, H. M., "Increasing the Relevance of QRA." Proc. Conf. Health, Safety and Loss Prevention in the Oil, Chemical and Process Industries. Butterworth–Heinemann, UK, 1993a.

Tweeddale, H. M., "Maximizing the Usefulness of Risk Assessment. Proc. Conf. Probabilistic Risk and Hazard Assessment, Newcastle, Australia. Balkema, Rotterdam, 1993b.

Tweeddale, H. M., Cameron, R. F., and Sylvester, S. S., "Some Experiences in Hazard Identification and Risk Short-Listing." *J. Loss Prev. Process Ind.* **5**(5), 279–288 (1992).

Wood, S., and Tweeddale, H. M. "Rosebank Peninsula Risk Assessment." Auckland City Council, April 1989.

Chapter 4

Risk and Reliability Criteria

When you have studied this chapter, you will have:

- some appreciation of the possible philosophical approaches to defining "How safe is safe enough?";
- an understanding of the principles behind typical statutory requirements; and
- some appreciation of the philosophical and moral issues involved.

This chapter addresses step 5, "Define Risk Targets," of risk management in Figure 1-4.

4.1 INTRODUCTION

Everyone lives with a degree of risk. This may result from some activity one undertakes voluntarily oneself, or from some external activity undertaken by others, or from some natural event.

By choice, people over the ages have come to live in communities, rather than by themselves. They do this because they perceive that the benefits of living in communities outweigh the disadvantages. For example, the majority of people live in cities, towns, or other settlements, because of the convenience of such life, such as closeness to employment, shops, transport, medical care, entertainment, and provision of services such as power, water, drainage, and garbage removal, in spite of the chance that their neighbors may be troublesome, that they may be involved in transport accidents, that they may have their houses burglarized, etc.

Different people perceive the advantages and disadvantages of community life differently. In particular, people have different perceptions of the acceptability of risk.

Much of the enjoyment of rock climbing comes from overcoming the sense of exposure to risk, in undertaking an activity which can result in serious injury or fatality if a mistake is made. One only needs to listen to someone talking about a climb the evening after it, hearing the details of individual movements of a hand or foot, to realize how awareness and enjoyment is heightened by the way the risk of falling focuses the attention. If the element of risk were removed, such as by having an elevating platform raised behind the climber so that it was never more than a meter below, the recreation would become tame, although the technical difficulty would be unchanged. On the other hand, if a person who enjoys rock climbing is exposed to a relatively low risk from an industrial activity near his or her home, the risk will probably be unacceptable unless it is perceived as being very much less than the risk when climbing. This illustrates the different perceptions of "voluntary" and "involuntary" risks.

Further, perceived risks are often very different from actual risks. A large stock tank could cause anxiety because of its proximity to housing, and because it is in a factory where large quantities of flammable material are known to be stored. Yet, if the tank is used for storage of water (and if the tank is below housing!), the risk to the housing would be very small indeed.

Anxiety about risk, or the perceived level of a risk, tends to be inflated if the exposed person does not understand the nature of the risk or cannot detect its cause. Someone may be prepared to cross a busy road several times each day, but may understandably be concerned about a demonstrably lower risk from some industrial activity that has the potential to cause an explosion or a gas escape, or from radioactivity, or from toxic substances or bacteria at very low levels in foodstuffs or the environment.

Risk implies uncertainty. If it is certain that some event will happen, then it is inappropriate to consider risk.

Risk has two dimensions: the severity of the consequences of the event, and the probability of its occurrence. So, when considering risk, it is necessary to nominate the effect and the probability (or frequency).

The concepts of risk criteria, risk and reliability engineering, and systematic risk reduction can be applied in numerous fields. The nature of the risk to be assessed will depend on whether the unwanted event would have an impact on:

- people
- the environment
- property
- production

It would also depend on how the effect of the event would be measured, such as injury or fatality to people, monetary value of damage to property or loss of its utility for a period, or duration and extent of damage to living species in the environment or damage to the visual aspects of the environment.

It is necessary that the criterion be defined in the same units as the assessed risk. Some types of criterion include:

- People

  risk of fatality to the member of the public most exposed (expressed as the probability of fatality per year caused by the activity being investigated)

  risk of fatality to an employee working on the hazardous plant or operation (sometimes expressed as the number of such fatalities expected per 100 million worked hours)

  frequency of events resulting in different numbers of fatalities (expressed as a curve drawn on a graph with axes of number killed and frequency of incidents with that impact)

- The environment

  frequency of a defined event with harmful consequences (e.g., frequency of release of a defined quantity of a known harmful material. This approach is particularly appropriate where the severity of the environmental effect of the release cannot be easily quantified, but it can be determined with reasonable confidence that a defined quantity of release would cause unacceptable damage)

- Property

  frequency of a defined level of harmful exposure (e.g., frequency of exposure to a defined high level of heat radiation or explosive blast overpressure) frequency of damage equivalent to a defined proportion of replacement cost of the exposed property (e.g., frequency of damage equivalent to 20% of the current replacement value)

- Loss of production or ability to supply a market

  frequency of damage or interruption to production resulting in a defined loss of profit (e.g., loss of production equivalent to 20% normal production for 4 weeks)

  frequency of damage or interruption to supply of a market for a defined period

- General: Frequency of "major incidents," which may be defined in various ways:

  several (e.g., 5) people killed

  many (e.g., 20) hospitalized

  severe property damage (e.g., $10 million)

  major, national media coverage

  public outcry sufficient to threaten continuation or recommencement of the hazardous activity.

Note that quantitative criteria can be developed for any type of loss, such as overspent contracts or time overrun on projects, provided that a measure can be defined for the magnitude of the loss, and the frequency or likelihood can be specified.

In many cases, the criteria for environmental effect, property damage, and resulting business interruption will need to be defined on a case-by-case basis, by discussion between members of the organization and other external organizations such as environmental authorities, insurers, and public interest groups. (For example, in considering the adequacy of safeguards designed into a toxic waste incinerator, it may be appropriate to define a maximum frequency for failure of the scrubbing system that cleans the combustion gases of irritant materials. Such a criterion could be used as a basis for determination of the extent of provision of standby emergency scrubbing equipment, and of the required reliability of the systems for detecting failure of the scrubber and actuating the standby equipment.)

In the case of risks to people, although discussion may be needed for particular cases, there is sufficient information available about other risks in the community, and risk criteria used in other cases, to guide selection of the appropriate criteria. (It is notable that a single criterion is often insufficient as a basis for determination of the acceptability of the risk from a proposed or existing plant or operation.)

4.2 THE PROBLEM WITH "ACCEPTABLE RISK"

A great deal of effort is often devoted to trying to define an "acceptable risk." This term is most commonly used by those generating the risk to justify their exposing others to it. In effect it is saying: "I determine that you will find this risk acceptable." This is both provocative and paternalistic. The level of risk that any individual will accept varies greatly between individuals, and from time to time.

Two terms can be used instead of "acceptable risk" to aid consideration and discussion of risk. These both direct discussion toward what is demonstrable rather than arguable. The terms are:

- "accepted risk" and
- "approved risk."

An accepted risk is one that is accepted by the person or people involved, wisely or unwisely, regardless of whether it is relatively high or low compared with everyday risks. It is an objective statement. Either they accept the risk or they do not.

An approved risk is one that has been approved by the appropriate statutory authority or regulator on behalf of the general community, regardless of whether it is accepted by those exposed. Again, it is an objective statement of fact. Such a risk may be relatively high or low compared with everyday risks, but the statutory authority or regulator has approved it because the balance of benefits against risks is valuable to the community. (It is then up to the statutory authority or regulator, as well as the generator of the risks and those exposed, to determine how best to respond to and manage those risks.)

It is very desirable not to speak of an "acceptable risk," but of an "accepted risk," or an "approved risk," or perhaps of "a risk that might be accepted by those exposed."

4.3 SOME EVERYDAY RISKS

4.3.1 Introduction

Criteria for risks to people are usually based partly on the level of risks already present in the community, and partly on community expectations

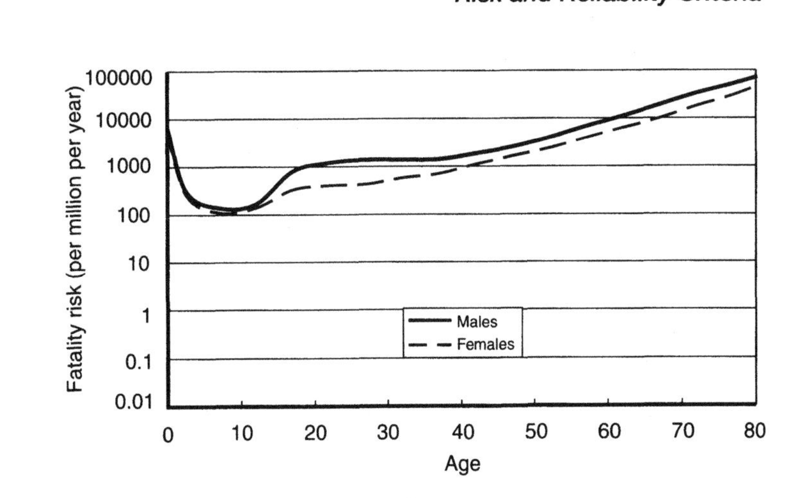

Figure 4-1. Risk of fatality—all causes—typical of a developed country. Source: ABS (1999), ABS (2000).

for the particular case. Typical risks from all causes, including health, are illustrated in Figure 4-1.

For comparison, typical risks to individuals in a developed country from a variety of specific activities are listed by Higson (1990) and others. A selection are shown in Figure 4-2 to the same scale as Figure 4-1.

4.4 RISKS TO MEMBERS OF THE PUBLIC FROM NEW PLANT

4.4.1 Individual Risk

There is widespread (but not unanimous) agreement in many countries that an additional risk from industrial sources of 1 chance in a million per year (i.e., 0.000001 per year) to the person most exposed is a very low level compared with risks which are accepted every day without question, and that such an additional risk may therefore be accepted by the wider community. Some countries or regions require a lower criterion. This is called an "Individual Risk Criterion."

A separate and very important issue is the degree to which such low levels of risk are able to be measured and quantified. That is discussed in a later chapter.

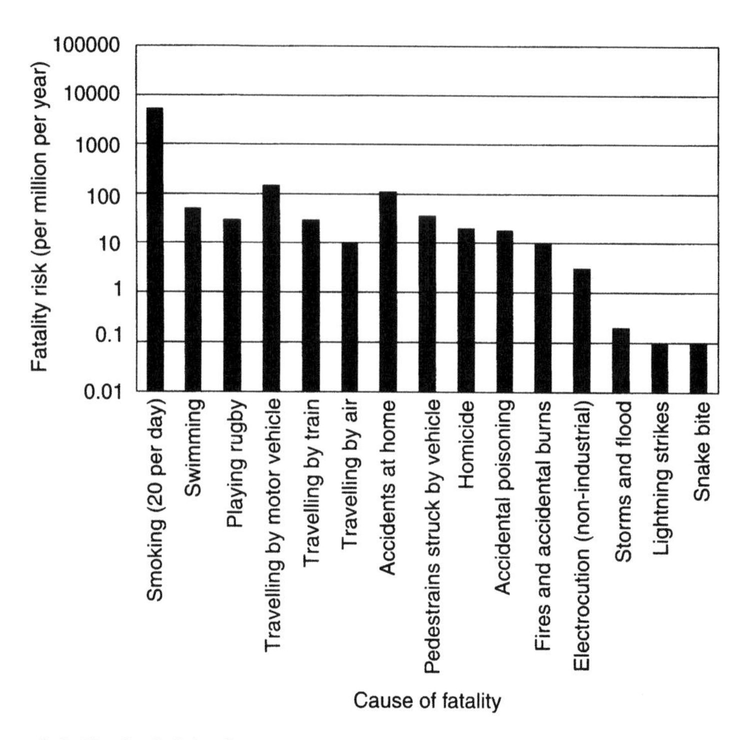

Figure 4-2. Typical risks from selected activities in a developed country (New South Wales, Australia, 1981–1986).

4.4.2 Societal Risk

A criterion that relates to the risk to any particular individual is insufficient on its own. There is an important difference between an incident that might occur with a probability of 1 in 1 million per year that could fatally injure 1 person and an incident with the same probability that could fatally injure 1000 people. For this reason, the "Societal Risk" (or "Group Risk") type of criterion is receiving attention in many studies.

A difficulty with this type of criterion is to decide on limits that do not exclude, as too high to be approved, many unlikely but major hazards which are accepted by the community at present. For example, the Thames Barrier in England has been consciously designed for the "1 in 1000 year high tide." It has been estimated that, in the event of the tide overtopping the barrier, hundreds of people would drown in spite of the emergency procedures. If such an event is plotted on the above chart, it is well into the "Too High" region. The same anomaly is found with major dams and other large structures in many countries.

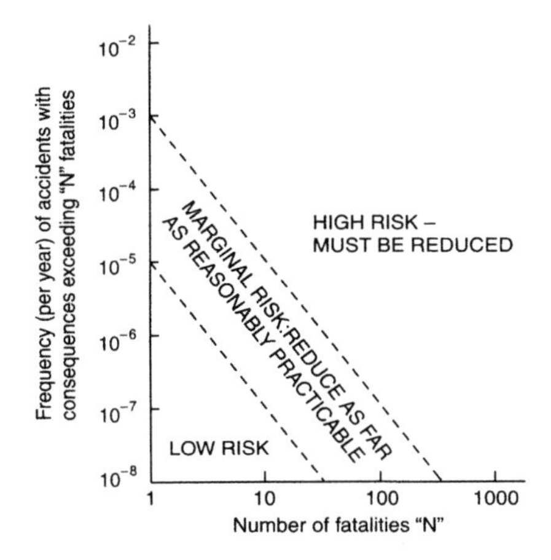

Figure 4-3. Example of approach to defining societal risk criteria.

There is a related difficulty with definition of criteria for societal risk for individual industrial facilities. Several small and independent facilities in the same region may each generate a low level of societal risk, whereas if the same facilities were all grouped under a single management, their combined societal risk may fall in the high-risk zone of the chart. So two process plants A and B, operating to the same safety standards and identical in all physical respects except that plant B comprises two parallel units of the same design as the single unit of plant A, will have different societal risks. The societal risk curve of plant B will be above that of plant A on the chart because the frequency of each postulated event on plant B will be twice that of plant A.

For that reason, the societal risk of traffic accidents for any region (e.g., city, state, country) is likely to fall in the high-risk zone; see Figure 4-3. Similarly, the societal risk for any major airport is likely to be in the high-risk zone, not because the airport is unsafe, but because of the large scale of the activity.

4.5 RISKS TO EMPLOYEES

4.5.1 "Fatal Accident Rate"

Employees are generally exposed to higher risks than members of the public, as employees are generally closer to the hazardous plant and operations than the public. One unit of risk sometimes used in assessing employee risks is the Fatal Accident Rate (FAR). The FAR is the number

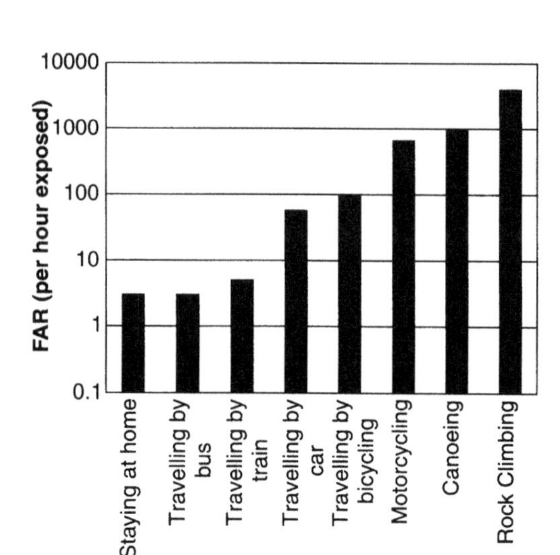

Figure 4-4. Fatal accident rates of nonindustrial activities (fatalities per 100 million hours worked or exposed.)

of fatalities per 100 million hours worked. It is sometimes calculated for specific activities an operator may be called upon to perform, but is more commonly averaged over the operator's total workload. The FAR is roughly equivalent to the number of fatalities per 1000 working lifetimes.

Typical values of FAR for nonindustrial and industrial activities are displayed in Figures 4-4 and 4-5 (to the same vertical scale).

Example:

A plant, which is operated by a team of shift workers, has a predicted frequency of generating a large explosion with a frequency of 1 in 10,000 per year. In the event of such an explosion, each worker on the plant has a probability of 40% being fatally injured. What is the FAR of the workers due to that plant?

Solution:

The plant is operated for 24 × 365 hours per year = 8760 hours. The explosion is predicted to occur with a frequency of 1 per 10,000 years, that is, 1 per 87,600,000 hours. Each operator faces a fatality risk of 0.4 per 87,600,000 hours.

Thus the FAR is 0.4 × 108/87,600,000 = 0.45 per 100 million worked hours.

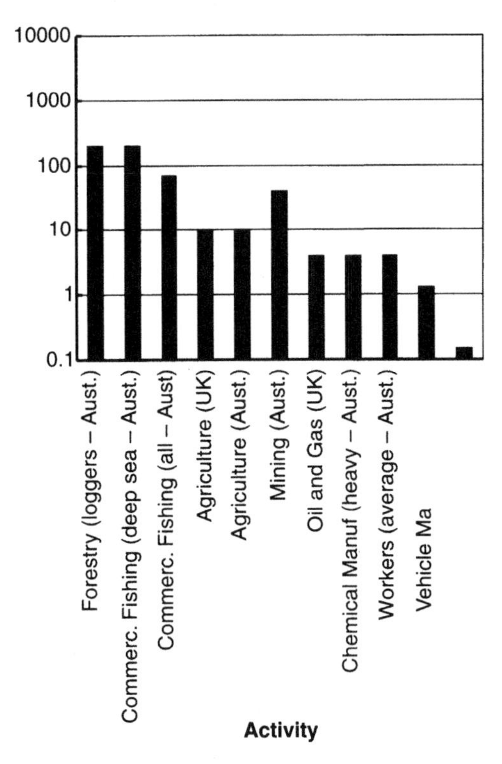

Figure 4-5. Fatal accident rates for typical industrial activities (fatalities per 100 million hours worked or exposed.)

4.5.2 Risk Criteria for Employees on Proposed New Plants

One approach that was taken in a section of the chemical industry is as follows:

Around half of the fatalities that have occurred have resulted from "process" causes, and around half from "occupational" causes. Thus the FAR from process causes would be around half of the total of 4, that is, around 2. The industry is at the safe end of the industrial spectrum, and the risks at work are similar to the risks at home, per hour exposed. There has not been any anxiety expressed by employees or statutory bodies that the industry is "unsafe," and in fact the contrary is the case.

If an attempt were made by any single company to reduce the risks to employees by a dramatic amount, then the costs of doing so could make that company uneconomic. Therefore, as the industry is already regarded as safe, the principle aim should be for continuing evolutionary improvement.

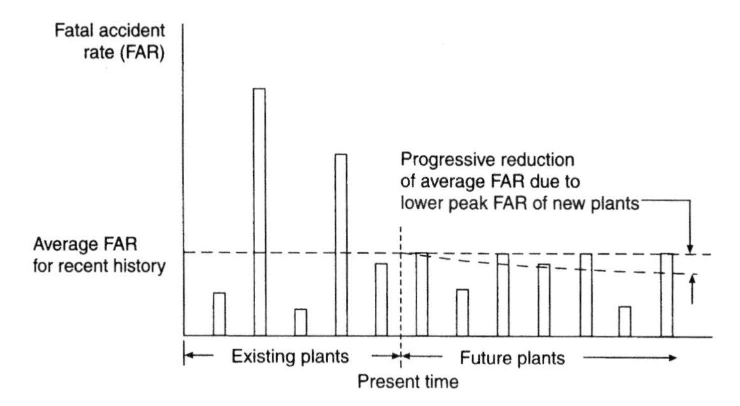

Figure 4-6. Reduction of average FAR by limiting the risks of new plants.

As the average historical FAR from process causes is around 2, and as this is the result obtained from existing plants, some of which will have higher FARs and some lower FARs, if the FAR of the employee most at risk on any new plant is limited to 2.0, then the average for the plant will be less than that, and as more new plants are commissioned as time passes, the average FAR will fall. The effect of this is illustrated in Figure 4-6. This is why, in some parts of the chemical industry, the criterion for FAR for a proposed new plant has been set (subject to statutory approval) at 2.0 for the employee most at risk.

In other industries, where the existing risk to employees is higher, it may be appropriate to adopt a different philosophical approach, and to aim for new facilities to have a lower FAR than the current average.

4.5.3 Risk Criteria for Employees on Existing Plants

With constantly rising community expectations for the safety of process plants, it is inevitable that an existing plant, designed to the standards of the time it was designed, and with the insights of that time, is less safe than a similar plant designed now.

It is often not practicable to upgrade an existing plant fully to current standards. In deciding the improvements to be made in an existing plant, it is necessary to take account of several factors that determine both the allocation of priority and the extent to which improvements should be made.

With limited resources (appropriately skilled staff, money), it is not possible to do everything at once (as discussed in Chapter 3). The factors to be considered include:

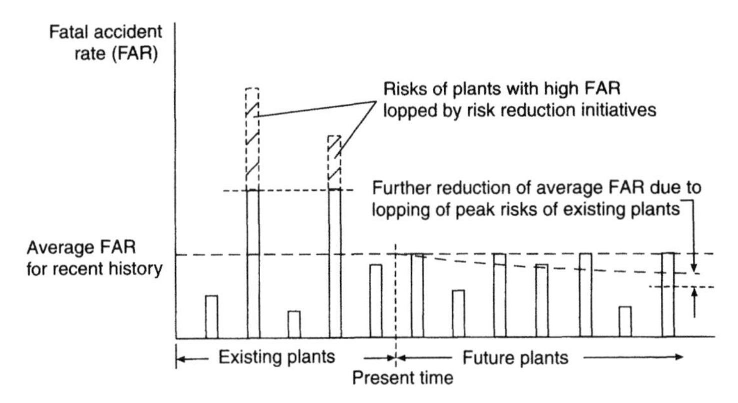

Figure 4-7. Reduction of average FAR by peak lopping of existing risks.

- the highest level of risk for any individual employee (no one person should be expected to be at high risk, even if the average for all operators is low);
- the number of people at risk (as an improvement that reduces the risk for many people would have more priority, other things being equal, than an improvement which produced the same improvement for few);
- the cost of achieving the improvement (both staff time and money); and
- the remaining productive life of the plant (as the benefit of risk reduction is really the product of the reduction by the number of people affected by the duration of the reduction).

Some organizations have set as a goal that no single operator on an existing plant should be exposed to a risk exceeding some defined level of FAR (e.g., 6) and have aimed to lop peak risks to that level, with the priority for that lopping depending on both (a) the extent to which the risk exceeds that level and the number of people exposed, and (b) on the elapsed time and resources required to implement the improvement.

The effect on average corporate FAR of lopping the peak risks of existing plants is shown diagrammatically in Figure 4-7.

4.6 ECONOMIC FACTORS IN RISK CRITERIA

Because it would be possible to spend an infinite amount of money in the quest for perfect safety, and because the supply of money (and other resources such as working time) is limited, it is necessary to take account of economics, at least to some degree. Various approaches have been suggested.

Attempts to put a value on human life have caused criticism from those who regard human life as priceless. Others point out that every human life has emotional value to friends and relatives, which money cannot compensate for. However, there are real limits to the amount that can be spent per life saved.

In principle, one absolute limit on the expenditure permissible per life saved can be indicated by dividing the annual gross national product by the annual number of births. The monetary figure obtained is the amount that could be spent to prolong the life of each baby born, assuming that no money was spent on any other activity in the community. In practice, there are many other demands for the wealth generated by the community annually, and so the amount able to be spent would be less. How that amount is defined will depend on personal priorities.

Another approach is to estimate how much money the person would have earned if he or she had not been killed in the accident. The next of kin can, in principle, be compensated financially by payment of this amount, but this does not eliminate the emotional damage and other intangible but real loss.

In seeking to allocate the limited financial and monetary resources of the community to risk reduction, it is, in principle, necessary to rank proposals in terms of the risk reduction achieved per dollar spent. Examination of these "cost effectiveness" levels achieved in practice (Morrall, 1986) shows a wide range, such as those shown below:

Steering column protection	$0.1 million per life saved
Alcohol and drug control	$0.5
Children's sleepwear fire prevention	$1.3
Benzene: control of fugitive emissions	$2.8

Very much higher values per life saved (e.g., thousands of millions of dollars) have been calculated for environmental cleanup projects where the aim has been to reduce the risks to people from residues of toxic chemicals.

A problem with using the value of life as the basis for setting the criterion is that, if it is decided that the monetary value of the lives saved is insufficient to justify the proposed expenditure on safeguards, the saving and the "costs" are directed to different groups: the saving being made by the organization which avoids the need to install the safeguards, and the costs (in the form of the risks) being borne by those exposed. This is regarded by the community as very dubious at least, and probably very unfair.

Example:

Consider the case for installation of undersea isolation valves to protect offshore platforms. A study undertaken by one international oil company concluded that the increase in safety did not justify the cost of installation, etc. The difficulties with this approach included:

- equity (the benefits of the decision not to install the valves would accrue to the company, which would avoid the need to outlay the expenditure, but the costs of not doing so—the risks—would be borne by the employees);
- professional ethics (the interests of the safety of the body of employees would be held to be of higher priority than the profitability of the corporation); and
- legal liability (if an accident were to occur which could have been prevented by installation of the valves, then a court may decide that the corporation had been negligent, and the individual(s) responsible for the decision may face personal prosecution, large fines, or even imprisonment).

If one company implements an expensive risk-reduction program, they suffer a competitive disadvantage compared with other organizations. Therefore there is a strong case for an industry-wide approach. However, one problem with such an approach is that its rate of development tends to be governed by the most reluctant member of the industry group. Thus there is a strong case for regulatory control of risk criteria.

4.7 REGULATORY APPROACHES TO SETTING RISK CRITERIA

4.7.1 Individual Fatality Risk

Where a regulatory jurisdiction has defined criteria for fatality risk, individual fatality risk is very commonly one of those chosen.

Sometimes different levels of fatality risk are defined for different types of location. For example, a level of 1 per million per year (pmpy) may be chosen for the fatality risk from an industrial facility at general residential areas, with a lower risk level being defined for areas where people may be more vulnerable (e.g., schools, retirement villages, hospitals) and higher levels for areas that are intermittently occupied or where people would be less vulnerable (such as commercial areas, parks and sports fields, and industrial areas).

4.7.2 Individual Risk of Dangerous Dose

A difficulty with the individual fatality risk is that the relationship between exposure and the probability of fatality is uncertain and undergoing frequent revision in the case of many types of exposure. In some jurisdictions it has been decided, instead of using the individual risk of fatality, to use the individual risk of being exposed to a "dangerous dose," where the dangerous dose of each type of exposure is defined separately.

Whereas there may be debate about the probability of fatality from a particular level of exposure (e.g., to heat radiation, explosion overpressure, toxic gas), whether a particular level is "dangerous" (i.e., has a significant probability of resulting in fatality, or at least serious injury) is much less debatable.

4.7.3 Societal Fatality Risk

When the Societal Fatality Risk approach was first postulated, it was adopted in some jurisdictions before its limitations were fully understood. As a result, in some jurisdictions there are rigidly defined boundaries on the societal risk chart between the zones of high, intermediate, and low risk.

In some other jurisdictions, the boundaries are provisional, or open to negotiation and interpretation. In yet other jurisdictions, there is a requirement that societal risks be calculated as a means of displaying the spectrum of magnitudes and likelihoods of the postulated incidents and to aid in forming a broad view of the risks and benefits of the industrial facilities, but not as a sole basis for a decision about the acceptability of those risks.

It is common, where a facility is calculated to generate risks in the intermediate zone between high and low, to require the risks to be reduced to a level that is "as low as is reasonably practicable," provided that the benefits of the activity that produces the risks are seen to outweigh the generated risks. The meaning of "as low as reasonably practicable" is discussed in Section 4.8.

4.7.4 Risk of Damage to Property or the Environment

In some jurisdictions, criteria are defined in relation to the frequency of exposure to damaging levels of heat radiation or explosion overpressure, with a view to limiting the risk of damage to adjacent facilities, whether these be housing, commercial, or industrial. Different frequency/exposure level combinations may be defined for different types of use to which the adjacent land is put.

4.7.5 Implicit Risk Criteria

In some jurisdictions the emphasis is on assessment of possible consequences, but not on assessment of the likelihood or frequency. Nevertheless, the likelihood is often implicit in the definition of the scenarios to be assessed, as in the case of the US EPA Chemical Accident Prevention Provisions (40 CFR 68) which require consequence assessment of both "worst-case" and "alternative case" scenarios, the latter being more probable but generally less severe.

4.7.6 Discretionary Powers

In view of the limitations of the methods available to quantify risks (see Chapter 7), many jurisdictions regard the criteria as general guidance only, with the regulatory authority holding discretionary powers to:

- allow proposed industrial developments even where the calculated risks exceed the criteria; and
- disallow proposed industrial developments even where the calculated risks meet the criteria.

4.8 THE MEANING AND USES OF "AS LOW AS REASONABLY PRACTICABLE"

Legislation in some countries requires risks to be reduced to a level that is "as low as reasonably practicable" (often abbreviated to ALARP).

It is not possible to define ALARP in purely objective and absolute terms. There will always be a need for experienced judgment and subjective opinion, and hence there will always be potential for debate. Ultimately whether specific cases meet ALARP standards may need to be decided in court.

However, a working definition of ALARP, in relation to risks from the activities of process industries on- and offshore, is:

A risk is ALARP if the facility or plant:

- uses the best available technology capable of being installed, operated, and maintained in the environment by the people prepared to work in that environment; and

- uses the best operable and maintainable management systems relevant to safety; and
- the equipment and the management systems are maintained to a high standard; and
- the level of risk to which people are exposed is not high by comparison with many risks accepted knowingly in the community without concern" (Tweeddale and Cameron, 1994).

Where risks are not so low as to be negligible, it is often a requirement that they be reduced to a level that is ALARP. An industrial organization may assert to a regulatory authority that the risks are ALARP. To demonstrate that the requirements of the above definition are met, it may be required that the organization demonstrate that it is meeting all the requirements for effective management of the risks postulated by Hawksley (see Chapter 10).

For a fuller discussion of ALARP, see Tweeddale and Cameron (1994).

4.9 CALCULATING AND DISPLAYING THE RISKS OF POTENTIAL LOSSES

4.9.1 Individual Fatality Risk

In principle, risks are calculated from the consequences and the frequency.

In many cases, the risks can be calculated by multiplying the two. For example: If the severity of an explosion is such that there is a 10% probability of killing a person at a particular location, and the explosion has a frequency of 0.001 per year, then the risk is simply:

Fatality risk = $0.001 \times 10\% = 0.0001$ per year

Where there is a range of incidents that expose the person at that point to risk, the total risk is determined by adding the risks of the separate incidents. This does *not* mean that the total risk is of all the incidents occurring at once!

It is often difficult to determine the total risk if some of the incidents could trigger some of the others. This is called the "domino" scenario and is less common than is perceived by the public. It is, however, possible, and needs to be checked for qualitatively at least, if quantification is difficult.

4.9.2 Societal Risks

Where there is a possibility of a range of incidents with different frequencies and severities, it is sometimes appropriate to express the risks

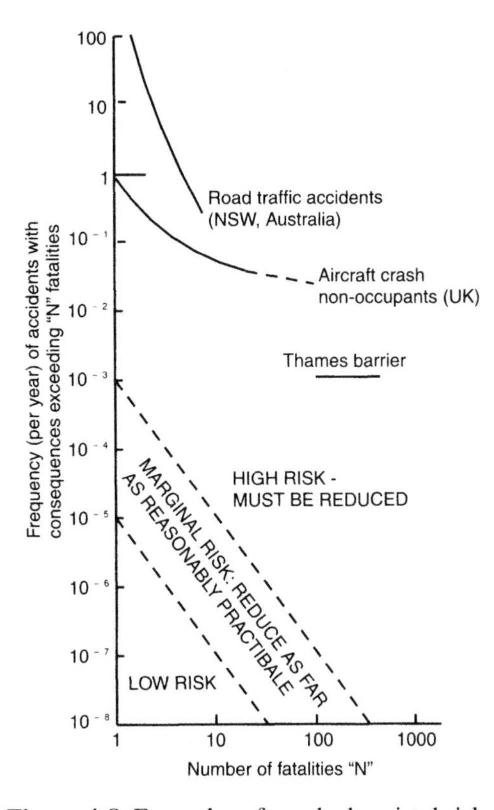

Figure 4-8. Examples of graphed societal risks.

as a "Group Risk" or "Societal Risk" graph. An example is shown in Figure 4-8.

These curves are found by sorting the incidents in descending order of severity, then determining the cumulative frequency, and plotting that on a log-log scale.

Example of Calculation of Societal Risks

Incident Severity (fatalities)	Incident Frequency (per year)	Cumulative Frequency (per year)
10	0.001	0.001
8	0.005	0.006
2	0.01	0.016
1	0.07	0.086

A graph is then drawn plotting cumulative frequency against incident severity. See Figure 4-9.

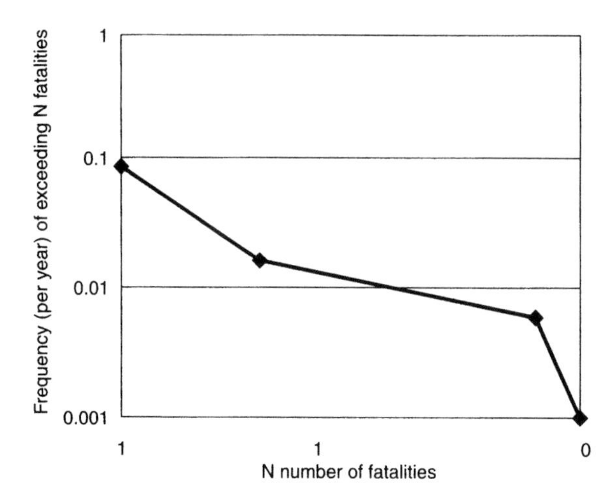

Figure 4-9. Illustration of constructed societal risk graph.

At first sight, societal risk may appear an ideal way of expressing risk criteria. But there are problems with it. For example, if criteria are defined for individual manufacturing sites, the larger the site, the more difficult it will be to meet the criteria. If a large site exceeds the criteria, it could be subdivided in such a way that each subdivision would meet the criteria, although nothing has really changed.

Again, as shown in Figure 4-8, major activities such as air transport and road transport commonly cannot, on a national or statewide scale, meet criteria set for industrial activities.

The Thames Barrier at London (UK) is another illustration. At times of unusually high tide, if the wind is blowing strongly from the east, the level of the sea in the Thames estuary rises substantially, with the potential to flood large areas of London. Because of long-term, gradual subsidence of the land, the risk of this has assessed as very high, perhaps as much as around 1 in 50 per year. So a barrier was built, able to be raised from the bed of the river at the onset of such times. It is estimated that this barrier, or the surrounding high ground, could now be flooded with a risk of around 1 in 1000 per year. This is clearly a major improvement. But it is estimated that, in the event of the barrier being overtopped, hundreds of people would drown, in spite of all the emergency procedures that might be taken. Thus, this installation is not able to reduce the risks from the natural hazard of flooding to a level meeting typical industrial societal risk criteria.

Thus, when attempting to define criteria using societal risk, it is necessary also to relate the criteria to the scale of the operation.

4.9.3 Other Types of Risks

Other types of risks can be calculated using the principles used for either individual risks or societal risks. In some cases, simple bar charts or pie charts may be appropriate displays. In other cases, innovative ways of displaying the results may be desirable.

4.9.4 Risk Contours

In some special cases, such as chemical plants, where there are numerous hazards with consequences of the same general type (such as risk of fatality) even though the hazards may be very different (e.g., fire, explosion, toxic gas escape), it is possible to use computer programs to calculate the risk levels on a topographical grid, and then to plot contours of risk on the grid (for an example, see Figure 4-10).

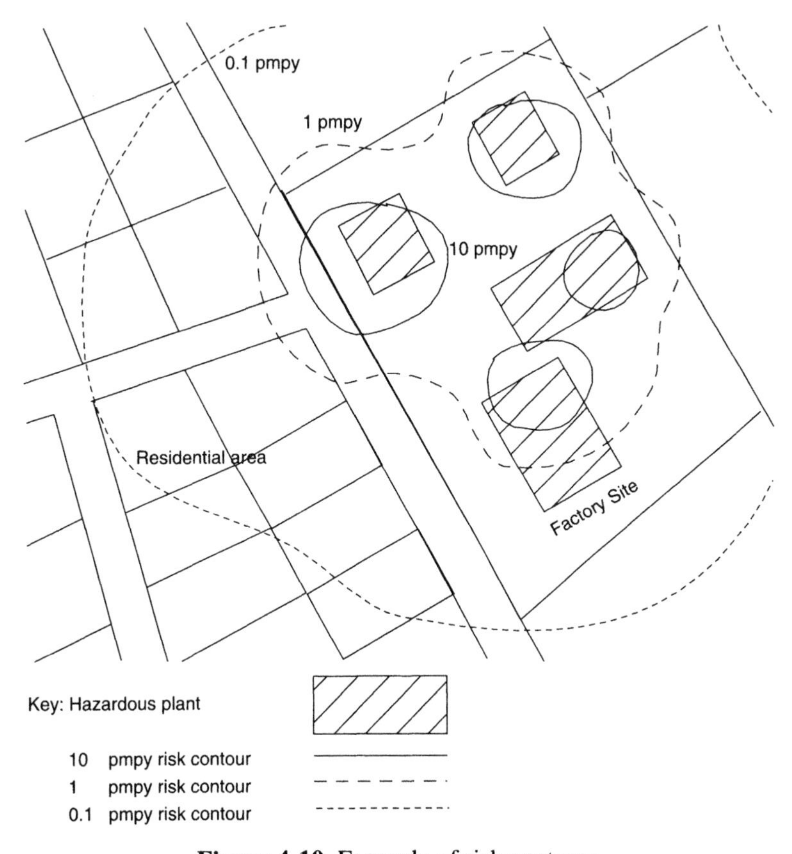

Figure 4-10. Example of risk contours.

Contours are also commonly used to display the frequency of exceeding excessive levels of hazardous exposure.

For example, there are computer programs which will prepare contours for the frequency of exposure to nominated levels of heat radiation, explosion overpressure, and toxic gas concentration. These are needed to assess compliance with guidelines in some jurisdictions. (See Chapter 9.)

REFERENCES

ABS, "Population by Age and Sex, Australian States and Territories." Australian Bureau of Statistics, Publication 3201.0, Canberra, June 1999.

ABS, "Deaths, Australia." Australian Bureau of Statistics, Publication 3302.0, Canberra, December 2000.

Higson, D. J., "Risks to Individuals in NSW and Australia as Whole." Australian Nuclear Science and Technology Organization, Lucas Heights NSW, 1990.

Morrall, J. F., "A Review of the Record." *Regulation*, Nov./Dec., 25–34 (1986).

Tweeddale, H. M., and Cameron, R. F., "Interpretation and Application of 'As Low as Reasonably Practicable' as a Measure of the Safety of Upstream Petroleum Facilities." Department of Primary Industries and Energy ALARP Review, DPIE, Canberra, 20 January 1994.

Chapter 5

Assessment of the Severity of the Consequences of Hazardous Incidents

A great deal of research continues into methods of estimating or calculating the severity of the consequences of hazardous incidents with increasing precision. The greater the precision, the greater the complexity of the method. The findings of the research often differ substantially.

This chapter aims to introduce the principles of those methods, to demonstrate simple but approximate methods, and to provide references for study of more detailed methods. When you have studied this chapter, you will be able to use mathematical methods to:

- estimate approximately the severity of the impact on the surroundings of a variety of types of process plant incidents, including fires, fireballs, explosions, and toxic gas escapes;
- estimate approximately the likelihood of any people being killed, or seriously injured, at varying ranges from the incident; and
- understand some of the limitations of the available methods.

You will also be able to review the available computer programs to select those that best meet your need for more rigorous assessment of consequences.

This chapter addresses step 6, "Assess Consequences," of risk management in Figure 1-4. (It also equips you to understand the principles and methods required to comply with the consequence assessment requirements of EPA 40 CFR 68 and OSHA 29 CFR 1910.119.)

5.1 INTRODUCTION: CALCULATION VERSUS ESTIMATION

In much risk assessment, it is not possible to calculate the magnitude of the consequences. For example, a risk engineer in a railroad company cannot calculate the number of people who would be killed if two trains collide. The consequences need to be estimated by experienced staff by examination of previous incidents.

In the process industries, it is possible to perform calculations of the impact at a distance of some kinds of incident. This is because the consequences are governed to some extent by physical and chemical laws. But it is important not to allow the rigor of these laws to blind us to the uncertainties about the circumstances of the incident, such as the location of a leak, the size of the hole, the presence or absence of ignition sources, the direction of the wind, the terrain over which a leaked burning liquid spreads, or the vulnerability of people to different levels of exposure. So, even where calculations are performed, there is a large element of estimation needed to convert the calculated answers into realistic conclusions about the likely consequences.

In particular, in the case of reliability, where the consequences are likely to be expressed in terms of the amount of production interruption (measured in days, or in physical quantity of a product), it is not possible to calculate the severity of the interruption, as it is determined by many factors such as the availability of spare parts, the extent of buffer storages, and the time required to effect repairs. Therefore the consequences in the case of a production interruption must be estimated. If such estimates are made by people with experience of the industry, they are likely to be of quite sufficient precision to be used effectively in risk management.

5.2 FIRES

5.2.1 Introduction

Fires in process plants can take many forms:

- pool fires,
- jet fires,

- fireballs, and
- flash fires.

The damage done by such fires depends on several effects:

- flame impingement,
- convection, and
- heat radiation.

Fireballs are described in Section 5.3. Flash fires are described in Section 5.4.

5.2.2 Pool Fires

A pool fire is the combustion of material evaporating from a layer of liquid at the base of a fire. It typically occurs when a pool of spilled flammable or combustible liquid is ignited.

The behavior of the resulting fire depends on a number of factors. These include:

- the nature of the liquid being burned,
- the ambient temperature,
- the effective diameter of the pool, and
- the extent of any breeze.

These affect the rate of vapor formation to provide fuel, and the supply of oxygen to the fire. The more volatile the liquid and the greater the ambient temperature (especially of the ground), the greater the rate of fuel supply. The greater the diameter of the pool, the less efficient is the supply of oxygen to the flame, resulting in a less radiant and more smoky flame. The greater the breeze, the greater the supply of oxygen, and the brighter and more radiant the flame.

In assessing the effects of a postulated fire, a number of assumptions are often made, relating to:

- the height of the flame above the pool,
- the inclination of the flame in the event of a breeze,
- the rate at which the pool burns away,
- the average surface emissivity, that is, the amount of heat energy radiated per unit area of the flame, and
- the proportion of the total heat of combustion that is radiated, as distinct from convected upward with the smoke.

Because of the imprecision of such assumptions, any assessments made which are based on them are necessarily approximate.

In assessment of the effects of pool fires, it is often assumed that the flame can be represented by a cylinder with a vertical axis of diameter equal to the average diameter of the pool, and height equal to twice the diameter. In practice, the flame of a large pool fire tends toward a cone, because of the inward flow of air at the base, replacing the air convected upward with the flame. The maximum height of the conical flame can be greater than twice the pool diameter. In the case of a very large and stable pool fire, the dominant impression is of smoke, with flashes of flame in places. Such a fire has a very low average surface emissivity of radiant heat.

5.2.2.1 Flame Impingement

The rate of heat input to vessels from flame impingement is used as the basis of sizing relief valves for fire duty.

The temperature of hydrocarbon flames is reported to vary typically between 1200°C and 1600°C, depending on the turbulence and aeration of the flame. For example, a jet flame of LPG is reported to reach a temperature of around 1600°C, whereas a calm pool fire with a dull red and smoky flame is reported sometimes to reach around 1200°C.

The US NFPA (1984) state that uninsulated vessels, under average plant conditions, when enveloped in flame, may be expected to absorb 63 kW/m^2 of exposed surface wetted internally by the contents.

Heat damage due to flame impingement is much greater than from convection of hot combustion products and from heat radiation. It is normal, in assessing the risks to people, to assume that anyone within the range of flame impingement will be fatally injured.

In assessing the effect of flame impingement on structures, it is necessary to consider the nature of the sections of plant exposed, the duration of exposure, any fire protection fitted, and the stresses on the members of any structures so exposed.

For example, electrical and instrument cables and instrumentation are very rapidly damaged or destroyed by flame impingement. Substantial steel sections are less rapidly affected than light sections, and the more heavily stressed the section, the greater the chance of structural failure. It is normal for plant layout to be designed to minimize the possibility of flame impingement on sensitive equipment or light structural steel sections. For example, flammable liquid pumps (which have a relatively high chance of fire following seal failure) are no longer normally located under pipe bridges, where there is a high likelihood of fire damaging cable runs, columns, beams, etc., and of starting additional leaks in flanged joints.

5.2.2.2 Convection

The potential for convected heat to damage sensitive plants located above potential sites of pool fires is generally covered by suitable layout. For example, it is normal for aluminum fin-fan heat exchangers to be located over the pipe bridges where there is the least chance of a leak, provided that plant vessels, pumps, etc., are sufficiently remote from the pipe bridges. Further, as the fin-fan exchangers are located above the pipe bridges, they are usually sufficiently high above the grade to be minimally exposed to any pool fire.

It is not common for the effects of convected heat to be formally estimated by calculation for the purposes or risk assessment.

5.2.2.3 Radiation

In a fire on a process plant, the main fire damage at a distance in the horizontal plane is due to heat radiation. Heat radiation intensity at a distance from a pool fire can be calculated using one or both of two methods:

- the *point source* method, and
- the *view factor* method.

They are both described below. Both have major inherent approximations.

Point Source Method

The steps in this calculation are as follows:

1. Estimate the location and diameter of the pool of fuel.
2. From the nature of the fuel and the diameter of the pool, using empirical data, estimate the height of the flame, and hence the location of the center of the flame.
3. From the diameter, using empirical data, estimate the rate at which the level of the fuel burns away (mm/min).
4. Calculate the rate of combustion of fuel (kg/sec).
5. Using the heat of combustion of the fuel, calculate the power generated by the fire (kW).
6. Using empirical data, estimate the proportion of the heat of combustion which is emitted as heat radiation.

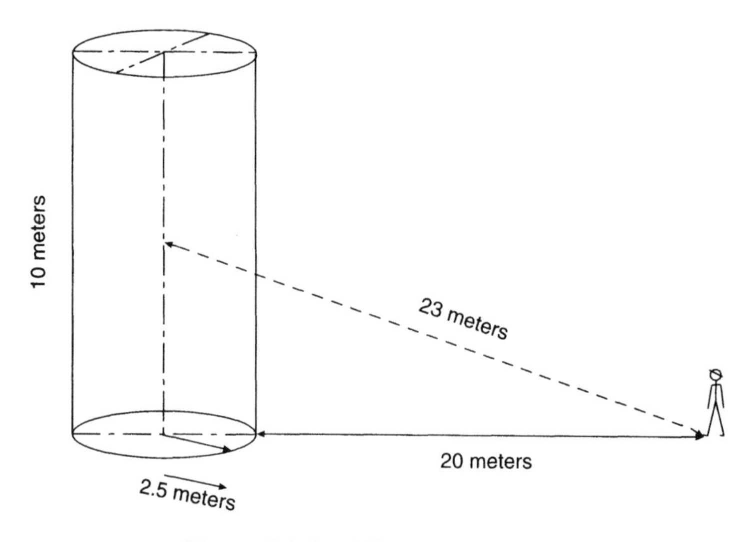

Figure 5-1. Pool fire of motor spirit.

7. Assuming that all the heat radiates from a single point at the effective center of the flame, calculate the distance from the point to the exposed person or equipment (m).
8. Assuming that the heat is radiated equally in all directions (untrue for all flames other than spherical ones), calculate the heat radiation intensity at that distance by dividing the radiated power by the surface area of a sphere of the radius to the exposed person or equipment.

The following example illustrates the basic approach.

Example 5.2.1: Pool Fire Heat Radiation: Point Source Method

A pool of motor spirit of 5 meters diameter is ignited. Calculate the heat radiation intensity at a distance of 20 meters from the edge of the pool.

Solution:

The configuration is as shown in Figure 5-1. The distance from the center of the pool to the observer is 22.5 meters. It is assumed here that the height of the flame is twice the diameter,[11] the height of the center of the flame is 5 meters above the pool. Thus the distance from the center of the flame to the observer is:

Radius $R = (5^2 + 22.5^2)^{1/2} = 23$ meters

[11] This is a common assumption, but formulae and empirical data are available for estimating the ratio of height to diameter for different materials and pool diameters. See CCPS (1999), Chapter 3.6.

Table 5-1
Suggested Burn-Down Rates for Pool Fires[2] (Various Sources)

Material	Rate of Burn-Down (mm/min)
Liquefied hydrocarbon gases (e.g., LPG)	8
Light hydrocarbon liquids (e.g., benzene, gasoline, xylene)	6
Heavy hydrocarbon liquids (e.g., fuel oil)	4
Flammable liquids with nonluminous flame (e.g., methanol)	4

The rate of loss of level in the pool is estimated from Table 5-1, assuming that motor spirit has a burn-down rate similar to that of benzene, which is 6 mm/min.[1]

With the density of motor spirit at 750 kg/m^3, the mass rate of combustion would be:

$$\text{Mass rate of combustion} = \text{pool area} \times \text{burn-down rate} \times \text{density}$$
$$= (3.142/4) \times 5 \times 5 \times (0.006/60) \times 750 \text{ kg/sec}$$
$$= 1.47 \text{ kg/sec}$$

With the heat of combustion of motor spirit at around 48,000 kJ/kg, and 35% of the heat radiated (Table 5-2), the rate of heat radiation (i.e., the radiant power of the flame) would be:

$$\text{Radiant power of flame} = 1.47 \times 48{,}000 \times 0.35 = 24{,}700 \text{ kW}$$

Table 5-2
Suggested Proportions of Heat Radiated from Flames[3] (Various Sources)

Material	Heat Radiated as % of Total Heat
Hydrogen	20
Hydrocarbons:	
C1 (e.g., methane)	20
C2 (e.g., ethylene)	25
C3 and C4	30
C5 and higher	40
Methanol	15
Gasoline	15

[12] These rates are greatly affected by the diameter of the pool and the wind speed at the time.

[3] Based on fires of around 1 to 3 meters diameter. These values reduce markedly as the diameter of the pool increases, due to the reduced efficiency of combustion and the resulting increase in soot formation, which obscures the flame from the observer. The values also depend markedly on the wind. See Lees p. 16/206, and Mudan and Croce in CCPS (1999), p. 217.

The surface area of a sphere of radius 23 meters is:

Area = $4 \times 3.142 \times 23 \times 23 = 6648\ m^2$

From this, the heat radiation intensity at a surface normal to the radius from the flame center is:

Heat radiation intensity = $24{,}700/6648 = 3.7\ kW/m^2$, that is, around $4\ kW/m^2$.

Inherent approximations in the point source method include:

- treatment of the flame as radiating from a single point, assuming uniform radiation in all directions (only true for a spherical flame);
- the estimate of the ratio of the height to the diameter of the flame;
- the rate of burn-down of liquid pools; and
- the average proportion of the total heat of combustion that is radiated.

Nevertheless, it is a commonly used approach for heat radiation in the far field. It is also more amenable to large-scale risk assessment computer programs where the risks from numerous postulated fires at various locations and with random orientations are assessed.

"View Factor" Method

The radiation intensity at a distance from a point source is inversely proportional to the square of the distance. The radiation intensity at a distance from an infinite line source is inversely proportional to the distance. The radiation intensity at a distance from an infinite plane source is constant.

The "view factor" method is based on knowing the radiation intensity at the surface of the flame, and reducing that in proportion to the ratio of the field of view from the exposed surface occupied by the flame, compared with the total field of view, adjusted for the angle of incidence of the radiation to the surface. This is illustrated by Figure 5-2.

By integrating the incremental cones of view from the point on the exposed plane surface to the flame, and adjusting for the angle of incidence of the center line of each cone onto the surface, the view factor is determined.

Tables and graphs of view factors have been published for a variety of shapes of flame and location of the exposed surface (Howell, 1982). An example is shown in Figure 5-4.

The intensity of heat radiation from flame surfaces (i.e. the surface emissivity) from the surface of flames of various materials is given in Table 5-3. These are average emissivities, including smoke obscuration, from pool fires. (Turbulent and more luminous flames would have higher emissivities.)

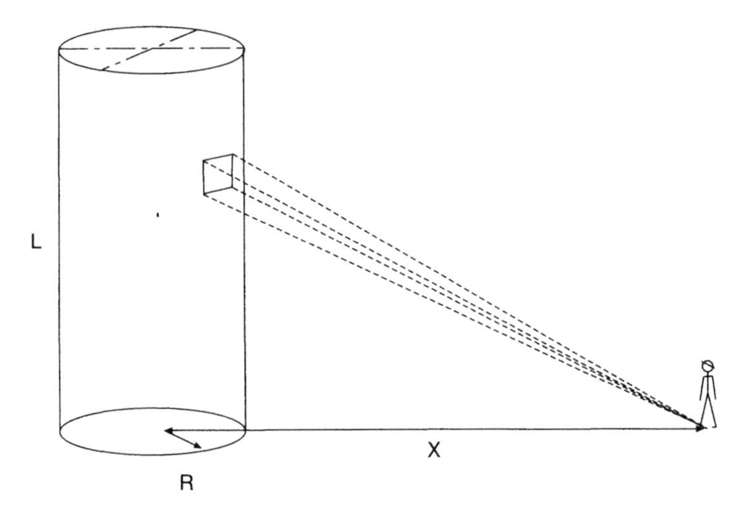

Figure 5-2. Illustration of view factor method.

Table 5-3
Suggested Average Surface Emissivity (i.e., Intensity of Radiation from the Surface) of Flames from Pool Fires (Various Sources)

Material	Surface Emissivity (kW/m^2)
Liquefied hydrocarbon gases	90
Light hydrocarbon liquids (e.g., benzene)	70
Ethylene oxide, propylene oxide	40
Vinyl chloride	25
Hydrogen sulfide	20
Organic liquids with low luminosity flame (e.g., methanol, carbon disulfide)	15
Carbon monoxide	15

Note: The surface emissivities of turbulent flames such as jet flames or fireballs can be much higher owing to more complete combustion. The emissivities can be in the range 150–300 kW/m^2.

Example 5.2.2: Heat Radiation from Pool Fire: View Factor Method

We use the same fire as in Example 5.2.1, with a pool of gasoline of 5 meters diameter. The exposed surface is at a distance of 20 meters from the edge of the pool. The configuration is as shown in Figure 5-3.

The radius X from the center of the pool to the exposed surface is 22.5 meters. The radius R of the pool is 2.5 meters. So the ratio $X/R = 9$. The height of the flame is assumed to be twice the diameter of the pool, so the ratio $L/R = 4$. Referring to Figure 5-4, the view factor is 0.03.

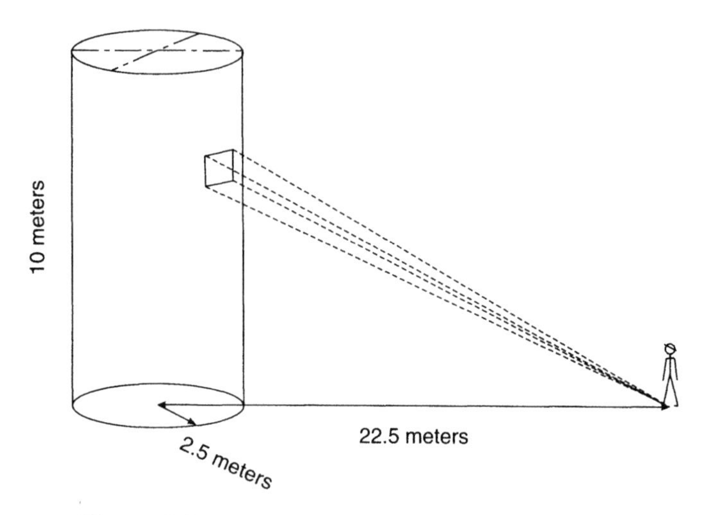

Figure 5-3. Pool fire example (view factor method).

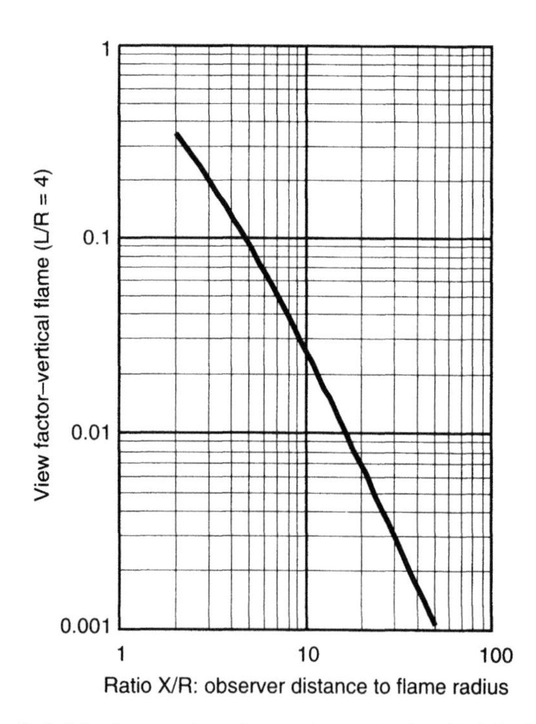

Figure 5-4. Maximum view factor for vertical cylindrical flame.

For gasoline, it is assumed here that the surface emissivity is the same as for benzene, that is, 70 kW/m^2 (Table 5-3).

Thus the maximum intensity of heat radiation at the exposed surface (perpendicular to the radiation from the flame) is:

$$\begin{aligned} \text{Intensity} &= 70 \times \text{view factor} \\ &= 70 \times 0.03 \\ &= 2.1 \end{aligned}$$

that is, around 2 kW/m^2. This can be compared with the value of 4 kW/m^2 obtained using the point source method.

The view factor method is said to be more reliable for short-range exposure, as it takes account of the finite size of the flame, which the point source method does not. However, there are approximations in the view factor method that affect its reliability. These approximations include the shape of the flame, the proportions of the flame, and the average surface emissivity. See Figure 5-5. (For more information, see Lees (1996), p. 16/207ff., and CCPS (1999), Chapter 3.6.)

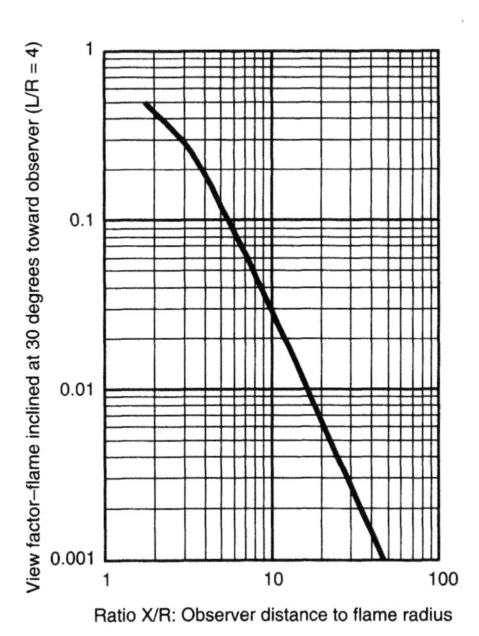

Figure 5-5. Maximum view factor for vertical cylindrical flame inclined at 30° toward observer.

5.2.3 Attenuation of Heat Radiation by Atmospheric Water Vapor

A method of calculating the attenuation of heat radiation by atmospheric water vapor (and carbon dioxide) is set out by Wayne (1991).[41] In summary, it is as follows:

Note that the constants in the formula below are selected for a flame temperature of 1500 K, which is stated to be midway between those of propane and LNG pool fires. It is also stated in the reference that the results are not very sensitive for different flame temperatures within a few hundred degrees of 1500 K.

$$\text{Transmissivity} = 1.006 - 0.01171 \log_{10} X_A - 0.02368(\log_{10} X_A)^2 - 0.03188 \log_{10} X_B + 0.001164(\log_{10} X_B)^2$$

where

$$X_A = X_{(H_2O)} = (RH \times L \times S_{mm} \times 2.88651 \times 10^2)/T$$

$$X_B = X_{(CO_2)} = L \times 273/T$$

and where

RH = fractional relative humidity (i.e., 0 to 1.0)

L = path length (meters)

S_{mm} = saturated water vapor pressure in millimeters of mercury at T kelvins

T = atmospheric temperature (kelvins)

Typical transmissivities are plotted in Figure 5-6. (Other approaches are set out by CCPS [1999], Chapter 3.4.)

5.2.4 Jet Fires

For more information, see CCPS (1999), Chapter 3.7, and Lees (1996), p. 16–221ff.

The following approaches can be used for assessing the radiation from a jet fire:

- Calculate the rate of escape of the fuel, the rate of heat generation due to combustion, and the heat radiated from the flame; then calculate the heat radiation using the point source method based on an estimate of the location of the center of the flame. (This method suffers from the assumption that the flame radiates equally in all directions, which in the case of a long, thin flame would not be correct.)

[14]Reprinted from Wayne, D. F. (1991): An economical formula for calculating atmospheric infrared transmissivities. J. Loss Prev. Process Ind., Vol 4, January with permission from Elsevier Science.

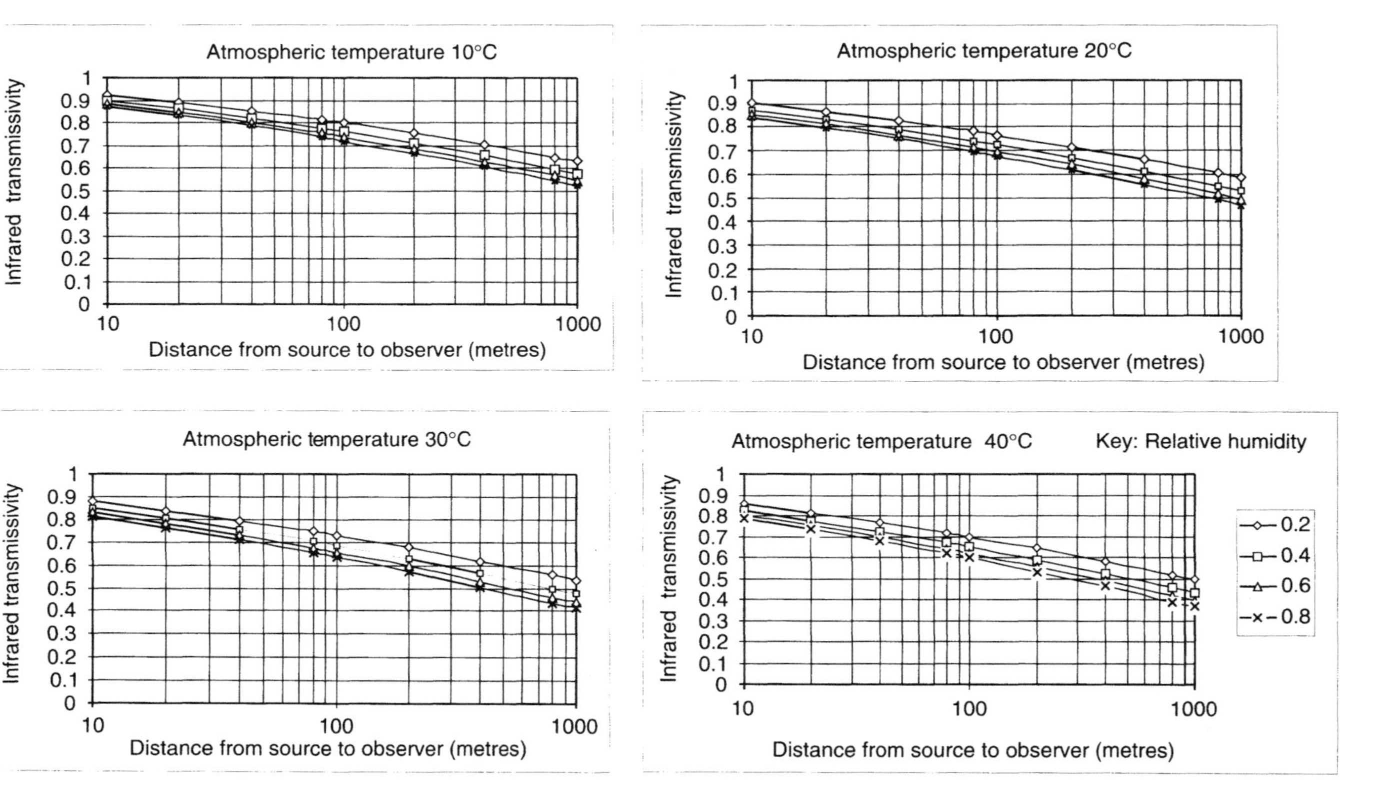

Figure 5-6. Atmospheric transmissivities for selected conditions.

- Estimate the shape and size of the flame, using an empirical relationship, then use the view factor approach.

These approaches are reviewed by Bagster and Pitblado (1988).

A review of the methods for estimating the length of a jet flame is set out by Bagster and Schubach (1995). CCPS (1999), Chapter 3.7, sets out approaches for estimating the length of the flame. Other methods have been determined by British Gas and Shell based on large-scale tests.

API-RP 521 (1996) suggests the following model:[51]

$$\text{Length} = 0.00326 \times (H_c \times \text{mass release rate})^{0.478}$$

where H_c is the heat of combustion of the escaping gas.

Methods for estimating the length of a flame from a flare and the tilt of the flame (commonly assumed to be 30° above the horizontal) are set out in API-RP 521 (1996). The flame length is estimated to be:

$$\text{Length} = 0.0177Q^{1.08}$$

where Q = heat release rate (MW)

The emissive power of a jet fire can be very high, for example, 350 kW/m^2, but is typically in the range 50–220 kW/m^2.

5.3 BLEVES

5.3.1 Introduction

BLEVE means "boiling liquid expanding vapor explosion" and is a rather unsatisfactory term for the sudden rupture of a vessel or system containing liquefied flammable gas (or flammable liquid held at pressure at a temperature above its atmospheric-pressure boiling point) due to fire impingement. The results are:

- a pressure blast due to the release of pressurized material in the vessel or system,
- missiles due to projected fragments of the vessel or system, and
- a fireball resulting from intense combustion of the turbulent mixture of escaped flammable vapor and liquid with the air.

Typically, the sequence of events is as follows:

1. A pressurized vessel of a liquefied flammable gas (such as an LPG stock tank) is exposed to fire.

[15] Reproduced by permission from API (1996): *Guide for Pressure Relief and Depressuring Systems*. API-RP 521, Edition 2, American Petroleum Institute, Division of Refining.

2. The vessel is heated by the fire. The pressure in the vessel rises as the contents are heated. Eventually the pressure relief valve operates, releasing the vapor evolved by the heat input to the vessel. The vapor is ignited by the fire and burns with a fierce, luminous, and highly radiant flame, but this flame may not cause any adverse effect of itself. The pressure is held at the relief valve set point.
3. The level of liquid in the tank falls as vapor escapes. Either because of flame impingement on the upper part of the tank above the internal liquid level, or because the level of liquid falls to below the level of the exterior flame, part of the vessel shell becomes overheated as a result of the lack of internal cooling by liquid. Its temperature rises to a point where it cannot withstand the stresses due to the internal pressure.
4. The tank ruptures. Note that this is not due to overpressure, but to overtemperature of the shell.
5. The fragments of the tank are propelled substantial distances. The contents instantaneously flash (partly) into vapor and ignite, producing a turbulent ball of very luminous flame which lasts for several seconds (up to around 15 depending on the mass of fuel involved), and which, after a few seconds touching the ground, rises into the air and becomes less luminous before burning out. There is a localized pressure blast wave due almost solely to the release of pressure in the tank, and barely to the combustion. The pressure blast is not damaging beyond the range of serious heat damage. (However, there may be a surge of low overpressure that rattles windows for some kilometers, without causing significant damage near the vessel.)

This scenario is shown in Figure 5-7.

A broadly similar scenario can occur with drums of solvent exposed to a fire for sufficient time for the contents to be heated above their atmospheric-pressure boiling point. A fire in a drum store can lead to large numbers of drums bursting, rocketing upward with a fireball of spray and vapor.

5.3.2 Calculation of the Diameter and Duration of a BLEVE

The two quantities that are calculated as a starting point for determining the effects of the heat radiated from a BLEVE are: the diameter and the duration. Various correlation studies have been carried out to determine the formulae appropriate to these quantities. Commonly used ones are:

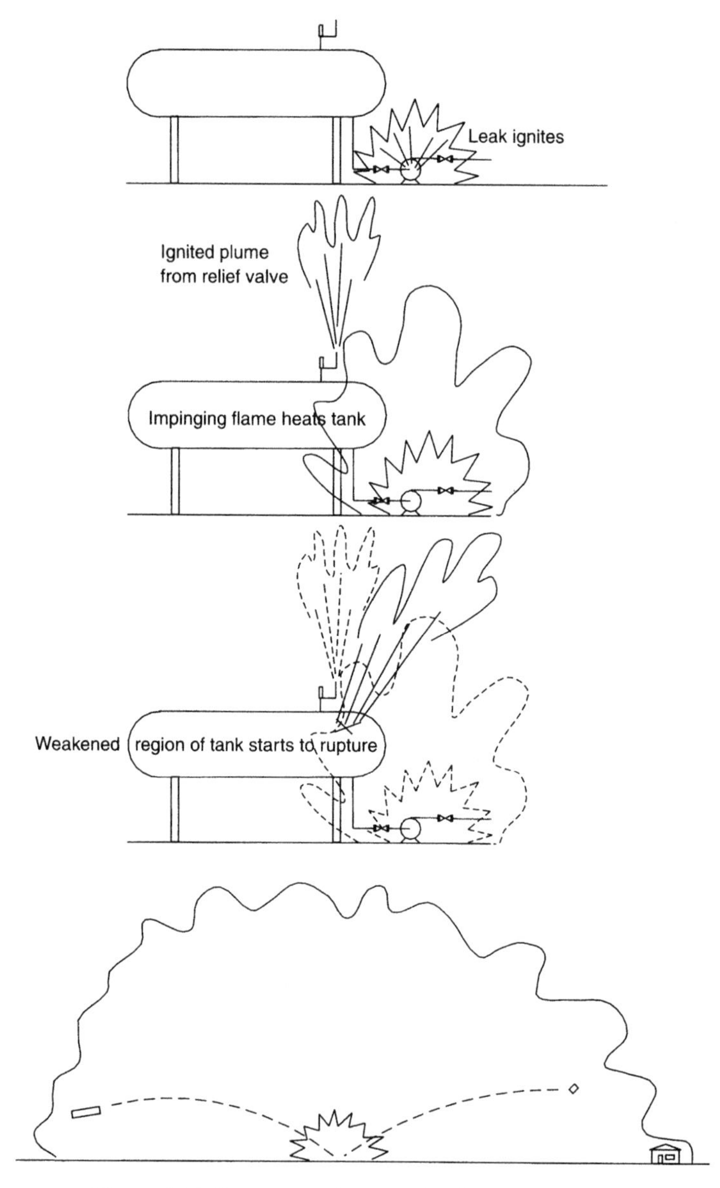

Figure 5-7. Sequence of events leading to a BLEVE.

$D = 6.48\,M^{0.325}$
$t = 0.852M^{0.26}$
where:
D = diameter (meters)
M = mass (kg) and
t = duration (sec)
(TNO, 1979)

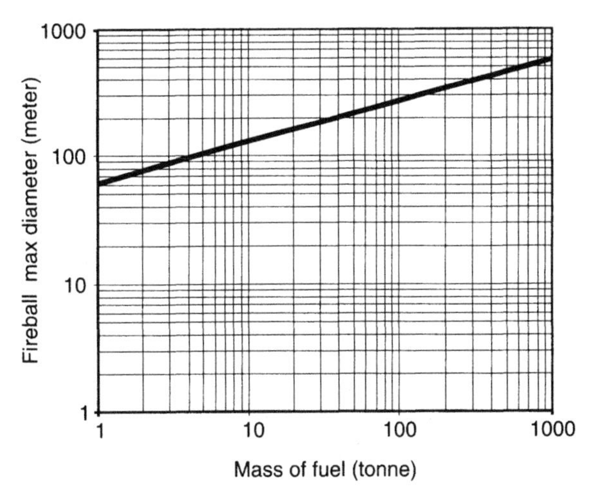

Figure 5-8. Maximum diameter of a BLEVE fireball.

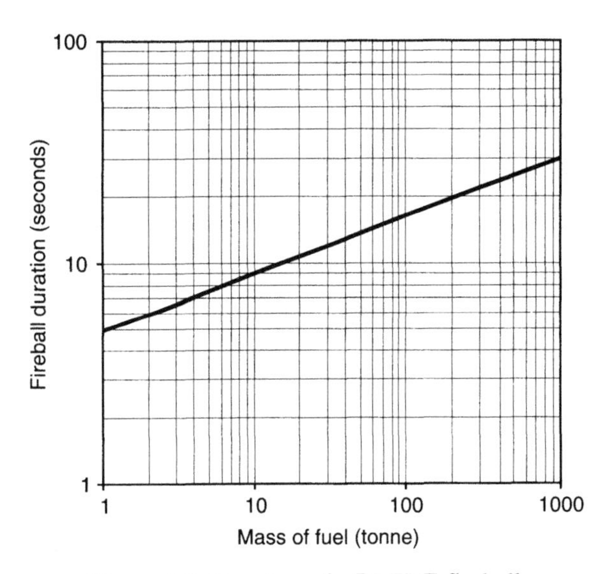

Figure 5-9. Duration of a BLEVE fireball.

Graphs of the TNO formula are displayed as Figures 5-8 and 5-9.

Details of tests undertaken by the UK Health and Safety Executive are set out by Roberts and others (2000), recording the diameters, durations, fireball emissivity, and heat exposures at various ranges. These results illustrate some of the uncertainty that exists about the fundamental properties of BLEVEs.

Example 5.3.1:

A 200-tonne stock tank of propane BLEVEs when one-third full. What is the diameter and duration of the fireball that results?

From the formulae quoted, or consulting the graphs, with $M=66{,}700$ kg, the maximum diameter could be around 240 meters, and the duration around 15 sec. But note that estimates of the durations and masses involved in actual BLEVEs show a wide scatter beyond the lines of "best fit." See Roberts *et al.* (2000).

5.3.3 Calculation of the Heat Radiation from a BLEVE

The heat radiation can be calculated either using the point source method, based on the mass of fuel being consumed and the duration of the fireball, or using the view factor method. Both are shown here.

5.3.3.1 Point Source Method

Note: Roberts *et al.* (2000) report that the fraction of the total heat of combustion that is radiated from a fireball ranges between 0.25 and 0.4. A suitable value for general consequence assessment is 0.35.

When calculating the heat radiation from a BLEVE, it is often assumed that the fireball is spherical and is just touching the ground. The configuration is as shown in Figure 5-10.

Example 5.3.2:

Calculate the heat radiation intensity at a distance of 300 meters from the center of a stock tank which BLEVEs when containing 67 tonnes of LPG.

From Example 5.3.1 above, the duration would be around 15 sec. With the heat of combustion of LPG being around 48,000 kJ/kg, and assuming

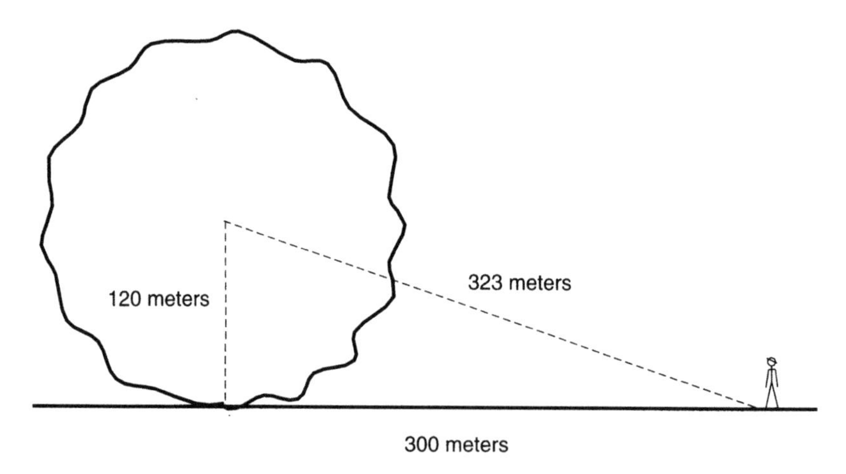

Figure 5-10. Typical configuration of BLEVE fireball used in consequence assessment.

a flame radiation efficiency of 35%[16] (i.e., assuming that 35% of the heat of combustion is radiated from the turbulent and luminous flame), the total heat radiated would be:

$$\text{Total heat radiated} = 66{,}700 \text{ kg} \times 48{,}000 \times 35/100 \times 1/15 \text{ kW}$$
$$= 74{,}704{,}000 \text{ kW}$$

If this is radiated evenly over the inside of a sphere of radius of 323 meters, the heat radiation intensity would be:

$$\text{Radiation intensity} = 74{,}704{,}000/(4 \times 3.142 \times 323 \times 323)$$
$$= 57 \text{ kW/m}^2 \text{ approx.}$$

The effect of attenuation of the heat radiation by atmospheric water vapor should also be calculated, as the distances to injurious levels of heat radiation are substantial, and attenuation may be significant.

Example 5.3.3:

Following the graphs shown in Figure 5-6, for a temperature of (for example) 20°C, a relative humidity of 50%, and an average distance from the fireball to the exposed person being around 250 meters, the transmission coefficient would be around 0.64.

Thus, the heat radiation intensity would be around $57 \times 0.64 = 36$ kW/m^2.

5.3.3.2 View Factor Method

Roberts *et al.* (2000) show that the shape of the fireball is influenced by the wind, and that the projected area of the fireball is greater when viewed crosswind than up- or downwind. Hence the view factor will vary accordingly. However, it is commonly assumed in consequence assessment that the fireball is spherical.

Roberts *et al.* (2000) note that the peak emissive power of the fireball, when averaged over the projected area at the time of peak emission, was in the range 270–333 kW/m^2 up- and downwind, and in the range 278–413 kW/m^2 crosswind. But this peak emission is of limited duration and does not apply at the time of greatest projected area. Hence it is suggested

[16] Alternatively, Crossthwaite *et al.* (1988) report that the radiation efficiency of the flame can be estimated from:

$$\text{Efficiency} = 0.27\ (P_s \times 10^{-6})^{0.32}$$

where P_s is the saturation vapor pressure of the liquefied gas (Pa, not kPa). Thus, for propane, the efficiency would be around $0.27\ (1{,}200{,}000 \times 10^{-6})^{0.32} = 29\%$.

Note that the intensity of the fireball ranges from a vividly luminous yellow-white flame to a dull red flame partly obscured by soot. The above efficiency of 29% would be the average. The most intense period of the flame would have a much higher efficiency.

that an emissive power of 250 kW/m^2 be used in consequence assessment, for the peak projected area and the full duration of the fireball. But it should be borne in mind that this is likely to be somewhat of an overestimate.

The maximum view factor of a sphere, not the view factor for a vertical or a horizontal plane, is:

$$F = \left(\frac{R_e}{R_r}\right)^2$$

where

R_e = emitter radius

R_r = radius to exposed point

F = view factor

In calculating the heat radiation, it is important to note that a BLEVE fireball, being highly turbulent and thus well supplied with air, is much more luminous than a flame from a pool fire, and for the early stages of the fireball the flame is not obscured by smoke. A value commonly used for the average surface emissivity over the duration of the flame is around 250 kW/m^2, although peak values ranging as high as around 400 kW/m^2 have been measured (Roberts *et al.*, 2000).

Example 5.3.4:

Calculate the heat radiation intensity at a distance of 300 meters from the center of a stock tank that BLEVEs when containing 67 tonnes of LPG. The configuration is as shown below.

Using the formula for the maximum view factor, and the results obtained for Example 5.3.1,

R_e = 120 meters; R_r = 323 meters

From this, the view factor is 0.138, and the heat radiation intensity would be:

$250 \times 0.138 = 34.5$ kW/m^2

The effect of attenuation of the radiation by water vapor should be applied as illustrated in Section 5.1 and is assumed to be as in Example 5.3.2 above, that is, a transmission coefficient of 0.64. Therefore the heat radiation intensity would be around:

$34.5 \times 0.64 = 22$ kW/m^2

The difference between the two results (36 and 22 kW/m^2) shows the uncertainty inherent in the methods.

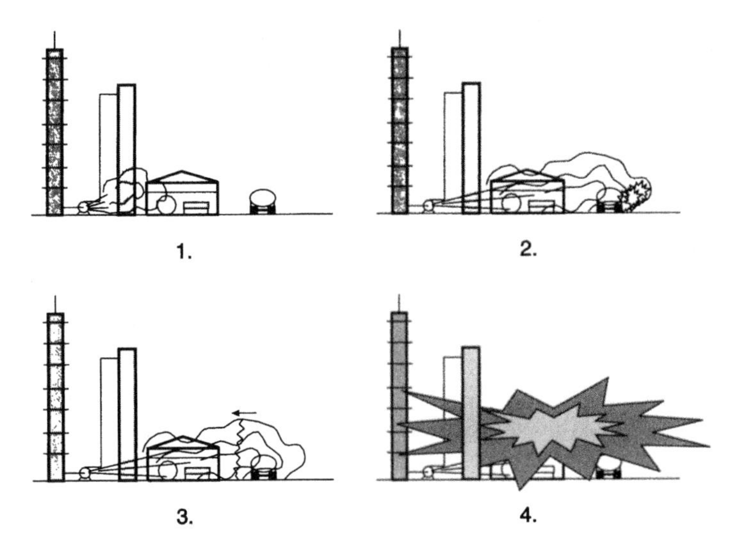

Figure 5-11. Typical initiation of a vapor cloud explosion.

A more precise but more complex approach is to calculate the total "thermal dose" in kJ/m^2 of an exposed surface, taking into account the time variation of emissive power of the fireball, and then comparing that with data linking thermal dose to the probability of fatality. Mudan (1984) sets out the effects of thermal doses.

5.4 VAPOR CLOUD EXPLOSIONS

5.4.1 Introduction

For many years it was believed that a cloud of flammable vapor, mixed with air, would not explode in the open air. It was believed that such a cloud would only flash, without a damaging blast wave.

Then attention was drawn to a number of incidents in which such a damaging blast wave appeared to have occurred.

Then, in 1974, the vapor cloud explosion at Flixborough in the United Kingdom removed all doubt. The control room of the plant was demolished by the blast (killing the 26 people inside), and most of the plant itself was badly damaged by both the blast and the subsequent fires. (See Chapter 13.)

A typical sequence of events leading to a vapor cloud explosion, shown in Figure 5-11, is described below.

1. A leak develops of a liquefied flammable gas, or a flammable liquid held at pressure at a temperature above its atmospheric-pressure boiling point. A proportion of the escaping liquid flashes instantaneously into

vapor, and much of the resulting spray of the unvaporized material absorbs sensible heat from the air and also evaporates effectively at once.

2. The vapor drifts with the breeze, mixing with the surrounding air, and the total mass of flammable vapor present in the cloud increases as the leak continues.
3. The vapor–air cloud reaches an ignition source, and in due course the concentration of the vapor rises above the lower flammable limit. The cloud ignites.
4. The flame front in the cloud accelerates (by a mechanism discussed below)...
5. ... to near sonic velocity, and generates a percussive shock wave.

Research on large vapor clouds during the 1980s suggested strongly that a cloud of most types of flammable vapor mixed with air will not explode if truly unconfined and unobstructed, no matter how large it is, but that the presence of obstacles leads to explosive rates of combustion in their vicinity. It is also suggested that the flame front slows down once it is clear of the obstacles. The obstacles create turbulence, which greatly increases the surface area of the flame front, and hence the mass rate of combustion and the rate of expansion of the burning cloud.

(Highly reactive gases, such as hydrogen, or vapor clouds which are highly turbulent, e.g., as a result of high-velocity escapes, can exhibit the same behavior as a congested cloud and can explode with a damaging blast wave.)

The effect of congestion, based on testing undertaken by British Gas Corporation, is illustrated by Figure 5-12.

This is a very important finding. The significance of it is that a leak from an isolated tank farm would be most unlikely to explode, and that an explosion of a cloud in a built-up plant area would involve only the mass of flammable vapor in the cloud within that built-up area. In that case, the energy released in the shock wave (the "power" of the explosion) would be related, not to the total mass of fuel released, or the total mass of fuel in the part of the cloud between the lower and upper flammable limits, but rather to the mass of fuel within the cloud within the lower and upper flammable limits within the bounds of the structure of the plant.

Although the pressures developed by a vapor cloud explosion in the open air do not usually rise sufficiently to be lethal to people directly, vapor cloud explosions cause fatalities in the following ways:

- by projecting missiles at people,
- by throwing people against solid objects,

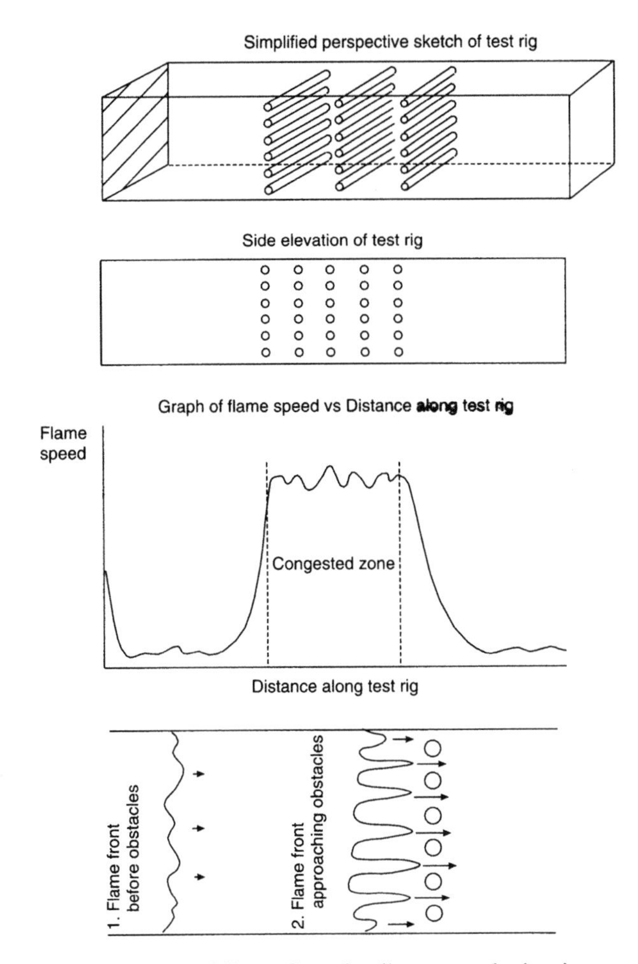

Figure 5-12. Acceleration of flame front leading to explosion in congested area.

- by collapsing buildings, and
- by enveloping people in the burning cloud.

5.4.2 Calculation of the Effect of an Unconfined Vapor Cloud Explosion

5.4.2.1 Outline of "TNT Equivalence Method"

There are various methods in use for calculating the effect of an unconfined vapor cloud explosion.

In essence, in the "TNT Equivalence Method" the mass of material in the cloud is converted to an equivalent mass of TNT, and the effect of the explosion determined from data from the extensive testing of TNT.

The main limitation of the TNT Equivalence Method is that, although it is a very satisfactory method for assessing the effects of a solid explosive, the behavior of an exploding gas cloud is very different, and the TNT Equivalence Method is not reliable in the near field. However, it is a simple method that is reasonably reliable in the far field.

The steps are as follows:

1. Calculate the likely mass of flammable vapor in the cloud in a confined or congested area:
   - calculate the mass of flashing liquid likely to escape, and
   - calculate the mass of vapor formed by flash or effectively immediate evaporation of that mass.
2. Calculate the equivalent mass of TNT.
3. By reference to a graph of overpressure for TNT, determine the explosive overpressure at the required radius.
4. From empirically determined tables of the effects of overpressure, determine the likely effect.

5.4.2.2 Calculation of the Mass of Vapor

The method illustrated below is for an escape of a flashing flammable liquid or liquefied flammable gas, such as LPG.

A similar approach is used for escapes of flammable gases, although different formulae would be used, and there is no need for the step of calculating the amount of vapor formed from escaping liquid.

1. Calculate the Mass of Liquid That Escapes

In the event of catastrophic failure of a vessel (rare), escape of the entire contents would probably be assumed.

In the event of a leak from a hole in a pipeline or a vessel, the rate of escape would be calculated using the usual formula for single-phase liquid flow:

$$G_L = C_d \cdot A\sqrt{2\rho \cdot \delta P}$$

where

G_L = mass flow rate (kg/sec)

C_d = discharge coefficient. (0.8 for smooth circular hole, 0.6 for irregular hole)

A = area of hole (square meters)

ρ = liquid density (kg/m^3)

δP = pressure difference across hole (Pa, not kPa!)

Example 5.4.1:

A propane stock tank holds 200 tonnes of propane at 25°C. The delivery pipeline of 100 mm dia fails totally at the nozzle of the vessel, and liquid propane escapes from the full diameter of the pipeline, under a pressure of 1250 kPag. At what rate does the liquid escape? (The density of the liquid is 492 kg/m^3.)

Solution:

The cross-sectional area of the pipe is:

$A = 0.1 \times 0.1 \times 3.142/4 = 0.00786\,\text{m}^2$

The rate of escape is:

$$G_L = 0.8 \times 0.00786\sqrt{2 \times 492 \times 1{,}250{,}000} = 220\,\text{kg/sec}$$

For methods of calculating the rate of escape of pressurized gases from holes in vessels, or from ruptured pipework, see Perry and Green (1984). For methods of calculating the rate of two-phase (liquid–vapor) escape of flashing or superheated liquids, see Morris (1990) and Leung (1990).

2. Calculate the Mass of Vapor Formed

When the pressure is removed from liquefied propane at atmospheric temperature, enough of it evaporates instantaneously and adiabatically ("flashes") for the latent heat of vaporization used in the evaporation to chill the remaining liquid to the atmospheric-pressure boiling point of the propane.

It is also commonly assumed that the spray of liquid thus formed partly evaporates as a result of absorption of sensible heat from the air, and that the proportion which evaporates is equal to the amount which flashes instantaneously. Thus the total amount of vapor formed is assumed to be twice the adiabatic flash.

An approximate formula for calculating the adiabatic flash is:

$$V = \frac{W \cdot C_{P(mean)} \cdot (t_1 - t_2)}{h_v}$$

where

V = weight of flash vapor produced (kg)

W = weight of liquid spilled (kg)

$C_{p(mean)}$ = geometric means of specific heats over range t_1 to t_2 (J/kg°C)

t_1 = temperature of the liquid in the process (°C)

t_2 = atmospheric-pressure boiling temperature of the liquid

h_v = latent heat of vaporization (J/kg)

Example 5.4.2:

At what rate is propylene vapor formed from the leak postulated in Example 5.4.1 above?

Assume that

$C_{p(mean)} = 2.3$ kJ/kg°C

$t_1 = 25$°C

$t_2 = -42$°C

$h_v = 420$ kJ/kg

$$V = \frac{220 \times 2.3 \times (25 + 42)}{420}$$

$$= 80.7 \text{ kg/sec}$$

This is doubled to include the effect of evaporation of spray, from which the rate of formation of vapor is estimated to be around 160 kg/sec.

In practice, it is reported that leaks of liquid propane, propylene, and other liquefied petroleum gases with lower atmospheric-pressure boiling temperatures rarely result in a pool of unevaporated liquid on the ground. Thus the above calculation is probably somewhat optimistic, that is, low.

5.4.2.3 Calculation of the Equivalent Mass of TNT

The first step is to determine the amount of vapor which is present in the cloud when it is ignited.

In the case of an instantaneous escape from a (rare) catastrophic failure of a vessel, the calculation is straightforward, being the mass of vapor formed by release of pressure on the total inventory of liquid in the vessel, determined as above.

In the case of a continuing escape, such as from a hole in a vessel or a pipeline, the task is to estimate the duration of the leak before the cloud reaches an ignition source.

In a production plant, it is often difficult to eliminate ignition sources entirely, because of the need for furnaces, etc. A common assumption, used where there is no better basis, is that if a cloud in an operating plant does not reach an ignition source within around 3 min, then ignition is unlikely. In the event of a breeze of (say) 3 m/sec, this is equivalent to an ignition source at a distance of 540 meters, which is a large distance in practical terms, in that most leaks would have dispersed to below the lower flammable limit before that distance in most atmospheric conditions.

The next step is to determine the equivalent mass of TNT. This is done using the following formula:

$$M_{TNT} = E\,\frac{M_{vap} \cdot H_c}{4600}$$

where

M_{TNT} = Equiv mass of TNT

M_{vap} = Mass of vapor

H_c = Heat of combustion of vapor (kJ/kg)

(The heat of combustion of TNT is around 4600 kJ/kg.)

E = Explosion efficiency compared with TNT

There are various approaches taken to the relative explosion efficiency. Some people prefer to use 10% or more, based on an estimate of the mass of vapor within the flammable range, rather than the lower efficiency for the entire mass of vapor. Where the vapor is confined—for example, in a building such as a compressor house—and a higher proportion than otherwise may be within the flammable range, then a higher relative explosion efficiency than 4% may be used, such as 10–20% of the entire mass. But it is difficult to be at all certain about the figure to be used in the case of total confinement.

As it is now widely accepted that only that part of a vapor–air cloud which is within a confined or congested area will contribute to the explosive blast effect, one method of estimating the equivalent mass of TNT is to base the assessment on a stoichiometric mixture fully occupying the congested area, and using an explosion efficiency of around 10–20%.

Example 5.4.3:

Propane vapor from a liquid leak forms a cloud at a rate of 160 kg/sec for 3 min before being ignited. Calculate the equivalent mass of TNT.

The total mass of vapor would be:

$160 \times 60 \times 3 = 28{,}800$ kg

The equivalent mass of TNT would be (assuming the heat of combustion of propane to be around 50,000 kJ/kg):

$$M_{TNT} = E \frac{M_{vap} \times H_c}{4600}$$

$$M_{TNT} = 0.04 \frac{28{,}800 \times 50{,}000}{4600}$$

$$= 12{,}521 \text{ kg}$$

That is, the equivalent mass of TNT is estimated to be around 12.5 tonnes, that is, probably somewhere between 10 and 15 tonnes.

IChemE (1994) suggest that the following efficiencies be used:

Material	Energy Ratio[a]	Efficiency E	TNT Equivalence
Acetylene oxide	6.9	0.06	0.4
Ethylene oxide	6	0.10	0.6
Hydrocarbons	10	0.04	0.4
Vinyl chloride	4.2	0.04	0.16

[a] Ratio of heat of combustion of the material to that of TNT.

They also recommend that an efficiency of 0.04 be used for other flammable gases except for highly reactive gases such as acetylene and hydrogen, where an (unspecified) higher efficiency would be preferred.

5.4.2.4 Determination of Explosive Overpressure

The explosive overpressure is determined from Figure 5-13. In using this graph, it is necessary first to determine the "scaled distance," where:

Scaled distance λ = Radius (meters)/(TNT mass (kg))$^{0.333}$

Then the overpressure is read off the graph.

Note that there is still debate about the overpressure inside the exploding cloud. Some feel that the overpressure does not generally exceed around

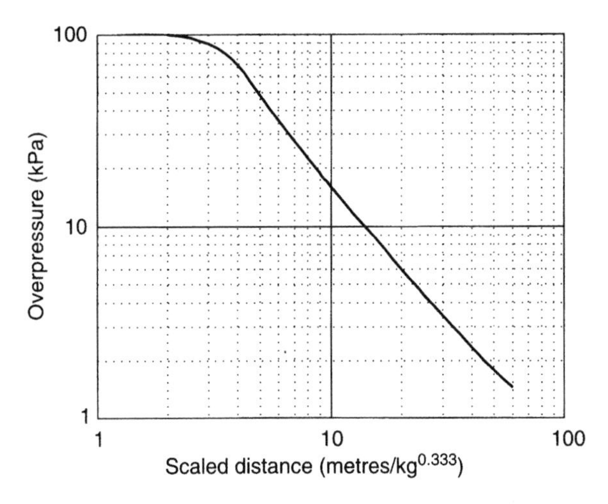

Figure 5-13. Incident overpressure for ground-level explosion.

100 kPa. Others believe that there are local points within the cloud, in regions of high turbulence, where the overpressure is much higher, and that the damage within the cloud is due largely to a large number of small zones where the cloud explodes very violently. To be conservative, it would be prudent to assume that the above curve continues upward following the general trend, rather than flattening off at around 100 kPa.

Example 5.4.4:

Determine the explosive overpressure at a distance of 200 meters from the center of a vapor cloud explosion with an equivalent TNT mass of 12.5 tonnes.

Solution:

$$\text{Scaled distance} = 200/(12{,}500)^{0.333}$$
$$= 200/23.2$$
$$= 8.6$$

Consulting Figure 5-13, the overpressure is found to be around 20 kPa.

5.4.2.5 Determination of Effect

Based on military experience, tables have been prepared of the degree of damage resulting from defined levels of overpressure. These tables are thus based on short-duration shock waves typical of TNT, rather than the longer duration pressure waves typical of vapor cloud explosions, but as the explosive efficiency used in calculating the equivalent mass of TNT is based on observation of effects from historical vapor cloud explosions, the table is a reasonable guide. See Table 5-4.

For use in quantitative assessment of risks to people, a graph (Figure 5-14) is given, indicating very approximately the risk to people exposed to an unconfined vapor cloud explosion in the open air or in a conventional building. The graph is necessarily very approximate. The functions are:

Risk in open (%) = $0.0003 \times$ (Overpressure kPa)^3

Risk in conventional building (%) = $0.04 \times$ (Overpressure kPa)^2

Note, however, that these graphs may overstate the risk of fatality at low overpressures. For example, the fatality risk in building at an overpressure of 10 kPa is calculated to be around 4%. This is probably excessive.

Table 5-4
Effects of Overpressure

Overpressure (kPa)	Effects
70	Regarded by some as a reasonable estimate of the overpressure at the edge of cloud. Anyone closer would probably be killed by direct flame envelopment. Outside the cloud, a person would be violently projected and may be fatally injured as a result. A person in a conventional masonry building would probably be killed by collapse of the walls and roof.
	There is severe structural damage, with steel structures severely distorted and displaced. It may be possible to salvage some large or solid equipment.
30	A person in open air may be killed by flying debris, etc., or by being thrown violently against a solid object. A person within a masonry building would have a substantial probability of being killed because of structural collapse. A person in the open would possibly suffer rupture of eardrums, but the overpressure as such would not cause any physiological damage such as crushed chest.
	There is serious structural damage around and above this overpressure. A conventional masonry building would be severely damaged and probably would need to be demolished.
15	A person in the open would not expect to be injured except by missiles. A person in a conventional building could be injured or conceivably killed by flying glass, collapsed ceilings, etc. A house would be rendered uninhabitable, but mostly could be repaired. Atmospheric-pressure storage tanks may be damaged, roofs pushed in, etc.
8	Glass fragments may be flung with sufficient force to injure. Roof tiles are damaged, and nearly all windows broken.
4	90% breakage of windows; little structural damage.

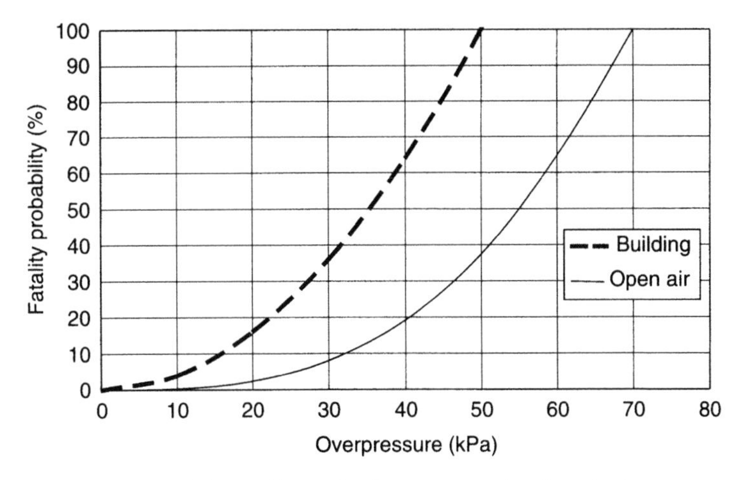

Figure 5-14. Fatality risks from vapor cloud explosion.

IChemE (1994) published a "survivability" graph relating the probability of fatality directly to the scaled distance. It makes no distinction between the probability of fatality in the open air and that in a building.

International Oil Insurers, London, published a guide (IOI, 1979) for estimating the damage to a plant from a postulated vapor cloud explosion. They define two damage circles: one representing an area within which the damage will average 80% of the replacement cost of the plant within that circle, and a larger circle such that the annulus between the two circles will suffer damage averaging 40% of the replacement cost of that plant, with the average of all damage within the larger circle being around 50% of the replacement cost of that plant.

They published tables of the radii, which corresponded to the following equations:

Radius (80% average damage) $= 33.4M^{0.333}$

Radius (40% average damage) $= 66.8M^{0.333}$

where radius is in meters, and M (tonnes) is the mass of the fuel in the cloud.

5.4.3 The British Gas VCE Method

This is described in IChemE (1994). In essence, it is a variation of the TNT Equivalence method, but it assumes that only the part of the flammable cloud that is in a congested area contributes to the blast. That part of the cloud is assumed to explode with a higher efficiency, typically

20%, than is assumed in the normal TNT Equivalence method, typically 4%. Thus, for a typical hydrocarbon with a heat of combustion around 10 times that of TNT, the TNT equivalent mass of the part of the cloud in the congested area becomes:

$$M_{TNT} = (0.2 \times 10) M_{CONGESTED} = 2 M_{CONGESTED}$$

5.4.4 The Multienergy Model

During the past 30 years there has been a great deal of research into the effects of vapor cloud explosions, particularly in the United States and Europe. Although it is widely agreed that the TNT equivalent method is reasonable for the explosion effects in the "far field," the fact that the effect of the combustion of a vapor cloud depends more on the manner of combustion (e.g., turbulent, confined) than on the mass involved has led to attempts to model the behavior to take account of the conditions in which the combustion occurs. Thus, instead of just one curve for blast effect versus distance, it is necessary to develop a family of curves, each for a particular type of condition.

One such approach, which has been widely accepted as reasonable, is the "multienergy model" (Van den Berg, 1985) developed by the TNO in The Netherlands for determining the overpressure at a distance from a vapor cloud explosion within a confined or congested area. (Since then, a variety of other models have been developed by other researchers, but, like the multienergy model, all are subject to further refinement and validation.)

In the multienergy model, the vapor cloud explosion is not regarded as a single event, but rather as a number of subexplosions, each derived from a source of turbulence in the burning cloud and separated slightly in time as the flame front travels from one source of turbulence to the next. In effect, each obstacle to free expansion of the burning cloud generates a wake of turbulence, which becomes the source of accelerated combustion and percussive overpressure. Thus the total overpressure generated by a vapor cloud explosion in a congested area is the sum of a number of small peaks of overpressure, spread over the period in which the combustion moves through the congested area. Van den Berg (1985) and Van den Berg *et al.* (1991) report that the blast effects are determined by the quantity and intensity of the combustion process, which are in turn dependent primarily on the size, shape, and nature of the partially confined and obstructed space, with the reactivity of the fuel–air mixture being less important.

The method is described below. It is important to note that the method relies on use of a family of curves of "Charge Strength Number," which is determined by the environment, that is, the extent of confinement and congestion. Research is continuing to clarify which curve to use for any particular application. While Charge Strength 6 or 7 may be appropriate for congested outdoor plant and Charge Strength 10 for full confinement, a conservative approach is to use Charge Strength 10 for all situations unless there is a clear case for using a lower value.

The method requires data on:

- the physical volume of the fuel–air cloud (enabling determination of the radius of a hemisphere of the same volume);
- the total combustion energy of the fuel in the cloud (E);
- the nature of the confinement or congestion (which determines the "charge strength number") from 1 to 10;
- the atmospheric pressure (P_o), typically 105 Pa; and
- the speed of sound in the atmosphere (c_o), typically around 335 m/sec.

The steps are:

1. Estimate the volume of the flammable vapor–air mixture that would be involved in the postulated explosion, for example, 1000 m^3.
2. Determine the radius of the hemisphere with the same volume as the postulated cloud. For example:

   $$R_{hemisphere} = (3 \times 1000/2 \times 3.142)^{0.333} = 7.8 \text{ meters}$$

3. Determine the "Combustion Energy Scaled Distance." For example: Assuming a stoichiometric composition of the vapor cloud, the typical combustion energy of a hydrocarbon-air mixture is around 3.5 MJ/m^3. Therefore the combustion energy is around $3.5 \times 1000 = 3500$ MJ $=$ 3.5×10^9 J.

   The Combustion Energy Scaled Distance for a radius to the exposed person or facility of, say, 500 meters, is:

   $$\bar{R} = R/(E/P_o)^{0.333}$$

   $$\bar{R} = 500/(3.5 \times 10^9/1 \times 10^5)^{0.333} = 500/32.6 = 15.4$$

4. Determine the "Dimensionless Maximum Static Overpressure." For example: Referring to the graph of scaled distance versus dimensionless

static overpressure (Figure 5-15), and using a "Charge Strength" of 6 or more (i.e., substantially congested), the dimensionless overpressure is around 0.013.

Thus, with an atmospheric pressure of 100 kPa, the blast overpressure would be around 1.3 kPa.

5. Determine the Dimensionless Positive Phase Duration. For example: Referring to the graph of scaled distance versus dimensionless positive phase duration (Figure 5-16), using any Charge Strength of 6 or more, the dimensionless duration is around 0.5.

Now, as:

$$\bar{t}_+ = \frac{(t_+ \times C_o)}{(E/P_o)^{0.333}}$$

then:

$$t_+ = [\bar{t}_+ \times (E/P_o)^{0.333}]/C_o$$

That is:

$$t_+ = [0.5 \times 32.6]/335$$
$$= 0.0486,$$

that is, around 0.05 sec

Summary of Equations

$$\Delta\bar{P}_S = \frac{\Delta P_S}{P_o}$$

$$\bar{t}_+ = \frac{t_+ \cdot C_o}{(E/P_o)^{0.333}}$$

$$\bar{R} = \frac{R}{(E/P_o)^{0.333}}$$

where:

P_o = atmospheric pressure (Pa)

C_o = atmospheric speed of sound (m/sec)

E = amount of combustion energy (J)

R_o = radius of hemispherical fuel–air cloud (m)

The Multienergy method is still subject to further research and validation. It is regarded by some authorities as being a better model than the TNT Equivalent method.

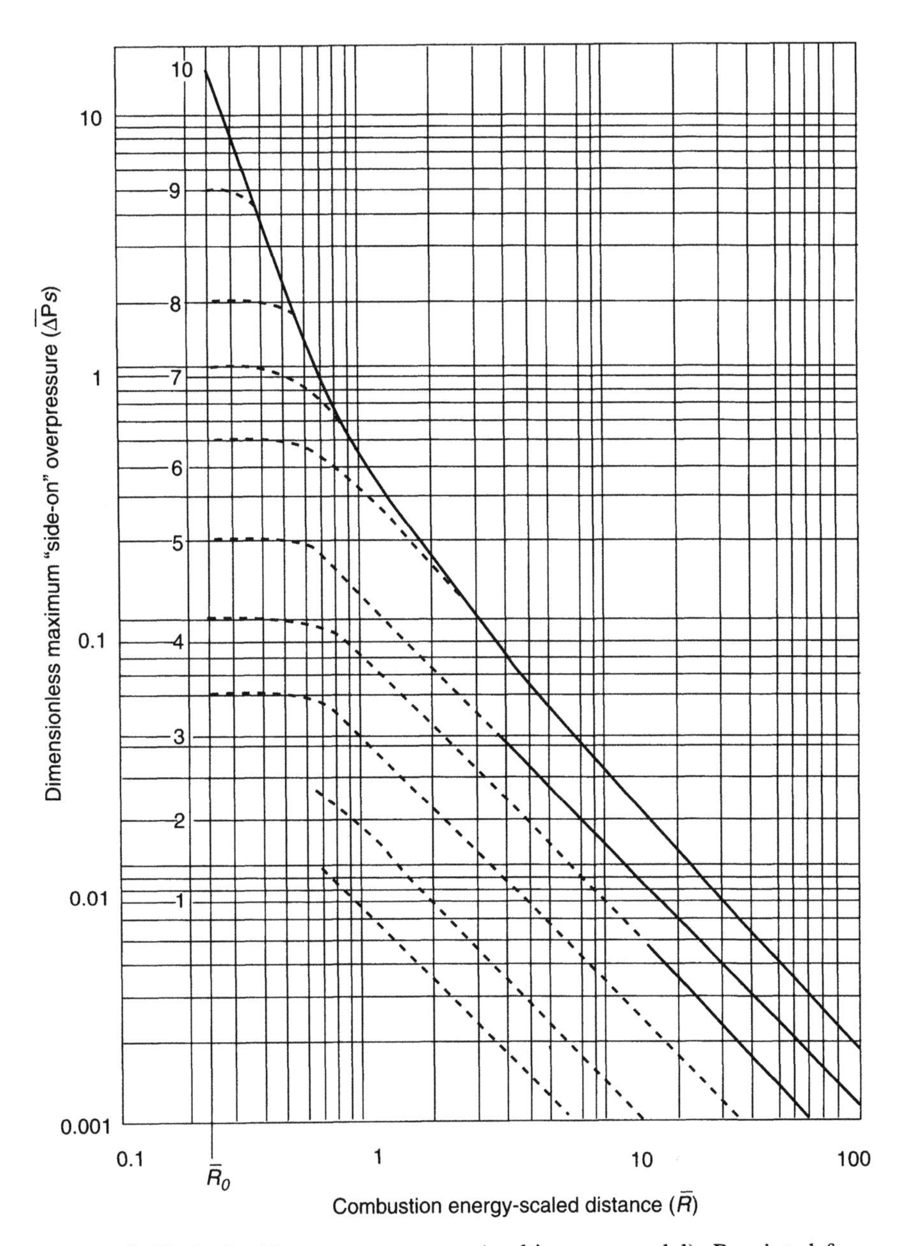

Figure 5-15. Peak side-on overpressure (multienergy model). Reprinted from Van den Berg, A. C. (1985): *The Multi-Energy Method—a framework for vapour cloud explosion blast prediction. J. of Hazardous Materials*, Vol. 12, pp. 1–10 (with permission from Elsevier Science and TNO).

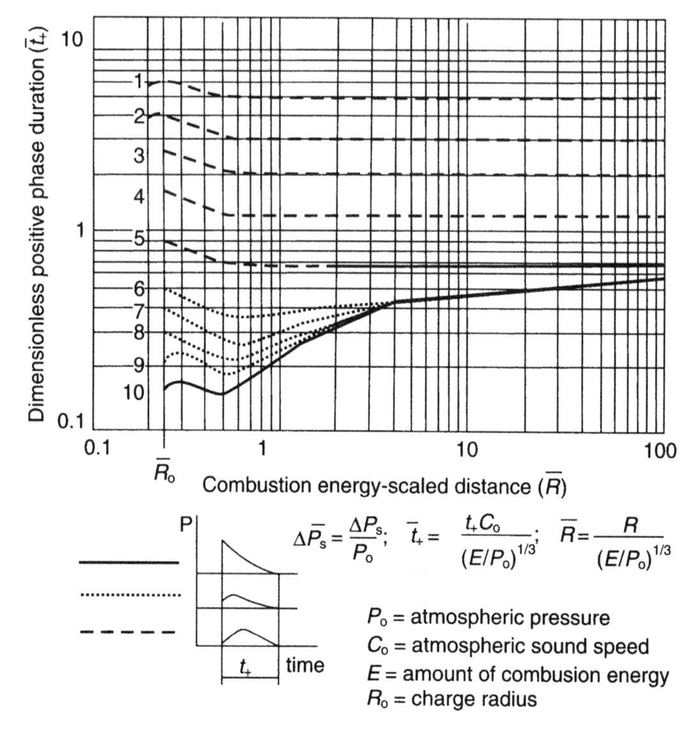

Figure 5-16. Positive phase duration (multienergy model). Reprinted from Van den Berg, A. C. (1985): *The Multi-Energy Method—a framework for vapour cloud explosion blast prediction. J. of Hazardous Materials*, Vol. 12, pp. 1–10 with permission from Elsevier Science and TNO.

5.4.5 Assessment of Frequency of Explosions, and Resulting Risks

The steps used in assessing the frequency of explosions are, in principle, as follows:

1. By analysis of the possible causes of leaks (e.g., pipe failures, operating errors) and use of data banks or by estimation using fault trees, etc., estimate the frequency of leaks of the postulated size.
2. Determine the size of the vapor cloud (as earlier).
3. Determine, for that size of cloud,
   - the probability of ignition, and
   - the probability of explosion if ignited.
4. Multiply the frequency of leak, the probability of ignition, and the probability of explosion if ignited, giving the frequency of explosions.

The first step is covered more fully in Chapter 6. The second step is covered earlier in this section. Several methods have been adopted for the third step.

(a) One method is to define ignition points on the plant by inspection, and to allot to each a probability that a flammable cloud will be ignited if it reaches the ignition point, then to determine the distance from the leak point to the lower flammable limit for that material for various weather conditions (see Section 5.6), then to use meteorological records to determine the probability of conditions that would result in the cloud reaching each ignition point in the various directions while still above the lower flammable limit. (This method assumes that it is possible to model the dispersion of gases within a built-up plant, which is not possible without CFD methods or wind-tunnel tests.)

(b) Another method is to use generic data based on analysis of historical incidents. A typical example is shown in Section 6.9.4.

(c) Another approach is entirely empirical. From reported estimates, graphs have been derived attempting to suggest the probability of ignition, and of explosion if ignited, of vapor clouds that have occurred in operating plants, storages, etc. One such set is shown as Figure 5-16 and is based on estimates from various sources:

- a cloud of zero tonnes has zero probability of ignition and of explosion;
- a cloud of 100 tonnes is certain to ignite;
- an ignited cloud of 50 tonnes has a 50% probability of exploding;
- a cloud of 1 tonne has around a 1 in 1000 chance of igniting and exploding;
- the probability of a cloud being ignited is proportional to the volume occupied by the cloud, that is, is proportional to the mass of the cloud;
- the probability of an ignited cloud exploding is proportional to its volume, as that determines the probability that it will contain congestion that will enable the flame front to accelerate to explosive speed.

From this the following indicative relationships may be derived:

Probability of ignition: $P_{ign}\ (\%) = M$ where M is mass of fuel in tonnes

Probability of exploding if ignited: $P_{exp}\ (\%) = 0.9M$

Probability of igniting and exploding: $P_{ign+exp}\ (\%) = P_{ign} \times P_{exp}$

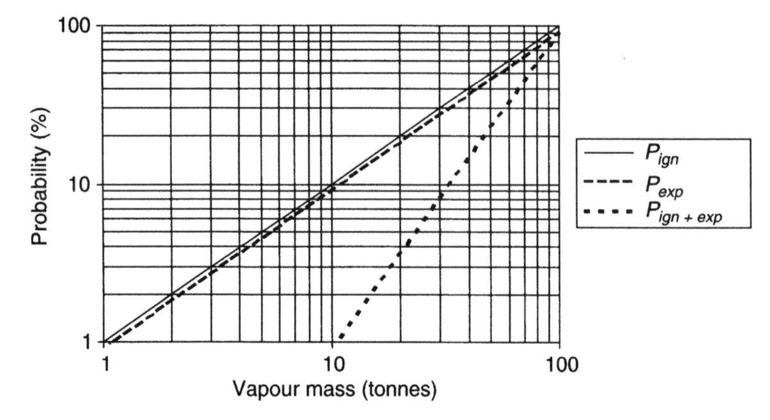

Figure 5-17. Suggested probability of ignition and explosion of unconfined vapor clouds in open-air operating plant.

See Figure 5-17. (Note: The graph is purely a rule-of-thumb estimate. It should be used only when better estimates based on local conditions are not available.)

Example 5.4.5:

It is estimated that a vapor cloud of 28,800 kg of propane would be formed in a production plant with a probability of 0.001 per year due to various causes. Given that it has a TNT equivalent mass of 12,500 kg, calculate:

(a) the probability of ignition and of explosion;

(b) the frequency of explosion;

(c) the risk to a person in a house at a distance of 200 meters, assuming that the cloud explodes at the point of release;

(d) the radius to the edge of the flash fire, assuming that the cloud does not explode if ignited; and

(e) the risk to a person in the open within the radius of the flash fire.

Solution:

(a) From Figure 5-16, the probability of ignition would be around 29%, and the probability of explosion if ignited would be around 26%.

(b) Thus, the probability of ignition and explosion would be around 7.5%. If the frequency of leaks is 0.001 per year, that is, around once per thousand per year on average, then the frequency of explosions would be:

$$0.001 \text{ per year} \times 0.075 = 0.000075 \text{ per year}$$

that is, around once per 13,000 years on average, or around 7.5 chances per 100,000 per year.

(c) Using the overpressure calculated in Example 5.4.4 (16 kPa), and using Figure 5-12, the probability of fatality for a person in a house exposed to 16 kPa would be around 12%. Therefore, with the explosion frequency of 0.000075 per year, the risk to the person would be:

$$0.000075 \text{ per year} \times 0.12 = 0.000009 \text{ per year},$$

that is, around 9 chances per million per year. (This can be compared with typical risk criteria and may suggest a need to reduce either the frequency of explosions, or their magnitude, or both.)

(d) Using the value of the TNT equivalent mass calculated in Example 5.4.3 (12,521 kg), and Figure 5-13, the radius of the flash fire can be very indicatively calculated assuming 70 kPa at the cloud edge.

From Figure 5-11, the radius of the flash fire corresponds with a scaled distance of around 3.8 $m.kg^{0.333}$. Therefore, the radius can be calculated from the following relationship:

$$\text{Radius (m)} = \text{Scaled distance} \times \text{TNTkg}^{0.333}$$

So,

$$\text{Radius} = 3.8 \times (12{,}521)^{0.333}$$
$$= 88 \text{ meters approx.}$$

(e) It is reasonable to assume that someone within a flash fire will be fatally injured. From (a) above, the probability of ignition is around 29%, and of explosion if ignited is around 26%. Therefore the probability of ignition but not explosion would be:

$$0.29 \times (1 - 0.26) = 0.21$$

With the frequency of leaks at 0.001 per year, the frequency of flash fires would be around 0.00021 per year, or 210 per million per year.

Note that this is possibly pessimistic, as many instances of ignition without explosion would occur too soon after the onset of the leak for a vapor cloud of the assumed size to have formed.

5.4.6 Confined Explosions

Ignition of a cloud of flammable vapor within a closed vessel, or a pipeline, can result in a detonation, where the blast or shock wave travels at sonic velocity and is characterized by a very sharp peak of pressure of short duration.

The resulting overpressures can be very high, typically up to around 7 to 8 times the initial pressure in the vessel before ignition. The higher the pressure before ignition, the higher the instantaneous peak pressure. If the gas is turbulent before ignition, the instantaneous peak pressure can be higher, around 9 to 10 times the pressure before ignition. In a detonation in a pipeline, in which the shockwave velocity can reach 3000 meters per second (compared with the velocity of sound at sea level of 330 meters per second), the pressure can reach 20 times the initial pressure in the pipeline. When it is considered that vessels are often designed such that the actual bursting pressure would be around 4 to 5 times the working pressure, it is evident that design to contain an internal deflagration would necessitate substantial extra cost, and design to contain a detonation is generally totally impracticable for economic or other reasons.

The peak pressure in a vessel is independent of the volume of the vessel, but the time taken to reach the peak pressure is a function of the cube root of the volume. This is the *cubic law.*

For any particular gas/air mixture, the value of the constant K_G, where

$$K_G = dP/dt\ V^{0.333}$$

is related to the flame speed and is an indication of the violence of the explosion.

Whereas the peak overpressure within a vapor cloud explosion in the open air is believed not to exceed 100 kPa, as the degree of confinement increases the instantaneous overpressure will increase toward the limiting condition of a closed vessel. Many references explore the venting volume required to limit the explosion pressure in a vessel to nominated levels, for example, Swift (1988).

Estimating the instantaneous overpressure that could occur within partially confined spaces such as between decks of an offshore oil platform is beyond the scope of this book. Various references exist that explore the pressure rise as a function of the venting area of enclosed volumes. This line of investigation continues.

Major international projects have been undertaken to develop computer programs to calculate the explosive overpressures in such structures. The models are extremely complex and require major computational time. The models need validation, which can be expensive. Scaled tests on physical models have produced markedly different results from those predicted with early drafts of such computer models.

5.5 OTHER EXPLOSIONS

5.5.1 Dust Explosions

5.5.1.1 Introduction

Wherever combustible dusts are handled, there is a chance of dust explosion. A large range of materials can give rise to dust explosions. More people (36) were killed in a dust explosion of a wheat silo (Westwego, Louisiana, 1977) than were killed in the Phillips Petrochemical plant explosion at Pasadena (24) in 1989 (see Chapter 13).

Because the combustible component of the fuel–air mixture (i.e., the dust) does not occupy significant volume, it is possible for a given volume of dust–air mixture to contain more fuel and oxygen than a similar volume of gas–air mixture. Thus a dust explosion in a closed vessel can generate higher pressures than a vapor explosion. Typically, dust explosions in closed vessels can generate peak pressures of up to 10 times (and more in some cases) the initial pressure in the vessel.

Typical scenarios for serious dust explosions are as follows:

• A combustible solid is being ground in a machine. Dust from previous operations has accumulated on all horizontal surfaces in the building, on ledges, girders, etc. Overheating or a spark between striking metal surfaces causes a small local explosion in the grinding machine. That local explosion shakes loose the accumulated dust in the building. Incandescent dust particles from the first explosion ignite the secondary cloud, and a very large secondary explosion occurs.

• A small dust explosion inside a machine ruptures the casing of the machine and propels the settled powder in the machine into the surrounding building, forming a cloud. This cloud is then ignited by incandescent particles or some other ignition source, resulting in a serious secondary explosion.

• An ignition source in a dusty environment inside a machine initiates a dust explosion that propagates through the ductwork of the plant to other vessels, etc., resulting in a very serious, damaging, and dangerous explosion.

Two important properties of dusts, relating to their explosion potential, are:

• minimum ignition energy, which determines the sensitivity to ignition, and

• flame speed, which determines the violence of the explosion.

Note: Eckhoff (1997) cautions that many of the documented properties of dust explosions and of explosible dusts have been determined in small-scale test equipment in which surface cooling effects are significant, and there is evidence that large-scale plant explosions are often much more severe than the documented data would suggest.

Tables are available for guidance on the sensitivity and explosive power of numerous dusts (e.g., Eckhoff, 1997).

However, it is now recognized that the minimum ignition energy for a particular fuel in a dust cloud depends on the shape and nature of the ignition source, spark, etc. Further, the power of a confined dust explosion is being found to depend on the shape of the confinement, because of the potential for creation of turbulence within the vessel. These are considered briefly by Eckhoff (1989, 1997).

Work has been done to determine the pressure rise in vented vessels (e.g., Lunn *et al.*, 1988).

5.5.1.2 Outline of Mechanism of Dust Explosion

The smaller the particles of a solid, the greater the ratio of surface area to mass. Therefore, the smaller the particle, the more rapidly it will heat up when exposed to heat radiation.

When a cloud of dust in the air is exposed to an ignition source, some of the particles may be heated to a sufficient temperature to ignite. Because of the high ratio of surface area to mass, the particle can burn fully in a very short period. The heat radiated from the particles initially ignited may be sufficient to heat adjacent particles to autoignition temperature, thus propagating the combustion process.

Depending on the circumstances, the combustion process can spread very rapidly through the dust cloud, propagated largely by heat radiation. If the rate of combustion is sufficient, the expansion rate of the air in the cloud is sufficient to generate a pressure rise of explosive magnitude and speed.

In a closed vessel, if a dust cloud is ignited, the pressure rises as the dust is consumed by the explosion. The rate of pressure rise depends on a variety of factors, including:

- the ignition energy,
- the speed of the flame front,
- the heat of combustion of the dust, and
- the size of the vessel.

The cubic law has been stated to apply to dust explosions:

$$(dp/dt)V^{0.333} = K_{St}$$

where dp/dt is the rate of pressure rise in the closed system; V is the volume of the closed system; and K_{St} is a constant specific to the material of which the dust is composed and defines the explosibility of the dust, that is, the power generated by such an explosion. (It should not be confused with the ease of ignition.)

Note: Eckhoff (1997) cautions that it has been found that K_{st} depends on a variety of factors such as the ignition energy and the shape of the container. Thus any industrial design based on K_{st} should have a substantial safety factor included. For this reason, tables of K_{st} are not included here.

5.5.1.3 The "Explosibility" of Dusts

Dust–air mixtures will only explode in a particular concentration range. Typically this ranges between a lower explosive limit of 0.02–0.09 kg/m^3 and an upper explosive limit of 0.7–3 kg/m^3. These vary substantially depending on the test method and so should be regarded as very tentative. The stoichiometric composition for many materials is in the range of 0.1–0.3 kg/m^3.

Dusts may be classified into different K_{st} classes. These classes do not reflect the probability of an explosion, or even its effects.

The particle size has a critical effect on the possibility of a dust explosion. Fine dusts explode more violently than coarse dusts. Dusts with a particle size above 700 micrometers cannot be made to explode even with a high ignition energy. A size of around 70 micrometers results in effectively maximum power for most materials.

The particle size distribution is very important. The median particle size is less important. Even a small proportion (e.g., 5–10%) of fines among a dust mostly comprising particles too large to explode is sufficient to propagate an explosion and to cause the larger particles to contribute. Mechanical handling of large particles can produce sufficient fines to render the dust explosive.

The best measure is not actually particle size, but particle surface per unit of mass, that is, the specific surface. There is a linear relationship between the specific surface and the violence of the explosion.

Typically in a closed vessel, the peak pressure (P_{max}) reached by a dust explosion can be 10–12 times the initial pressure. (This is higher than with a flammable gas–air mixture, which is typically 8 times.)

Typical values of P_{max} are listed below in Table 5-5 (from Eckhoff, 1997).

The ignition energy can have a large effect on the K_{St} factor (i.e., the violence of the explosion) and some effect on the final pressure reached.

The minimum ignition energy depends on the type of the dust and on its concentration. For each type of dust (nature of material, particle size distribution) there is a most easily ignited concentration. The minimum

Table 5-5
Values of P_{max} (Reprinted from Eckhoff, R. K. (1997): *Dust Explosions in the Process Industries.* Second Edition. Butterworth Heinemann, Oxford, UK with permission from Elsevier Science and the author.)

Material	P_{max} (barg)
PVC	8.5
Milk powder	8.1
Maize starch	10.1
Polyethylene	7.5
Coal dust	9.0
Resin dust	7.6
Brown coal	9.0
Wood dusts	9.2
Cellulose	9.3
Aluminum	12.5

Table 5-6
Minimum Ignition Energy (Reprinted from Eckhoff, R. K. (1997): *Dust Explosions in the Process Industries.* Second Edition. Butterworth Heinemann, Oxford, UK with permission from Elsevier Science and the author.)

Material	Minimum Ignition Energy (mJ)
Aluminum (fine)	<1
Barley grain dust	100
Cellulose	250
Maize starch	300
Wheat flour	100–540
Brown coal	160
Aluminum ("paint fine")	10
Methyl cellulose	12–105
Sulfur	<1–5
Toner (copier)	<1
Wood dust (coarse)	100
Wood dust (fine)	7

ignition energy at the most easily ignited concentration varies from less than 1 mJ to more than 1000 mJ.

Typical values of the lowest minimum ignition energy are listed in Table 5-6 (from Eckhoff, 1997). Note that these are highly dependent on the particle size distribution.

5.5.1.4 Predicting the External Consequences of a Dust Explosion

Most references cover the effects of a dust explosion within a closed vessel (tank, silo, ducting, etc.). There is little quantitative information available on the external effects of a dust explosion within a building.

However, it is suggested here that an estimate may be made by using a variation of the TNT Equivalent approach. This estimate will be very approximate and does not take account of the different combustion rates of different materials.

The steps are:

- determine the heat of combustion of the dust (kJ/kg);
- determine the volume of the likely dust cloud within the explosive limits (m^3);
- determine the likely mass of dust per cubic meter (kg);
- determine the proportion of the dust that will be able to contribute to the explosion (i.e., the quantity cannot exceed the stoichiometric quantity and may be limited by some of the particles being above 400 micrometers);
- determine the total heat of combustion of the mass of the dust contributing to the explosion;
- convert this to an equivalent mass of TNT using an appropriate relative efficiency factor (about which there is little empirical data, so a value of between 10% and 100% could be used—say 30%); and
- estimate overpressure radii as for TNT.

Note that this method would only apply for the external effects of an explosion within a weak containment vessel, for example, a very thin vessel, or a building with weak cladding, such that the restraint imposed on the explosively expanding cloud is insufficient to affect the overpressures. The method is not suitable for assessing the effect within the containment.

There is extensive literature available for determining the explosive pressure developed within a closed but vented enclosure, but the influence of factors such as the ignition energy on the venting area required has only been recognized fairly recently. The general rule appears to be to use the standard sizing methods, being conservative, then to add a "good margin"!

5.5.2 Solid-Phase Explosions

The explosive effect of unstable materials can be determined from the explosive power related to that of TNT. These are tabulated by Blatt. Examples are listed below as an illustration of the range.

Material	Explosive Power (TNT = 100)
Ammonium nitrate	56
Nitroglycerin	169
Nitromethane	134
Picric acid	106
Sodium chlorate	15

5.6 TOXIC GAS ESCAPES

5.6.1 Introduction

A variety of toxic gases are produced, processed, stored, and transported. Examples are chlorine for water purification and manufacture of plastics and solvents, hydrogen fluoride and hydrogen sulfide in refinery operations, ammonia in refrigeration and manufacture of fertilizers, hydrogen chloride in manufacture of a range of chemicals and pharmaceutical products.

The physiological effects of exposure to a toxic gas depend on both the concentration and the duration of exposure. The concentration of a toxic gas at a particular location remote from the escape point depends on the weather at the time: the wind direction, wind speed, and the "atmospheric stability" (explained later) all are important.

In calculating the risks to people from a postulated toxic gas escape, the steps are:

- calculate the concentration at the required distance for a range of wind and weather conditions;
- for each calculated concentration, and with the estimated duration of exposure, estimate the physiological effect and the probability of fatality per such exposure;

- using the probability of occurrence of each of the selected wind and weather combinations, calculate the probability of fatality per leak; and
- using the estimated frequency of the leak, calculate the risk of fatality per year.

5.6.2 Calculation of Concentration of Gas Downwind of Leak

5.6.2.1 Form of Gas Plume Downwind of Leak

If gas escapes continuously from a single point, it moves with the wind, following the continually varying path of the wind and gradually dispersing in both the horizontal and vertical planes due to turbulence.

Figure 5-18 illustrates the plan view of a gas plume from an escape from a single point. Figure 5-19 illustrates the gas concentration in a cross section perpendicular to the wind, illustrating the usual assumption that the concentration follows a Gaussian or normal distribution. (This is only the case with a gas of roughly the same density as that of air, or of a dense

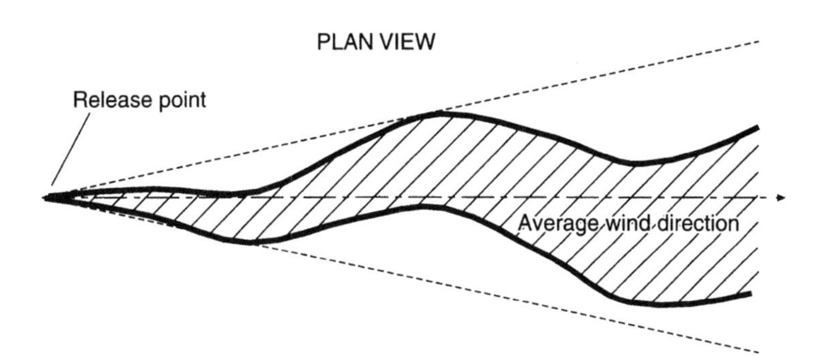

Figure 5-18. Plan view of gas plume.

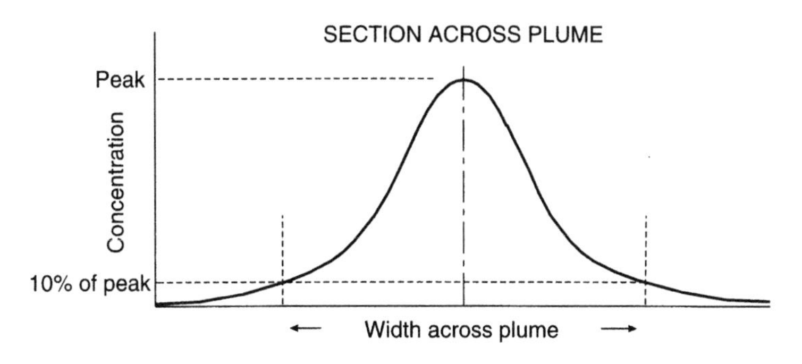

Figure 5-19. Cross section of gas plume.

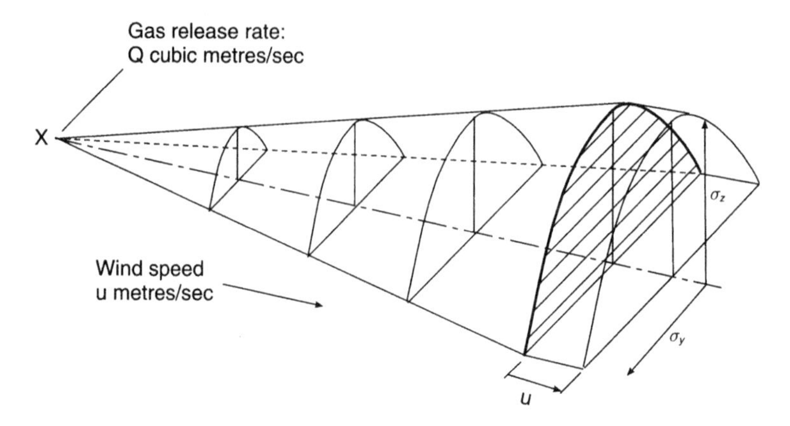

Figure 5-20. Principle of calculation of concentration.

gas sufficiently far downwind for the gas–air mixture to have around the same density as air.) For a person on the center line of the average wind direction, the gas concentration will vary with time, as the center line of the instantaneous wind direction varies.

In this assessment of the effects of toxic gases, the average concentration of the toxic gas on the average center line of the wind direction is used.

5.6.2.2 Principle of Gas Dispersion Calculation

Figure 5-20 illustrates the principle of calculation of the concentration of a gas downwind of the escape point as set out by Pasquill (1961, 1962).

The escape from point X gradually disperses in both the horizontal and vertical planes in the downwind direction. The cross-sectional area occupied by the plume at the point of interest is a function of the extent of horizontal dispersion and of vertical dispersion. If the gas is escaping with a rate of Q m^3/sec, and if the wind speed is u m/sec, then the volume of air into which the gas is dispersed is proportional to the horizontal dispersion, the vertical dispersion, and the wind speed. Thus the concentration at that distance is proportional to the rate of leakage divided by the volume into which the gas is mixed per second.

5.6.2.3 Atmospheric Stability

On a normal day, the temperature of the air near the ground is higher than that at an altitude. This is because of the sun heating the ground, which in turn heats the air above it. Thus the density of the warm air near

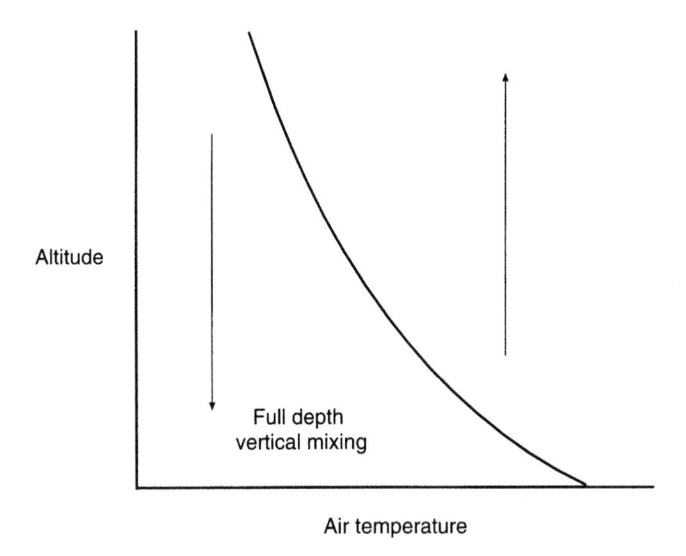

Figure 5-21. Temperature profile (unstable).

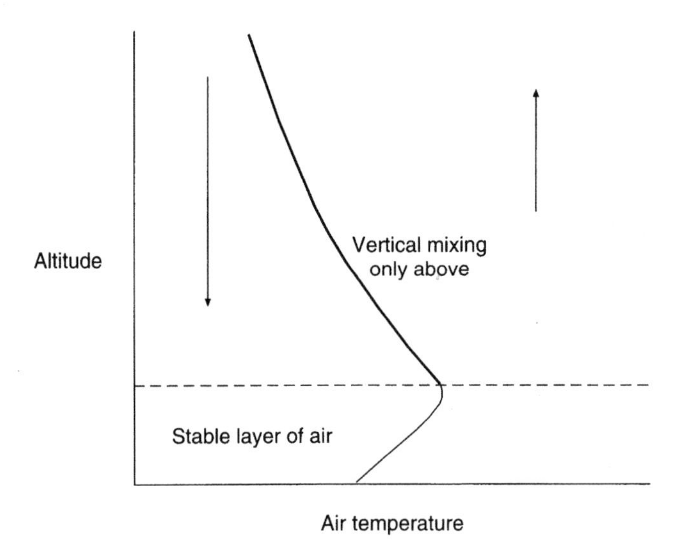

Figure 5-22. Temperature profile (stable).

the ground becomes less than the air above it, and so the warm air rises, being replaced by cooler air from above. See Figures 5-21 and 5-22.

This is described as "unstable" or "lapse" atmospheric conditions. Gas escaping at a time of unstable air will be rapidly dispersed in the vertical direction, thus reducing the ground-level concentration at a distance from the escape.

When the sky is clear and the ground is not being heated by the sun, such as on a clear night or early on a frosty morning, the temperature of the ground falls because of radiation of heat to space. Thus the temperature of the air at ground level falls to less than that of the air above it. In such conditions, the air near the ground is more dense than that above it, and there is little vertical mixing of the air. See Figure 5-22. A gas escape at a time of "stable" atmospheric conditions will hardly disperse in the vertical direction and can drift for long distances with little reduction of concentration.

5.6.2.4 Downwind Concentration from a Continuous Escape of Neutral Buoyancy Gas

Pasquill (1961) defined a range of atmospheric stabilities, from A (hot sunny day with a high degree instability and vertical mixing) through to F (inversion, with stable air layering and little vertical mixing other than due to turbulence caused by the air movement with any breeze), and then developed a graph showing the extent of horizontal and vertical dispersion for different atmospheric stabilities. See Table 5-7.

Figures 5-23 and 5-24, developed from the work of Pasquill, Gifford, and others by Turner (1970), display the horizontal and vertical dispersion coefficients. Note that they were derived from tests in a rural setting, in a flat, open terrain.

Pasquill (1961) proposed the following formula, which enables calculation of the time-weighted average concentration (χ parts per part) of a gas of neutral buoyancy at a point X meters in the downwind direction, Y meters off the center line, and Z meters aboveground downwind of a release from an effective height H meters aboveground.

Table 5-7
Key to Stability Categories © Crown Copyright Met Office Reproduced under Licence No. MetO/IPP/2/2003 0002

Surface Wind Speed (at 10 m) (m/sec)	Heating from Sun			Nighttime	
	Strong	Moderate	Weak	Thin Overcast or More Than Half Low Cloud	Clear to Half Low Cloud
<2	A	A–B	B	–	–
2–3	A–B	B	C	E	F
3–5	B	B–C	C	D	E
5–6	C	C–D	D	D	D
6+	C	D	D	D	D

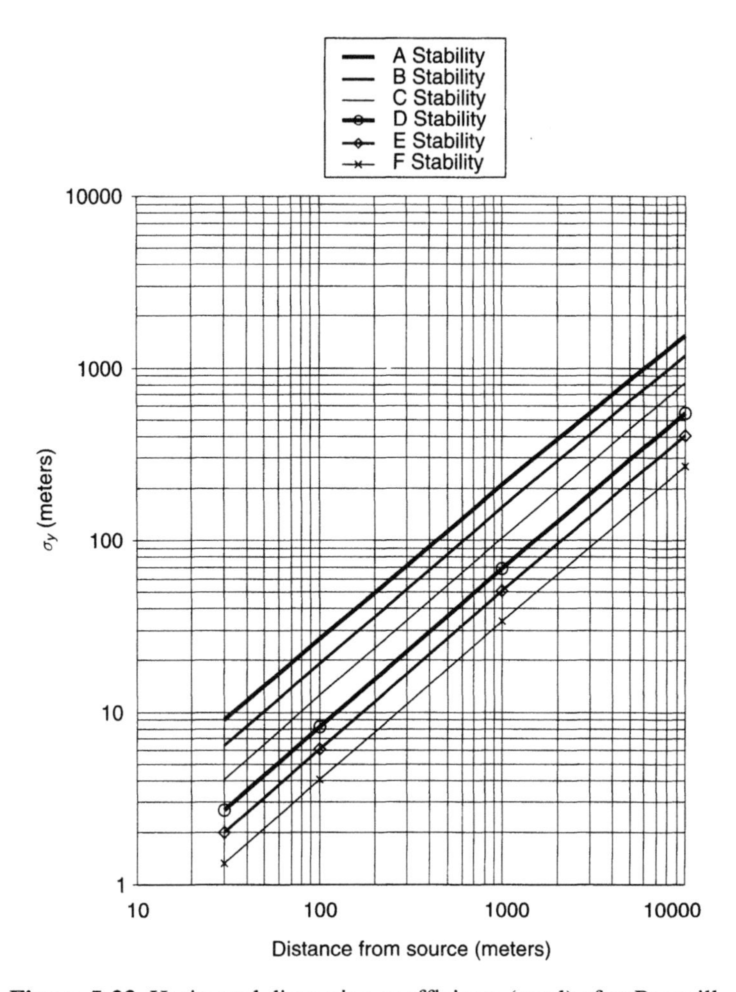

Figure 5-23. Horizontal dispersion coefficients (rural) after Pasquill.

crosswind direction term

$$\chi(x,y,z,H) = Q\frac{1}{u}\frac{1}{\sqrt{2\pi\sigma_y}}\exp\left[-0.5\left(\frac{y}{\sigma_y}\right)^2\right]$$

$$\times\frac{1}{\sqrt{2\pi\sigma_z}}\left[\exp\left[-0.5\left(\frac{H-z}{\sigma_z}\right)^2\right]+\exp\left[-0.5\left(\frac{H+z}{\sigma_z}\right)^2\right]\right]$$

vertical direction ground reflection term

For a center-line concentration from a release at ground level or from a stack at height H, this reduces to:

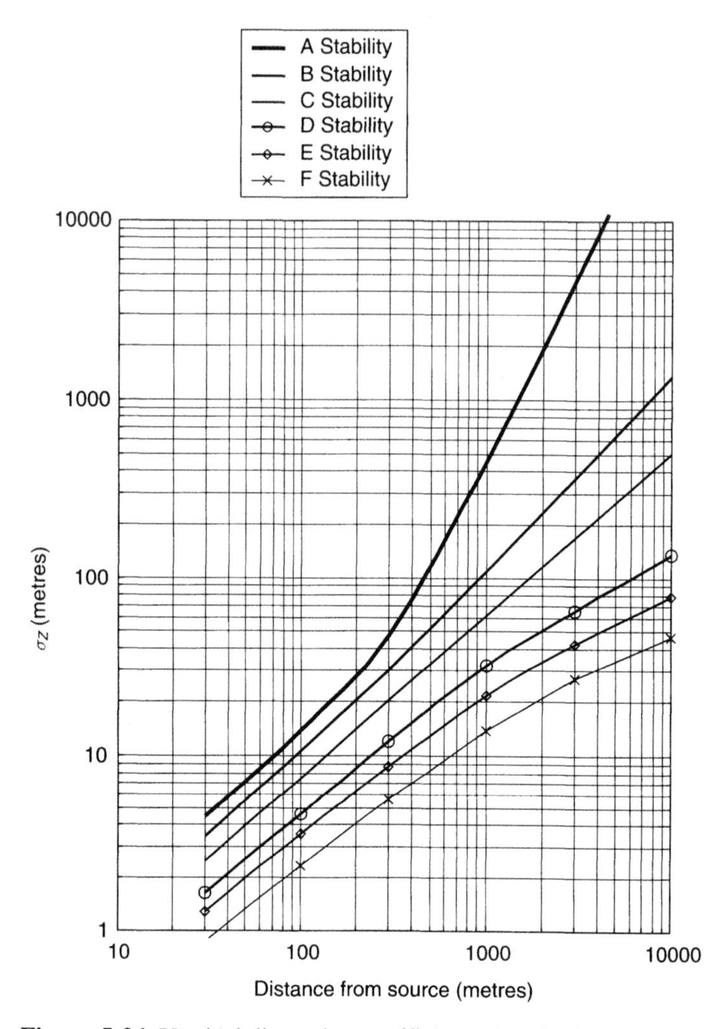

Figure 5-24. Vertical dispersion coefficients (rural) after Pasquill.

$$\chi(x, 0, 0, H) = \frac{Q \cdot F}{\pi \sigma_y \sigma_z u}$$

where

χ = average concentration on the downwind center line (dimensionless) (e.g., 0.0001 = 100 ppm) and

Q = gas release rate (m^3 per second)

F = stack correction factor for elevated releases = $\exp(-H^2/2\sigma_z^2)$

H = height of stack above the grade (m)

σ_y = horizontal dispersion coefficient (m) (see Figure 5-23)

σ_z = vertical dispersion coefficient (m) (see Figure 5-24)

U = wind speed (meters per second)

Example 5.6.2:

A gas escapes at a rate of 2 m^3 per second from a ground level source on a sunny but mild day when the wind is blowing at 3 m/sec. What is the approximate average concentration at a point 500 meters downwind?

Step 1. Determine the atmospheric stability. The most appropriate stability category is B.

Step 2. Determine the horizontal and vertical dispersion coefficients.

By reference to Figures 5-23 and 5-24, for B stability and a range of 500 meters, the horizontal dispersion coefficient is around 80 meters, and the vertical dispersion coefficient is around 50 meters.

Step 3. Determine the stack factor. As the release is at ground level, the stack factor reduces to 1.0.

Step 4. Calculate the concentration.

Substituting in the Pasquill formula, the concentration is:

$$\chi = \frac{2 \times 1}{3.142 \times 80 \times 50 \times 3}$$

$$= 0.000053 = 53 \text{ ppm}$$

In practice, it is much preferable to use one of the many computer programs available for all gas dispersion assessments. The purpose of illustrating the method above is to illustrate some of the principles of gas dispersion, to provide an understanding of the basis of such computer programs.

5.6.2.5 Rural and Urban Dispersion

Note that the graphs of dispersion coefficients (Figures 5-23 and 5-24) were derived in a flat, open rural setting, without major sources of turbulence. It is reported that the "roughness factor" for the terrain used by Pasquill would have been around 0.03 meters, that is, the obstacles in the area would have had a typical height of around 0.3 meters. (The roughness factor used in dispersion calculations is often around 10% of the height of the typical obstacles.) Thus, the Pasquill–Gifford coefficients should be used only for such surroundings. Further, note that Pasquill's work was based on short-term average concentrations, typically 3 min. So the concentrations are nominally 3-min averages.

Other graphs similar in form to those in Figures 5-23 and 5-24 have been produced for urban regions, but these should not be used now that computer programs are readily available.

5.6.3 Special Cases

5.6.3.1 Introduction

Other factors which affect the dispersion include:

- whether the release is continuous or a brief burst or puff;
- roughness of the ground or obstacles in the path of the dispersing gas;
- momentum of gas released vertically from a stack, causing the plume to rise above the top of the stack;
- jet mixing of gas in the vicinity of the escape point;
- releases from the tops of buildings; and
- density and buoyancy effects.

These are discussed in turn below.

5.6.3.2 Instantaneous Escapes

In the event of an effectively instantaneous escape (e.g., from the bursting of a container, or a short puff of gas), the gas is sometimes treated as a hemispheroid (i.e., a flattened or elongated hemisphere) of increasing size and dilution the further it travels downwind. See Figure 5-25.

Calculation of the concentration downwind of an "instantaneous" escape is best undertaken using a computer model.

Deciding whether to treat an escape as continuous or instantaneous can be difficult. A short continuous escape can, in the far field, resemble an instantaneous escape, in that the peak concentration may not reach a stable

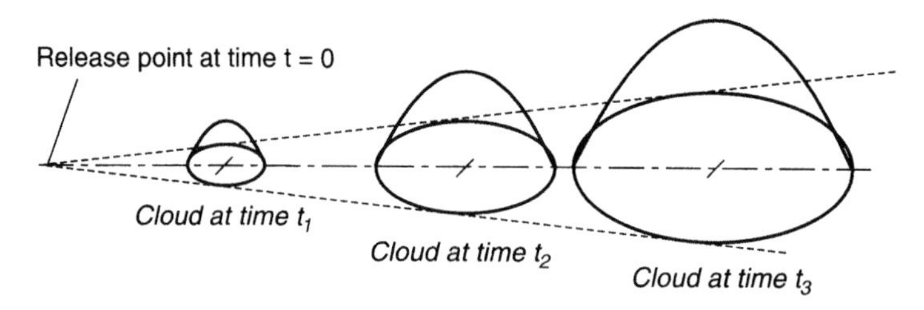

Figure 5-25. Illustration of instantaneous release.

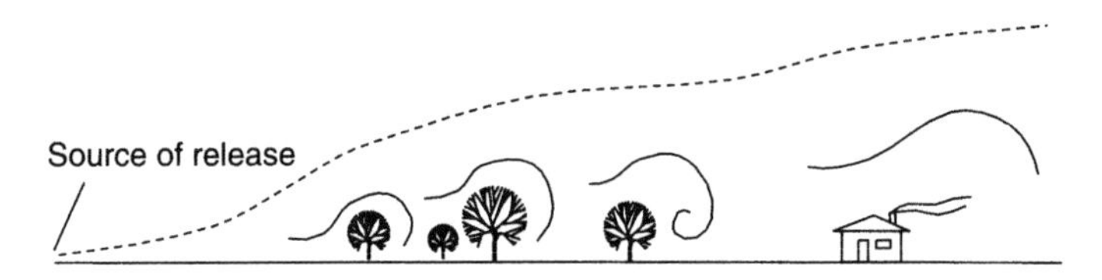

Figure 5-26. Effect of surface roughness.

level such as would be attained in the case of a continuous escape. One test is to regard any escape as continuous which, at the observer, is substantially longer than it is wide.

5.6.3.3 Surface Roughness

The effect of surface roughness or obstacles is to increase turbulence, particularly but not only in the vertical plane, and thus to improve dispersion and reduce the downwind concentrations. See Figure 5-26.

5.6.3.4 Momentum of Released Gas

Gas released from the top of a vertical stack will continue upward initially, progressively being directed horizontally by the wind. The effect is to raise the effective height of the stack. This effect is sometimes called the "plume rise." See Figure 5-27.

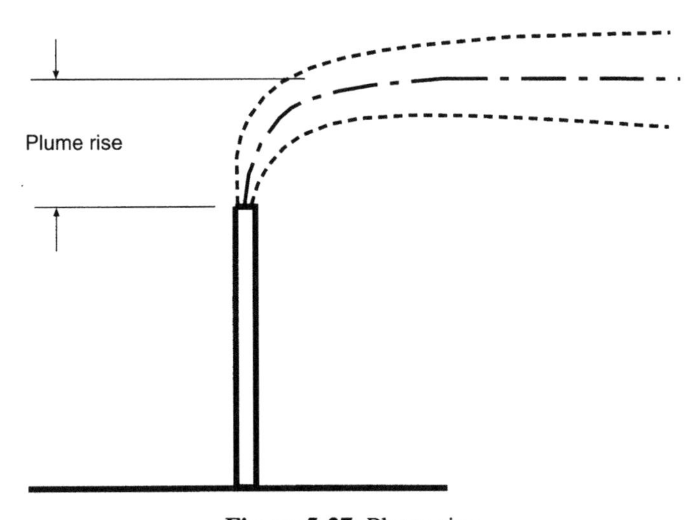

Figure 5-27. Plume rise.

5.6.3.5 Jet Mixing

Gas released with significant velocity will mix with the surrounding air, greatly reducing the concentration close to the point of release. The effect is to produce a notional release point at a distance upwind of the actual release point. See Figure 5-28.

5.6.3.6 Releases from the Tops of Buildings

A release from the top of a building, or a stack on the top of a building, is affected by the downdraft on the downwind side of the building, and the effective height of the release may be much less than the actual height. A common assumption is that the effective height is the actual height of the release above the top of the building. See Figure 5-29.

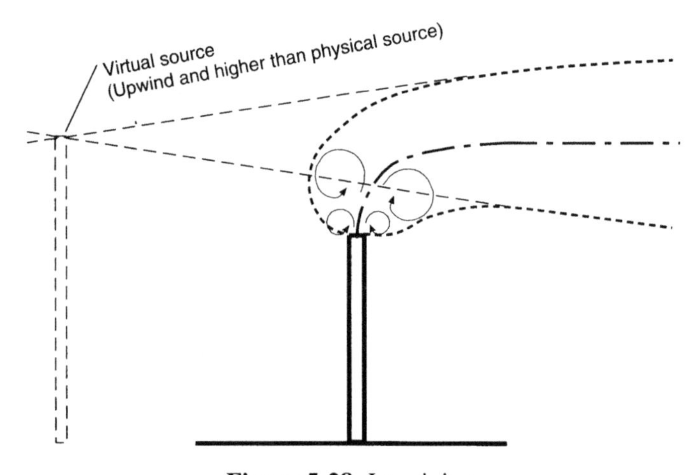

Figure 5-28. Jet mixing.

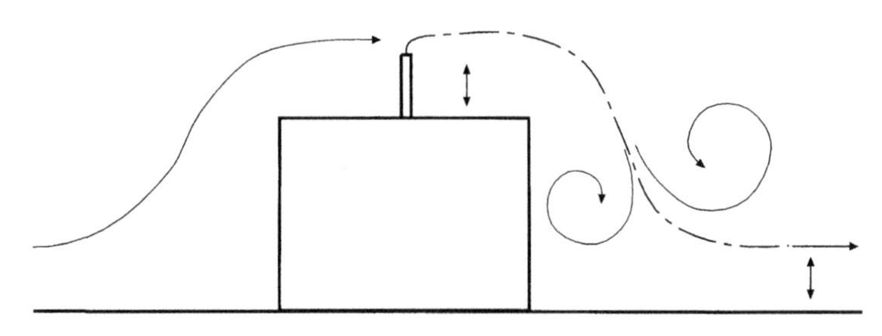

Figure 5-29. Release from the top of a building.

5.6.3.7 Density and Buoyancy Effects

A release of a buoyant or dense gas will initially be affected by momentum effects such as plume rise and jet mixing, just the same as gases of the same density as air. Once the initial momentum effects have diminished, the buoyancy or density can have a marked effect on the path and dispersion of the gas, until the dispersion has reached a degree where the density of the air/gas mixture is effectively the same as the surrounding air. From that point the dispersion is effectively as set out by Pasquill.

A dense gas released from a stack will slump toward the ground, and then spread sideways (while moving longitudinally with the wind) with a rolling action which entrains air and aids dilution. See Figure 5-30.

The mathematics of handling all these cases is normally handled by computer packages. In recent years substantial full-scale testing has been done to attempt to validate the mathematical models used, and to calibrate them by determining the values of incorporated constants.

The various models appear to be highly dependent on the "source terms" used, that is, the assumptions made about the initial release; whether evaporation from a pool of chilled liquid (rate of evaporation, size of pool, etc.) or leak from a pipe (orientation of the leak, the size of the hole, etc.)

5.6.4 Effects of Toxic Gas

The effects of a toxic gas depend on both the concentration and the duration of exposure. Probit mathematics can be used to predict the

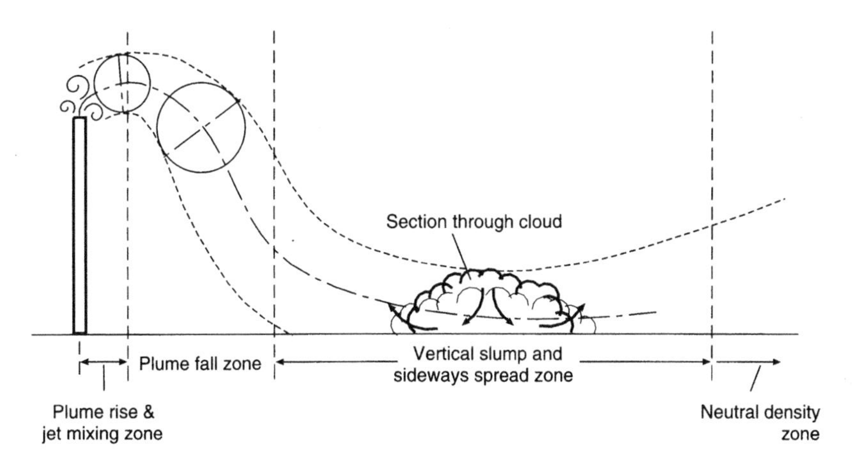

Figure 5-30. Density effect.

probability of fatality for a defined exposure (concentration, duration). See Section 5.8.

5.7 ENVIRONMENTALLY DAMAGING ESCAPES

Many chemicals have the potential to damage the environment. Particular examples are those chemicals which are produced because of their biological activity, such as pharmaceuticals, pesticides, herbicides, fertilizers, and hormones.

It is often difficult to define the extent of damage which will be caused by postulated specific escapes, such as *X* tonnes of material M in the event of a fire and the material being washed to the drainage system by firewater.

A preferable approach is to define a quantity of material which could be expected to have an unacceptable effect on the environment (including the drainage system in some cases, such as where effluent is treated biologically before release), then to examine routes by which that quantity could escape. This avoids continuing debate about the area in the environment which would be affected, the nature of the effect (e.g., plants, microorganisms, effects on the food chain) and the duration of the effect.

Assessment of the extent, nature, and duration of the environmental effect of various chemicals is a specialist task requiring a good understanding of the specific local environment and estimation, possibly with some calculation as an aid. This is beyond the scope of this text.

5.8 ASSESSMENT OF PROBABILITY OF FATALITY USING PROBIT MATHEMATICS

5.8.1 Introduction

If a population is exposed to a hazardous event, such as a defined level of heat radiation for a particular duration, or a concentration of toxic gas for a particular duration, the effects on all the people would not be identical. Some would be more susceptible, and some very much less. This is accounted for using *probit mathematics*. The application of the method to major hazards is described by Eisenberg *et al.* (1975), although many of the probit relationships quoted by them are now regarded as very pessimistic. For a more recent review of the basis and the limitations of the probit approach, see Lees (1996), p. 9–72ff. The approach is equally applicable, in principle at least, to damage to the living environment and to property.

The general form of the probit function is:

$$Y = k_1 + k_2 \log_e V$$

where Y is a measure of the percentage of the exposed population, property, or environment to sustain the defined degree of injury or damage; k_1 and k_2 are constants that depend on the nature of the exposure (examples are shown later in Table 5-9); and V is a variable determined from the severity of the exposure, sometimes a function of both intensity and duration (as with toxic gases), sometimes just the intensity (as with explosions). In the former case, V is expressed as $V = C^n T$.

The logarithmic relationship is used because the response to exposure roughly follows a skewed normal distribution, roughly represented by a log-normal distribution.

Probit equations are valid only within the range of data from which they have been derived. Outside that range, they have the potential for grossly incorrect results.

The relationship between the calculated probit Y and the percentage of the exposed population to be affected to the defined extent is shown in Figure 5-31.

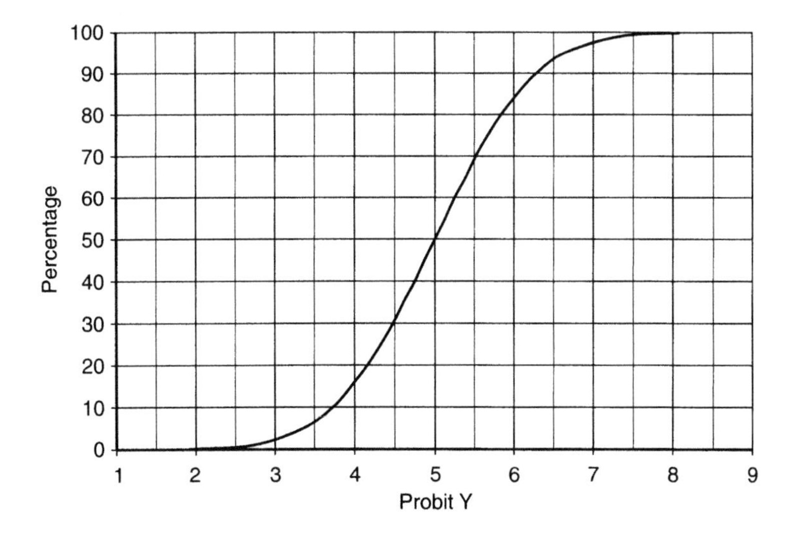

Figure 5-31. Relationship between probit and percentage.

5.8.2 Application to Fatality Risks from Heat Radiation from a Fire

For heat radiation fatality risks, for a person wearing normal clothing, the probit relationship is:

$$Y = -37.23 + 2.56 \log_e (I^{1.333}\ t)$$

where I is heat radiation intensity (W/m^2) and t is exposure time (sec). (See Lees, 1996, p. 16–257ff.)

Example 5.8.1:

A person is exposed to heat radiation of 15 kW/m^2 for 1 min. What is the probability of fatal injury?

$$Y = -37.23 + 2.56 \log_e (15{,}000^{1.333} \times 60)$$
$$= 6.0$$

Referring to Figure 5-31, the percentage of fatalities in a population so exposed would be around 85%. Because of the imprecision of the method, all that can be said with confidence is that there would be a very high probability of fatality for people so exposed.

The physical and physiological effects of exposure to heat radiation are reported by various sources to be as shown in Table 5-8.

Lees (1996, p. 16–258) presents a graph showing the relationship between heat flux, exposure time, and first- and second-degree burns.

Table 5-8
Effects of Heat Radiation

Heat Radiation (kW/m^2)	Effect
1.2	Heat from the Sun at noon
1.5	Minimum for pain
5–10	Will cause pain in 10–30 sec and second-degree burns in around 30 sec
12.5	In time may heat timber to a temperature at which it will ignite with pilot ignition
25	Spontaneous ignition of timber; thin uninsulated steel structures weaken

5.8.3 Application to Fatality Risks from Heat Radiation from a BLEVE

The probability of fatality of someone exposed to a BLEVE can be estimated using the probit method. The probit equation for exposure to a BLEVE is different from that for steady-state fire (Section 5.8.2) because of the shorter duration of exposure and the (often) much higher heat flux.

The probit equation is:

$$Y = -10.7 + 1.99 \log_e\{(I^{1.333}\, t)/10{,}000\}$$

For example, using the results for the BLEVE considered in Example 5.3.1, where t = 16 sec and I = 16,000 and 36,000 W/m^2 (depending on the assessment method used), the corresponding probits are 2.2 and 4.3.

Thus the probability of fatality, according to the two separate methods of assessment, would be either extremely small, or around 25%.

The physiological effect of exposure to a BLEVE, according to Brasie (1974), is that up to one fireball diameter from the edge of the fireball, an exposed person is likely to suffer second- to third-degree burns, and up to two fireball diameters is likely to suffer first-degree burns. This, however, is only a broad guide and should be not be used in isolation from calculations and understanding of the local situation (topography, layout, etc.).

Based on this empirical rule, it would be expected that an exposed person would suffer second- and third-degree burns. This suggests that the probability of fatality would be significant.

A more recent method of evaluating the physiological effect is to refer to the graph shown in Lees (1996), pp. 16–258.

5.8.4 Application to Fatality and Injury Risks from Toxic Gas Exposure

5.8.4.1 Probit Method

For toxic gas fatality risks, the probit relationship depends on the particular gas. The form of the relationship is:

$$Y = k_1 + k_2 \log_e(C^n\, t)$$

where k_1, k_2, and n are constants depending on the gas, C is the concentration in parts per million, and t is the exposure time in minutes.

Typical values of the constants are as shown in Table 5-9.

Note that, although the constants in Table 5-9 suggest precision, they are derived by attempting to fit curves to limited and dubious records of

Table 5-9
Typical Probit Constants for Toxic Gases (Sources as shown)

Substance	k_1	k_2	n
Ammonia (Opschoor et al., 1992)	−15.8	1	2.0
Ammonia (World Bank, 1988)	−9.82	0.71	2.0
Chlorine (Withers and Lees, 1985)	−8.29	0.92	2.0
Chlorine (Withers and Lees, 1986)	−8.29	0.92	2.0
Hydrogen chloride (World Bank, 1988 cited by CCPS 1999)	−16.85	2.00	1.0
Hydrogen sulfide (GASCON2, 1990)	−36.2	2.366	2.5

past incidents. As a result there are several sets of probit constants in use for many common toxic materials, which produce markedly different results. See, for example, Schubach (1995), and Ferguson and Hendershott (2000). Further, the actual toxic effect of exposure depends greatly on the level of activity of the exposed people. Very active people will be more greatly affected. This is just one cause of uncertainty about toxic gas effects on people. (For more comprehensive listing of probit constants, see CCPS, 1999.)

Example 5.8.2:

A population is exposed to chlorine at 200 ppm for 10 min. What is the probability of fatality per person on average?

Solution:

Using the relationship:

$$Y = -8.29 + 0.92 \log_e(200^2 \times 10)$$

$$Y = 3.6$$

Thus, using Figure 5-31, the probability of fatality would be around 6% on average. In practice, because of the uncertainty of the dose/response relationship, all that can be said with confidence is that the exposure would be very serious, with a significant chance of fatality.

5.8.4.2 HSE Dangerous Toxic Load/Dangerous Dose Approach

Because of continuing uncertainty about the dose/response relationship, and the resulting continuing debate about the probit relationships, Turner and Fairhurst (1989) developed an approach that defines a "dangerous toxic

Table 5-10
Typical Constants for Determining Whether the "Toxic Dose" Is Excessive

Substance	Index n	Toxic Dose Limit A
Acrylonitrile	1.0	9600 ppm min
Ammonia	2.0	3.76×10^8 ppm^2 min
Chlorine	2.0	1.08×10^5 ppm^2 min
Hydrogen fluoride	2	2.4×10^6 ppm^2 min
Hydrogen sulfide	4	2.0×10^{12} ppm^4 min
Sulfuric acid mist	1	2.16×10^5 (mg/m^3) min

(Reprinted from Turner and Fairhall [1989] by permission)

load" or "dangerous dose" in terms of the concentration (in ppm) raised to some power and the duration of exposure. That is, $C^n\,t$ must be less than a defined toxic load limit A. This can be used in setting risk criteria, by setting a maximum frequency of exposure of a population to that dose. By this means the debate about calculation of the fatality risk is avoided, as the dangerous dose is defined such that there is no doubt that it would have a very severe effect on any exposed population.[17]

Tentative toxicological relationships being considered by the HSE include those shown in Table 5-10.

Example 5.8.3:

Consider the same chlorine incident postulated in Example 5.8.2 above, with exposure to a concentration of 200 ppm for 10 min.

For chlorine, $n = 2$, $A = 1.08 \times 10^5$ ppm^2 min. The toxic dose would thus be $200^2 \times 10 = 400{,}000$ ppm^2 min.

This substantially exceeds the toxic dose limit of 108,000 ppm^2 min. Thus the exposure meets the definition of a dangerous toxic load and must not exceed a frequency of 10^{-6} p.a.

REFERENCES

API, "Guide for Pressure Relief and Depressuring Systems. API-RP 521," 2nd ed. American Petroleum Institute, Division of Refining, 1996.

[17] A dangerous toxic load, or dangerous dose, typically has the following general characteristic impact: almost every exposed person would suffer severe distress; a substantial fraction of those exposed would need medical attention; some would be severely affected and need continuing long-term medical treatment; and the most highly susceptible people (e.g., aged, sick) would be killed (typically 1% to 5%).

Bagster, D. F., and Pitblado, R. M., "Thermal Radiation Hazards in the Process Industry." Proc Chemeca 88, Sydney, August 1988.

Bagster, D. F., and Schubach, S. A., "A Tentative Linear Jet Fire Model." Proc. Conf. Loss Prevention in the Oil, Chemical and Process Industries. Society of Loss Prevention in the Oil, Chemical and Process Industries (Singapore), 99–105, December 1995.

Blatt, A. H. "Compilation of Data on Organic Explosives. Service Directives, NO-b10, OD-01 OSRD No. 2014." Office of Scientific Research and Development.

Brasie, W., *Loss Prevention*, Vol. 10, p. 135. American Institute of Chemical Engineers, 1974.

CCPS, "Guidelines for Consequence Analysis of Chemical Releases." Center for Chemical Process Safety, AIChE, New York, 1999.

Crossthwaite, P. J., Fitzpatrick, R. D., and Hurst, N. W., "Risk Assessment for the Siting of Developments Near Liquefied Petroleum Gas Installations." *IChemE Symposium Series 1988*, No. 110, pp. 373–400 (1988).

Eckhoff, R. K., "Differentiation and Tailor Making—The Rational Approach to Prevention and Control of Gas and Dust Explosions. *J. Loss Prev. Process Ind.* **2**, 122 (1989).

Eckhoff, R. K., *Dust Explosions in the Process Industries*, 2nd ed. Butterworth–Heinemann, Oxford, UK, 1997.

Eisenberg, N. A., *et al.*, "Vulnerability Model: A Simulation System for Assessing Damage Resulting from Marine Spills." USCG Report CG-D-137-175, available as NTIS Report AD-A015-245 (1975).

Ferguson, J. S., and Hendershot, D. C., "The Impact of Toxicity Dose-Response Relationships on Quantitative Risk Analysis Results." *Process Safety Prog.* **19**(2), 91–97, Summer (2000).

GASCON2 (1990), Gascon2, A Model of Estimate Ground Level H_2S and SO_2 Concentrations and Consequences from Uncontrolled Sour Gas Releases (volume 5), E. Alp, M. J. E. Davies, R. G. Huget, L. H. Lam, M. J. Zelensky. Energy Resources Conservation Board, Calgary, Alberta, Canada, October 1990.

Howell, J. R., *A Catalog of Radiation Configuration Factors*. McGraw-Hill, New York, 1982.

IChemE, *Explosions in the Process Industry*. Major Hazards Monograph, 2nd ed. The Institution of Chemical Engineers, Rugby, UK, 1994.

Institute of Petroleum (UK), *Model Code for Safe Practice in the Petroleum Industry*, Part 9: *Liquefied Petroleum Gas Safety Code*, Vol. 1: *Large Bulk Pressure Storage and Refrigerated LPG*, 2nd ed. 1987.

IOI, "The Evaluation of Estimated Maximum Loss from Fire or Explosion in Oil, Gas and Petrochemical Industries with Reference to Percussive Unconfined Vapor Cloud Explosion." International Oil Insurers, London, 1979.

Lees, F. P., *Loss Prevention in the Process Industries*, 2nd ed. Butterworth–Heinemann, Oxford, UK, 1996 (3 volumes).

Lunn, G. A., Brookes, D. E., and Nicol, A., "Using the K_{st} Nomographs to Estimate the Venting Requirements in Weak Dust-Handling Equipment." *J. Loss Prev. Process Ind.* **1**(3), 123–133 (1988).

Leung, J. C., "Two Phase Flow Discharge in Nozzles and Pipes—A Unified Approach." *J. Loss Prev. Process Ind.* **3** (1990).

Morris, S. D., "Flashing Flow through Relief Lines, Pipe Breaks and Cracks." *J. Loss Prev. Process Ind.* **3** (1990).

Mudan, K. S., "Thermal Radiation Hazards from Hydrocarbon Pool Fires." *Proc. Energy Combust. Sci.* **10**, 59–80 (1984).

NFPA, "Flammable and Combustible Liquids Code." National Fire Protection Association, 1984.

Opschoor, G., van Loo, R. O. M., and Pasman, H. J. (1992), "Methods for Calculation of Damage Resulting from Physical Effects of the Accidental Release of Dangerous Materials." International Conference on Hazard Identification and Risk Analysis, Human Factors, and Human Reliability in Process Safety, Orlando, Florida, January 15–17, 1992.

Pasquill, F. "The Estimation of the Dispersion of Windborne Material." *Meteorol. Mag.* **90**(1063), 33–49 (1961).

Pasquill, F., *Atmospheric Diffusion: the Dispersion of Windborne Material from Industrial and Other Sources*. Van Nostrand, London, 1962.

Pasquill, F., and Smith, F. B., *Atmospheric Diffusion*, 2nd ed. Halsted, New York, 1983.

Perry, R. H., and Green, D. (Eds.), *Perry's Chemical Engineers Handbook*, 6th ed. McGraw-Hill, New York, 1984.

Roberts, T., Gosse, A., and Hawksworth, S., "Thermal Radiation from Fireballs on Failure of Liquefied Petroleum Gas Storage Vessels." *Trans. IChemE* **78, Part B** (2000).

Schubach, S., "Comparison of Probit Expressions for the Prediction of Lethality Due to Toxic Exposure." *J. Loss Prev. Process Ind.* **8**(4), 197–204 (1995).

Swift, I., "Design of Deflagration Protection Systems." *J. Loss Prev. Process Ind.* **1**(1), 5–15 (1988).

TNO, "Methods of the Calculation of the Physical Effects of the Escape of Dangerous Material." Report of the Committee for Prevention of Disasters, Directorate General of Labour, Ministry of Social Affairs, Balenvan Andelplein 2, 2273 KH Voorburg, The Netherlands, 1979. (Referred to as the TNO Yellow Book.)

Turner, D. B., "Workbook of Atmospheric Dispersion Estimates." EPA Office of Air Programs Publication No. AP 26, *Quart. J. R. Met. Soc.* **93**, 383–384 (1970).

Turner, R. M., and Fairhurst, S., "Assessment of the Toxicity of Major Hazard Substances." Specialist Inspectors Report 21. *Health Safety Executive* HMSO (1989).

Van den Berg, A. C., "The Multi-Energy Method—a Framework for Vapour Cloud Explosion Blast Prediction." *J. Hazardous Mater.* **12**, 1–10 (1985).

Van den Berg, A. C., van Wingerden, C. J. M., and The, H. G., "Vapor Cloud Explosion Blast Modelling." TNO Prins Maurits Laboratory, PO Box 45, 2280 AA Rijsijk, The Netherlands; proceedings of CCPS Conference and Workshop on Mitigating the Consequences of Accidental Releases of Hazardous Materials, Louisiana, May 20–24, 1991.

Wayne, D. F., "An Economical Formula for Calculating Atmospheric Infrared Transmissivities." *J. Loss Prev. Process Ind.* **4**, 86–92 (1991).

Withers, R. M. J., and Lees, F. P., "The Assessment of Major Hazards: The Lethal Toxicity of Chlorine." Parts 1 and 2. Journal of Hazardous Materials 12, 3 (December), 231–282 and 283–302 (1985).

Chapter 6

Assessing the Frequency or Likelihood of Potential Hazardous Incidents or Losses

When you have studied this chapter, you will be able to analyze the possible contributory factors or causes which can lead to a possible plant incident; estimate the probability of such incidents occurring; and recognize some of the common weaknesses in design or operation of instrumented control and protective systems that increase the likelihood of mishap.

This chapter addresses risk management step 7, "Assess Frequency," in Figure 1-4.

6.1 ANALYSIS OF CAUSES OF INCIDENTS USING FAULT TREES

6.1.1 Introduction

Where a postulated hazardous incident, or failure of equipment to perform as required, would result from a combination of events, it is very helpful to analyze the structure of the causes by use of a "fault tree." This analysis is also used when calculating the expected likelihood or frequency of the incident or failure.

As with calculation/estimation of consequences (Chapter 5), it is important to recognize the limitations of any analysis and calculation of the likelihood or frequency of incidents. While analysis and calculations can help us understand the cause–effect structure of possible incidents, and their likelihood, there are many unquantifiable factors that can have a major effect on the likelihood. These must be taken into account in any proper assessment, but they can only be estimated, and so have wide limits on their accuracy.

This chapter describes first a systematic approach to analysis of the causes as a basis for quantification of likelihood, and then the mathematical methods for quantifying those elements that can be quantified. It also provides some broadly indicative data to use in preliminary assessments. Later chapters discuss the limitations of these methods, and the other important factors to be considered.

6.1.2 Distinguishing "Demands" from "Protection Failure"

There are two concepts which are important for construction of fault trees and especially for calculating the frequency of incidents or failures. They are *demand* and *protection failure.*

In the case of a boiler for generating steam, the pressure may be controlled by a pressure sensor sending a signal to an automatic controller, which in turn sends a signal to a control valve in the fuel line, such that if the pressure starts to fall below the set point then the valve increases the fuel flow; and if the pressure starts to rise above the set point then the valve closes partly to reduce the fuel flow.

The control system is needed continually for normal operation. Any failure of any component of the control system causes a problem.

This can be contrasted with the pressure relief valves fitted to the boiler. They are designed to release excess pressure to prevent the boiler from bursting if the pressure control system fails. It is important to note that the pressure relief valves are not needed continually, since they are inactive as long as the pressure control system is operating correctly. But the moment that the pressure control system fails and the pressure starts to rise to an excessive level, then the pressure relief valves are needed, and only then.

A failure of the normal method of operation (e.g., a failure of the boiler pressure control system, or failure of an operator to perform a routine operation correctly) is called a *demand.*

A failure of the response to a demand (e.g., a failure of the pressure relief valves) is called a *protection failure*.

A protective system is one that is needed only when the normal means of operation fails, that is, when a demand occurs.

This distinction is important for two reasons:

- when constructing a fault tree, it is helpful to show the undesired hazardous incident resulting from a combination of a demand AND a protection failure; and
- when assessing the frequency of hazardous incidents, it is essential for demands to be shown as frequencies, and for protection failures to be shown as probabilities. The reason for this is described more fully later.

More examples of demands and protection failures are as follows:

- If a car breaks down on a freeway, and the nearest telephone has been damaged by vandals, then the breakdown is the demand, and the telephone being in a vandalized condition is the protection failure.
- If a fuel stock tank leaks, and the leaking material is held within the bund, then the demand is the leak, and the protective system is the bund (which has not failed in this case).

It is common for hazardous incidents to arise from a much more complex situation than simply one possible type of demand, and one possible type of protection failure. It is when analyzing such situations that fault trees are most useful.

6.1.3 Constructing Fault Trees

There are two forms of logical connection between events that are commonly used in construction of fault trees of the possible causes of accidents. They are *AND* and *OR*. These two forms of logical connection (or *gate*) are often represented as in Figure 6-1.

- In the case of an AND gate, event C will only occur if both events A and B occur.
- In the case of an OR gate, C will occur if either event A or event B occurs.

These gates can be used to assemble a fault tree as shown in the following example of a hot oil heating system. Refer to the simplified flowsheet in Figure 6-2.

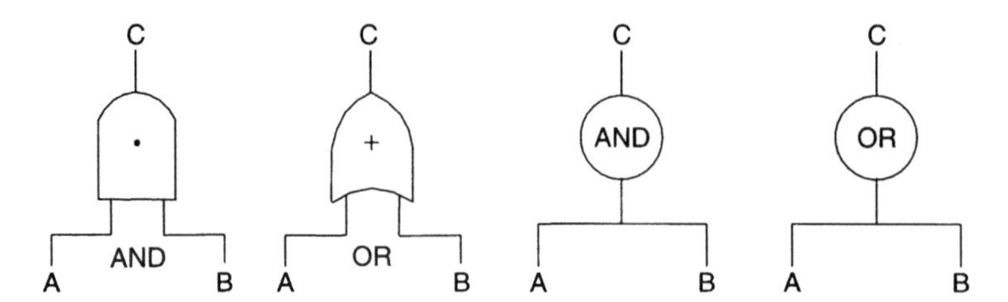

Figure 6-1. Logic gates—alternative representations.

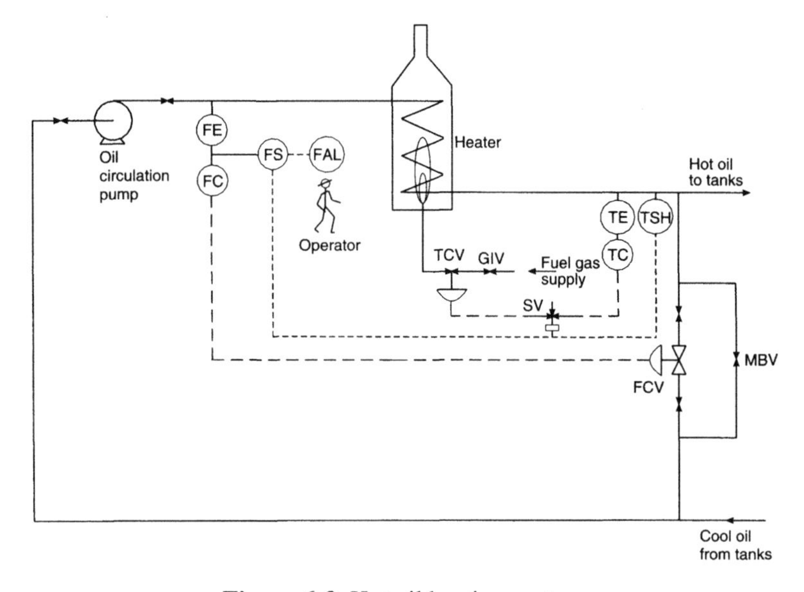

Figure 6-2. Hot oil heating system.

The flowsheet of Figure 6-3, from which much irrelevant detail has been omitted, shows the heating section of a plant supplying hot oil to heating coils in bitumen tanks. Oil returning from the bitumen tank farm is circulated by the heating oil circulating pump through a gas-fired heater back to the bitumen tanks.

It is important that flow be maintained through the heater, or the heater coils may overheat, rupture, and cause a large and damaging fire. The flow rate to the heater is sensed by a transducer FE, which sends a signal to a flow controller FC, which in turn sends a signal to a flow control valve FCV, which opens progressively if the flow to the heater falls for some reason (e.g., due to oil flow to a bitumen tank being closed down) and recirculates heating oil back to the pump. There is a manual bypass valve, MBV, which is normally closed, and is intended for use principally if FCV is undergoing maintenance.

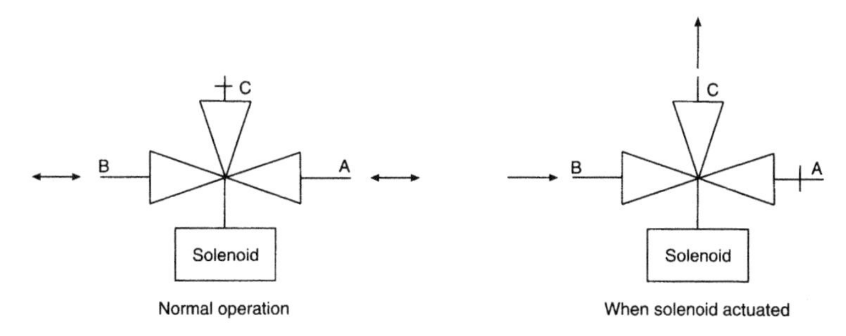

Figure 6-3. Explanation of solenoid valve operation. In normal plant operation, the solenoid valve permits free flow of the control air between the temperature controller TC and the temperature control valve TCV, i.e., between A and B in the above diagrams. The port C is normally closed. When the solenoid is actuated by a signal from the low flow switch FS or the high temperature switch TC, port A is closed and port C opened. This releases the air pressure in the pneumatic line to the diaphragm actuating the temperature control valve TCV, which is fitted with a spring that causes the TCV to close.

If the flow of heating oil to the heater is too low (e.g., due to the pump stopping, a failure of the flow control system, or a major leak), the signal from the flow transducer FE actuates the flow switch FS, which in turn actuates the solenoid valve SV, which releases the air pressure in the pneumatic line to the temperature control valve TCV, which then closes under spring pressure. The flow switch also actuates the low-flow alarm FAL, alerting the operator and enabling him or her to take appropriate action to prevent overheating of the heater coils, such as opening the manual bypass valve MBV or closing the manual gas isolation valve GIV.

There is also a high temperature switch TSH in the oil delivery line from the heater. If it detects excessive temperature due to some cause such as failure of the temperature control system TE, TC, and TCV, it actuates the solenoid valve SV and the temperature control valve closes under spring pressure.

Fault Tree Analysis can be used to display the logical structure of the complex combinations of equipment failures that can lead to the heater burning out at a time of very low or zero flow and the gas burner continuing to operate.

The starting point is to identify the possible triggering events, or demands, that can potentially lead to the coils burning out. They include:

- Zero or low oil flow:

  pump failure

  flow control system failure

large leak of oil

partial or full blockage of a pipeline

valve closed

- Excess gas flow (e.g., at a time of low oil flow)

Gas flow control system failure

Each of these demands should be analyzed separately, as the feasible protective response may vary between the demands. The total frequency of the coils burning out would be the sum of the results obtained from each of the analyses (as the coil burnout would result from "pump failure OR flow control system failure OR large leak of oil OR partial of full blockage of a pipeline OR, etc."). The analyses can be combined into a single fault tree, but there is no need to do so.

For brevity, only one demand will be analyzed here: pump failure. For simplicity, some of the actions that an alert operator would probably take are omitted here (e.g., starting the standby pump, manually actuating the temperature control valve). The fault tree is shown as Figure 6-4.

This fault tree can be expressed in words as follows:

- The heater coils will burn out if both the pump fails AND the protective response fails.
- The protective response fails if both the automatic response fails AND the manual response fails.
- The automatic response fails if either FE fails OR FS fails OR SV fails OR TCV fails.
- The manual response fails if either FE fails OR FS fails OR the operator fails OR GIV fails.

(Note that the over-temperature protection system [TSH, SV, and TCV] will not operate if the pump stops, as there is no flow to convey the overheated oil out of the heater to the temperature sensor TSH.)

Inspection of the above fault tree shows that two elements (FE and FS) appear more than once. It is therefore necessary to examine whether the fault tree can be expressed in a manner that avoids this repetition, as it often leads to double counting and can lead to the calculated frequency of the unwanted "top event" being either grossly overstated or understated.

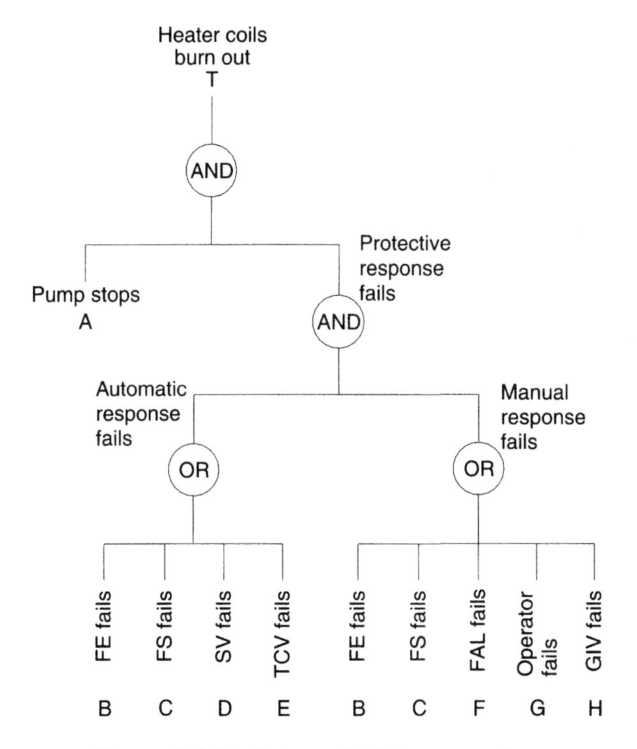

Figure 6-4. Fault tree—initial construction.

This examination can sometimes be done by inspection, but where the result is not obvious, it is prudent to use Boolean algebra. At first sight, this may be daunting, but it is actually very simple after a little practice.

6.1.4 Reduction of Fault Trees with Boolean Algebra

The above tree can be expressed as below, with the top event of the fault tree ("Heater coil burns out") represented by the symbol T. For simplicity, the element names are replaced by single letters as shown below.

Pump	A	TCV	E
FE	B	FAH	F
FS	C	Operator	G
SV	D	GIV	H

The logic displayed in Figure 6-4 can be expressed in Boolean algebra as:

$$T = A \cdot (B + C + D + E) \cdot (B + C + F + G + H)$$

This is expanded using Boolean algebra, which in this step is identical with conventional algebra:

$$\begin{aligned} T = A \cdot (&B \cdot B + B \cdot C + B \cdot F + B \cdot G + B \cdot H \\ &+ C \cdot B + C \cdot C + C \cdot F + C \cdot G + C \cdot H \\ &+ D \cdot B + D \cdot C + D \cdot F + D \cdot G + D \cdot H \\ &+ E \cdot B + E \cdot C + E \cdot F + E \cdot G + E \cdot H) \end{aligned}$$

These can be reduced using the rather strange-looking Boolean identities set out below (see also Figure 6-5).

Important Boolean Identities

1.	$A \cdot A = A$	"A and A equals A"
2.	$A + A = A$	"A or A equals A"
3.	$A + A \cdot B = A$	"A or (A and B) equals A"

Explanation of the last identity: The large outer ellipse in Figure 6-5 represents a number of objects of various shapes and colors. The small ellipse A represents the objects that are square. The small ellipse B represents the objects that are red in color. The overlapping area "A and B" (or $A \cdot B$) represents the square red objects. Clearly the square red objects are a subset of the square objects, so counting the square objects "A" includes the square red objects "$A \cdot B$."

The above expression can be simplified.

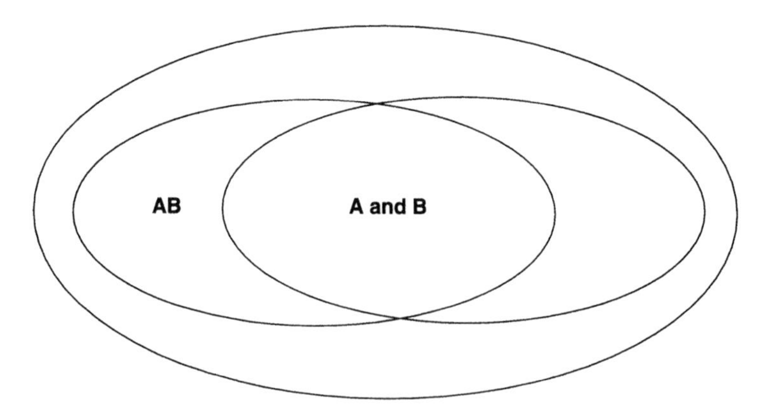

Figure 6-5. Explanation of Boolean identity.

- First, the term "BB" reduces to "B" (using the first identity);
- Then every other term including "B" can be struck out (using the third identity), that is, eliminating every other term in that row and column, such as $B+B\cdot C=B$, and $B+B\cdot F=B$.

Thus:

$$\begin{aligned} T=A\cdot(&\overset{\mathbf{B}}{\cancel{B\cdot B}}+\cancel{B\cdot C}+\cancel{B\cdot F}+\cancel{B\cdot G}+\cancel{B\cdot H}\\ &+\cancel{C\cdot B}+C\cdot C+C\cdot F+C\cdot G+C\cdot H\\ &+\cancel{D\cdot B}+D\cdot C+D\cdot F+D\cdot G+D\cdot H\\ &+\cancel{E\cdot B}+E\cdot C+E\cdot F+E\cdot G+E\cdot H) \end{aligned}$$

Similarly "CC" in the second row reduces to "C," and all the other terms including "C" can be struck out.

Thus:

$$\begin{aligned} T=A\cdot(&\overset{\mathbf{B}}{\cancel{B\cdot B}}+\cancel{B\cdot C}+\cancel{B\cdot F}+\cancel{B\cdot G}+\cancel{B\cdot H}\\ &+\cancel{C\cdot B}+\overset{\mathbf{C}}{\cancel{C\cdot C}}+\cancel{C\cdot F}+\cancel{C\cdot G}+\cancel{C\cdot H}\\ &+\cancel{D\cdot B}+\cancel{D\cdot C}+D\cdot F+D\cdot G+D\cdot H\\ &+\cancel{E\cdot B}+\cancel{E\cdot C}+E\cdot F+E\cdot G+E\cdot H) \end{aligned}$$

Rewriting the expression:

$$T=A\cdot(B+C+D\cdot F+D\cdot G+D\cdot H+E\cdot F+E\cdot G+E\cdot H)$$

This is then factorized, when it reduces to:

$$T=A\cdot\{B+C+(D+E)\cdot(F+G+H)\}$$

Converting back to the names of the components, and redrawing the fault tree, this becomes as shown in Figure 6-6.

This fault tree can also be expressed in words as follows.

- The heater coils will burn out if both the pump fails AND the protective response fails.
- The protective response fails if either FE fails OR FS fails OR a combination of failures occurs.
- A combination of failures that will lead to failure of the protective response occurs if there is (a failure of either SV OR TCV) AND (a failure of either FAL OR the operator OR GIV).

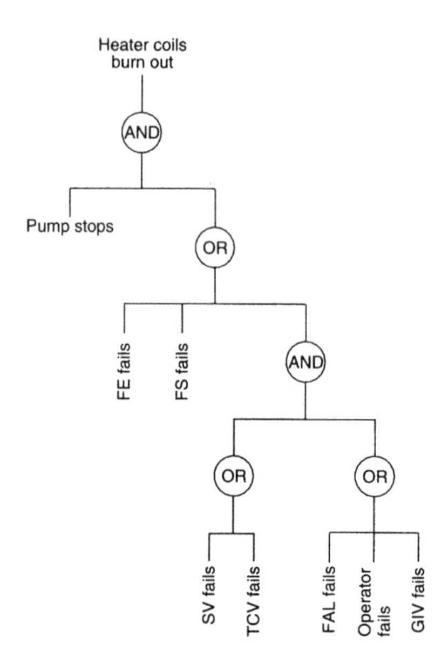

Figure 6-6. Reduced fault tree.

This can be checked by inspection, comparing the above logic with the flowsheet (see Figure 6-3).

The reduced fault tree illustrates the importance of simplifying a fault tree to avoid double counting and inclusion of unnecessary data. Note that it is possible, in a fully reduced fault tree, for some elements still to appear more than once. However, if the earlier fault tree were quantified (using methods set out later), a quite different result would have been obtained from what the fault tree in Figure 6-6 will produce.

6.2 INTRODUCTION TO RELIABILITY MATHEMATICS

6.2.1 Introduction

6.2.1.1 Scope of this Section

In hazard analysis and probabilistic risk assessment, we are concerned more with *unreliability* than with reliability. We are concerned with the frequency of failures, or with the probability that an item of equipment will not operate as required when called upon to do so. We are principally concerned with the frequency of incidents that have a severe physical impact as a result of the failure.

In production management, reliability has an extra dimension. Not only are we concerned with the frequency of mishaps (e.g., equipment breakdown) which cause loss of production, but also we are concerned with the duration of the unavailability of the equipment because of the breakdown and the effect that breakdown will have on production of the plant as a whole.

6.2.1.2 Important Concepts

There are a number of concepts that need to be understood when assessing the reliability of a system (which may comprise equipment, or people, or both).

Capability or *functionality* is the ability of a system to do what it is designed to do; that is, "Can it work?" That is primarily the responsibility of the designer, rather than the reliability analyst. Nevertheless, it is important for it to be verified that equipment is capable of performing as planned. The answer to the question about the capability is found by technical investigation.

Reliability of a component or system is the probability that it will perform a prescribed function for a specified period, or on a specific occasion, given a specified range of conditions—in other words, "Will it work?" The answer to this question is found using basic probability theory.

When an assessment is to be made of a system, it is important to define the boundaries of the system, for example, in assessing the reliability of a control and protective system, it must be decided whether to include assessment of the reliability of the power supply to that system.

Availability of equipment is the time the equipment is in operable condition as a proportion of the total time.

6.2.2 Component Behavior

It is important to define what is meant by "failure" of a component or a system. *Failure* is performance outside the upper and lower specified acceptable levels.

If a large number of nominally identical components are put into service at the same time, the frequency of failures will commonly be fairly high initially as the parts are "run in," or as initial faults are detected. Then there is a period of relatively low failure frequency, followed eventually by a rising frequency as the components start to wear out. This may be graphed as the "bathtub curve" in Figure 6-7.

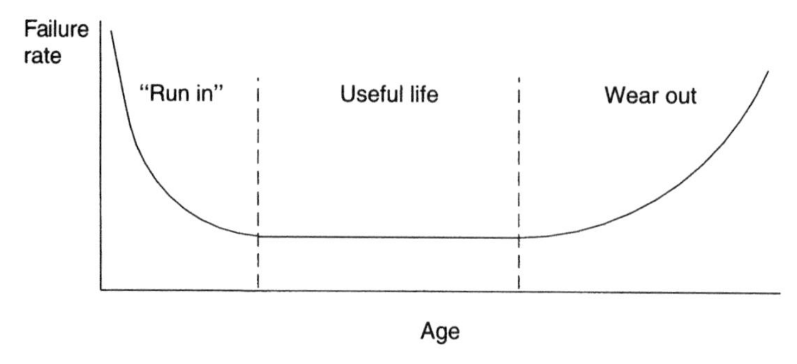

Figure 6-7. The "bathtub" curve.

The reliability mathematics covered here applies only to the "useful life" period, which is assumed to have a constant frequency of failure with time, that is, the probability of a failure occurring in any particular period of time (e.g., 1 month, 1 year) remains constant as the months pass. Although this is an approximation of varying precision for any particular situation, the principles established with this assumption are very important and widely applicable in practice.

If a number of components are in working order at time $t=0$ in the useful life phase, and if the failure rate is constant at f (e.g., f failures per year; very possibly a small decimal), then the probability of failure by time t will be:

Reliability over time $t=e^{-ft}$

For example, a component with a failure rate of 0.05 per year would have a reliability over a period of 10 years of:

$R(t)=e^{-ft}=e^{-0.05\times 10}=1/e^{0.05\times 10}=0.606$ (i.e., 61%)

and the probability of failure in 10 years is:

$1-R(t)=0.394$ (or 39%)

6.2.3 Probability of Failure of Protective Systems

A tank has water drawn from it intermittently, and at varying rates. It is fitted with a level control system which supplies water from an external supply until the tank is full again. There is also a high-level trip system, which actuates if the level rises higher than the "full" level, and shuts down the external water supply to prevent overflow.

Example 6.2.1: Level Control and Protection

Consider the following trap question. If a level control system fails once per 10 years, and the high-level alarm fails once per 10 years, how often will the tank overflow?

It is tempting to say the frequency is $0.1 \times 0.1 = 0.01$ per year, or a chance of 1 in 100 per year. However, this would be quite wrong. Consider the same arithmetic if the frequencies were 2 times per year in each case. The answer would be found to be 4 times per year, which is impossible if the controller fails only twice per year.

Checking the arithmetic, it will be realized that the answer obtained at first is actually 0.01 per square year. This dimension of square years does not make sense.

To be able to answer Example 6.2.1, it is necessary to convert the frequency of failure of the protective system to a probability of being in a failed state when called upon to act. To do this, it is necessary to know how often the protective system is tested.

If the protective system is tested once every T years, then the probability of the protective system having failed within any one test period will be:

Probability (of failure in time T) $= 1 - e^{-fT}$

where f is the failure frequency of the protective system.

This is illustrated in Figure 6-8 as the broken line. As the time between tests increases, the probability of the protective system being found, on test, to have failed approaches 100%.

Now, e^{-fT} can be expanded to give:

$$1 - fT + \frac{(fT)^2}{2!} - \frac{(fT)^3}{3!} + \ldots \text{ etc.}$$

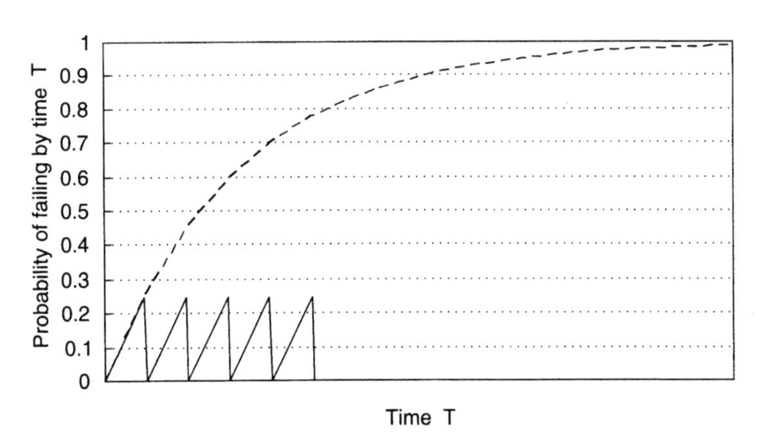

Figure 6-8. Probability of failing by time T (with and without testing).

If the probability of a failure occurring in any one particular period between tests is very low, that is, T is very short, $f \cdot T \ll 1$, then the higher order terms become negligible, and so:

Probability of failure in time $T = f \cdot T$ (approx.)

This is shown diagrammatically in Figure 6-8. The protective system is frequently tested, such that the probability of failing in any one time interval (T) between tests is low (shown above as around 15%). Thus the sloping part of the sawtooth-shaped curve, which is actually part of an exponential curve, is a reasonable approximation to a straight line. Each time the protective system is tested, either the system is found to be operational, or it is repaired and restored to operational condition. Thus at that time, the probability of being in a failed state either is or becomes zero.

In Figure 6-9, tests 1, 2, and 5 find the protective system in operable condition, while tests 3, 4, and 6 find the system to be in a failed state. At the time of the test there is no way of determining how long the system has been a failed state, that is, "dead." But, assuming that the system is in the "normal life" part of the bathtub curve, the probability of failing at any time between tests is the same. So the system is equally likely to have been in a failed state for a short time as for a long time.

Thus the average time, per failure, for which a protective system will have been in a failed state, that is, "dead," will be one-half of the time between the tests. The time for which the system has been inoperable is called *dead time*.

The fraction of time for which the high-level alarm will be in a failed state (i.e., "dead") will be the average dead time per failure, multiplied by the failure frequency.

Fractional dead time (FDT) $= \frac{1}{2} fT$

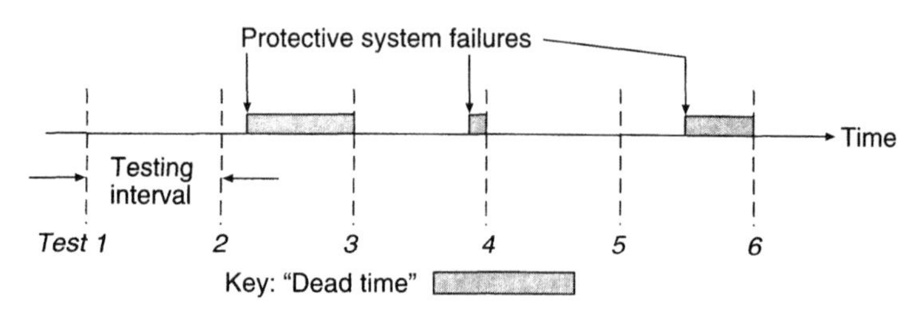

Figure 6-9. Tests of protective systems showing "dead" periods.

where

f = protective system failure frequency (the units are time^{-1})

T = interval between tests of the protective system (using consistent units of time)

Note that the fractional dead time (or FDT) is dimensionless and is the probability that the protective system will be in a failed state at the moment when it is called upon to act.

Where there is a demand with a frequency D, and the protective system has a fractional dead time of FDT, then the frequency of demands when the protective system is in a failed condition (i.e., the "Hazard Rate") is:

Hazard rate (HR) $= D \times \text{FDT} = D \cdot \tfrac{1}{2} fT$

Example 6.2.2:

Assume that the high level alarm in Example 6.2.1 is tested every 3 months. It is found, on consulting records, that it has failed around once every 10 years. The average time for which it is in a failed state on each occasion of failure will be half of the time between tests, that is, half the test interval—averaging 1.5 months or 0.125 of a year on each occasion.

Thus Example 6.2.1 would be solved as follows:

The frequency of failure of the high level alarm is 0.1 per year. Assume that it is tested every 3 months, that is, every 0.25 years.

Therefore its fractional dead time is:

$\text{FDT} = 0.5 \times f \times T = 0.5 \times 0.1 \times 0.25 = 0.0125$

The demand rate, that is, the frequency of failure of level control, is 0.1 per year.

HR = Demand rate $(D) \times \text{FDT} = 0.1 \times 0.0125 = 0.00125$ per year

This is equivalent to a 1 in 800 chance per year, or (more loosely and misleadingly) once per 800 years.

Note the terms:

Demand rate is the frequency of situations requiring protective response (e.g., frequency of failure of control systems, frequency of failure of whatever systems are needed in normal, routine operation).

Fractional dead time (sometimes called the *unavailability*) is the probability that a system will be in a failed state at any particular (randomly selected) moment.

Hazard rate is the frequency of the undesired (top) event.

6.2.4 Effect of Testing Time, Repair Time and Failure to Rearm

In practice, whenever a protective system is being tested, it is unavailable for the duration of the test. This adds to the FDT by an amount t/T, where t is the time taken to undertake the test.

Where the time taken to repair the protective system whenever it fails is R, then an additional fR is added to the fractional dead time.

Also, where a system has to be deactivated for the test, there is a probability that the tester will forget to rearm the system. If the probability of failing to rearm the system is P, then the full equation for fractional dead time becomes

Fractional dead time $= \frac{1}{2} fT + t/T + fR + P$

Note that the above formulae apply only if $fT \ll 1$ and if $DT \ll 1$, that is, if there is a very low probability of a demand occurring, or of the protective system failing, in any particular period between protective tests. (This is not an onerous restriction, as any instrument system, etc., that had a high probability of failing between tests would attract attention and action would be taken to reduce the probability.)

6.2.5 Hazard Rate with an Untested Protective System

It is bad practice to install a protective system and not to test it. An approximate formula for calculating the hazard rate for a system comprising a demand and an untested protective system is:

$$HR \text{ (untested protection)} = \frac{Df}{D+f}$$

This can be compared with Example 6.2.2 used above to illustrate fractional dead time. If the level controller fails with a frequency of 0.1 per year, and the (untested) high-level alarm is of a type that typically fails with a frequency of 0.1 per year, then the hazard rate would be:

$HR = (0.1 \times 0.1)/(0.1 + 0.1) = 0.01/0.2 = 0.05$ per year

Thus the overflow frequency of once per 10 years is reduced by a factor of only 2 by using an untested high-level alarm, compared with a factor of 80 if the alarm is tested quarterly.

The usefulness of an untested protective system is even more dramatically limited when the demand frequency is very low compared with the protection failure frequency. For example: An electrical switch room is located where it is just conceivable that a leak of flammable gas could enter it through its ventilation system. A flammable gas detector is installed in the air intake, to shut down the ventilation system in the event of flammable gas being detected.

It is estimated that:

- the frequency of gas leaks reaching the ventilation air intake is 0.001 per year;
- the frequency of failure of the gas detector is 0.2 per year.

It is to be decided whether, in view of the low likelihood of the gas leak reaching the switch room, it is really necessary to test the gas detector at the normal frequency of once per 3 months, or whether it would be reasonable to leave it off the testing schedule altogether.

There are three cases to be considered.

Case A: If the gas detector is not installed at all, then the hazard rate will be the same as the demand frequency: 0.001 per year, that is, 1 in 1000 per year.

Case B: If the gas detector is installed, but not tested, then the hazard rate will be:

$HR = Df / (D+f) = 0.0002/0.201 = 0.000995$ per year

that is, 1 in 1005 per year. This is effectively the same result as Case A: the untested gas detector contributes almost nothing to the safety of the switch room.

Case C: If the gas detector is installed and tested every 3 months, the hazard rate will be:

$HR = D \times \text{FDT} = 0.001 \times 0.5 \times 0.2 \times 0.25 = 0.000025$ per year

that is, 1 in 40,000 per year. This is an improvement by a factor of 40.

Thus a two-part principle emerges:

- If a protective system is installed, it must be tested.
- If it cannot be tested, it should be removed to prevent a false sense of security.

6.3 QUANTIFYING INCIDENT FREQUENCY ON FAULT TREES

6.3.1 Introduction

Earlier in the chapter, two forms of logical connection were shown: the AND gate and the OR gate.

Where there are two independent events, the probability that both will occur is:

$$P(\mathrm{A} \cdot \mathrm{B}) = P(\mathrm{A}) \times P(\mathrm{B})$$

(but watch out for hidden linkages between the two events, i.e., make sure that they are really independent of each other).

The probability that one or the other will occur (i.e., A or B) is:

$$P(\mathrm{A} + \mathrm{B}) = P(\mathrm{A}) + P(\mathrm{B}) - P(\mathrm{A}) \times P(\mathrm{B})$$

As $P(\mathrm{A})$ and $P(\mathrm{B})$ are usually small, the third term above is usually negligible compared with the sum of the first two terms (but care should be taken that it is not ignored when the probabilities are not small and the third term is thus not negligible).

6.3.2 Rules for Quantifying Events on a Fault Tree

Watch out for the trap of multiplying frequencies, and thus getting possibly attractive results which are totally wrong.

If the components which are needed continuously (i.e., those which will generate a demand when they fail) are specified in frequency units, and all the other components are specified in probability units, then there is usually no problem. But at each gate, check that the following rules are not being broken:

- Frequencies are added at an OR gate (getting a frequency result).
- Probabilities are added at an OR gate (getting a probability result).
- Frequencies and probabilities cannot be added (mixed units: meaningless).
- Frequencies cannot be multiplied (frequency squared units: meaningless).
- One frequency can be multiplied with probabilities at an AND gate (frequency result).

6.3.3 Quantification of a Fault Tree

The fault tree shown in Figure 6-6 is quantified at the top of the next page using the following indicative (fail-to-danger) data:

Protective systems are tested four times per year.

Note that the pump failure is the only demand. The other elements are all part of the protective response system. Therefore the pump failure

Pump	1.5 per year
FE	0.02 per year
FS	0.1 per year
SV	0.1 per year
TCV	0.05 per year
FAL	0.05 per year
Operator	0.1 probability
GIV	0.005 per operation (i.e., a probability of 0.005 of failing when needed)

must be expressed as a frequency, and all the other elements must be expressed as dimensionless numbers—FDT, probability, etc.

Thus the fractional dead times are:

FE: $FDT = 0.5 \times 0.02 \times 0.25 = 0.0025$

FS: $FDT = 0.5 \times 0.1 \times 0.25 = 0.0125$

SV: $FDT = 0.5 \times 0.1 \times 0.25 = 0.0125$

TCV: $FDT = 0.5 \times 0.05 \times 0.25 = 0.00625$

FAL: $FDT = 0.5 \times 0.05 \times 0.25 = 0.00625$

The data given for the operator and for the GIV are already in probability form.

These are inserted into the fault tree and evaluated as shown in Figure 6-10. The total frequency is 0.0256 per year, that is, around 1 in 40 per year.

It is evident, looking at the fault tree, that the reliability of the system depends principally on the reliability of FE and FS, which between them contribute $(0.0025 + 0.0125) = 0.015$ to the total probability of 0.017 of the response failing; that is, they contribute around 88% of the total, even though they are not the most unreliable components. This is because they are needed by both the automatic and the manual protective response systems. If either of them fails, the whole protective response fails.

There is a further point to be made about this example. In the design, the components appear independent, and so mathematically their failures have been treated as independent events. But this overlooks a situation known as *common-mode failure.* In common-mode failure, nominally independent failures occur because of some common factor. In this case, for example, it is possible that the instrumentation has been maintained by an unskilled tradesman who was recently hired. Thus several components may have been rendered inoperable by the same chance event. So the combined probability of several components being in a failed state is

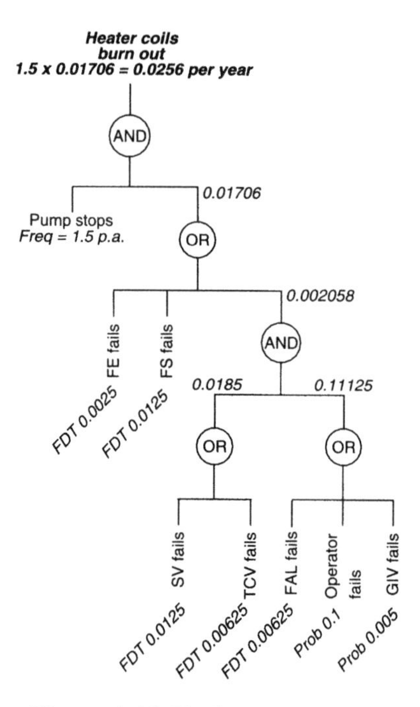

Figure 6-10. Evaluated fault tree.

likely to be much greater than the calculated value. Common-mode failure is discussed more fully later.

Approaches to definition of cost-effective improvement of the reliability of a system, such as that above, using the insights from the fault tree, are discussed later.

6.4 ALTERNATIVE APPROACH TO ASSESSING THE FAILURE FREQUENCY OF A SYSTEM: THE CUTSET APPROACH

A *cutset* is a combination of component failures which will lead to failure of the whole system.

6.4.1 Cutset Method

The approach takes the following steps.

1. By inspection of the control and protection system, one can set out the *success paths* that will prevent overheating of the oil heater

coils. Each success path is a series combination of components or activities which together will prevent the unwanted event. Provided any one of the paths is operating, the overheating will not occur.

Referring to the oil heater example set out earlier, they are:

Path 1	PU(mp)
Path 2	FE, FS, SV, TCV
Path 3	FE, FS, FAL, OP, GIV

2. Identify the combinations of component failures which will inactivate all of the paths.

First, one looks for single components that, if they failed, would inactivate all of the paths. There are none.

Next, one looks for combinations of two components which would inactivate all paths:

Path 1	PU
Path 2	FE, FS, SV, TCV
Path 3	FE, FS, FAL, OP, GIV

The appropriate combinations of two components are:

PU · FE

PU ; FS

These are called cutsets, that is, sets of components that would cut all the paths. (PU would inactivate path 1, and either FE or FS would inactivate both path 2 and path 3.)

Next, one looks for combinations of three components. One does not consider any combinations which include the combinations already noted as cutsets, as to do so would violate the Boolean identity ($A + AB = A$).

The combinations with three components are:

PU, SV, FAL

PU, SV, OP

PU, SV, GIV

PU, TCV, FAL

PU, TCV, OP

PU, TCV, GIV

Table 6-1
Evaluation of Cutsets

Cutsets	Frequency FDT/ Probabilities	Cutset Frequency
PU, FE	*1.5*× 0.0025	*0.00375* per year
PU, FS	*1.5*× 0.0125	*0.01875*
PU, SV, FAL	*1.5*× 0.0125 × 0.00625	*0.000117*
PU, SV, OP	*1.5*× 0.0125 × 0.1	*0.001875*
PU, SV, GIV	*1.5*× 0.0125 × 0.005	*0.000094*
PU, TCV, FAL	*1.5*× 0.00625 × 0.00625	*0.000059*
PU, TCV, OP	*1.5*× 0.00625 × 0.1	*0.00094*
PU, TCV, GIV	*1.5*× 0.00625 × 0.005	*0.000047*
	TOTAL	= *0.0256* per year

These, too, are cutsets.

In this example there is no need to look for combinations of four components, as there are only three paths.

The full list of cutsets can then be evaluated, using the frequency, FDT, and probability data as before. Note that only one element in any cutset can be a frequency, and that if any cutset contains a frequency then they all must (to ensure that all cutsets have the same units, either frequency or dimensionless). In Table 6-1, frequency values are shown in italics. These values are the same as those calculated using the fault tree method.

6.4.2 Discussion

When quantifying the frequency of the top event using the cutset approach, note that only one component of each cutset can be a frequency (shown in italics in Table 6-1). All the others must be probabilities. In the tabulated case, the first component in each case is a frequency, and the others are probabilities or fractional dead times (i.e., dimensionless).

In the same way as the quantified fault tree can be used in diagnosis of the principal contributors to unreliability, so the cutsets can be inspected to show the main contributors. In the case of this example, they are clearly PU · FS and PU · FE; together they contribute around 88% of the total.

Although one could aim for higher reliability by selecting more reliable components for FE, FS, and the pump, it would be better to restructure the automatic and manual protective systems to be independent of each other, so that there are no one-component or two-component cutsets. This is evident by considering the magnitude of the three-component cutsets in the example, where their total contribution is only around 12% of the total.

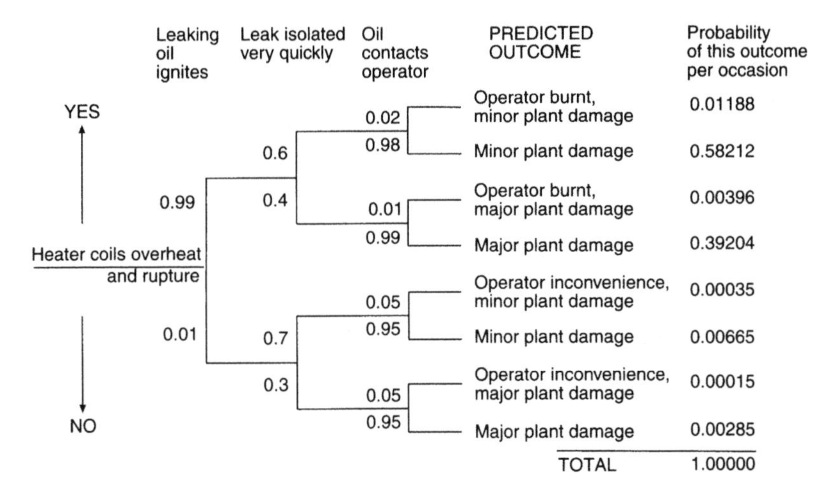

Figure 6-11. Event tree.

In a complex system, both fault tree analysis and cutset analysis can become very laborious if undertaken manually and are open to error. There are computer packages available which handle fault tree analysis, cutset analysis, and calculation of the total failure frequency.

6.5 ASSESSING THE PROBABILITIES OF VARIOUS OUTCOMES USING EVENT TREES

It has been estimated using the fault tree above that the heater coils would overheat with a frequency of around 0.0256 per year, that is, around 1 in 40 per year. The outcome of such a coil rupture and fire could depend on a variety of circumstances, such as whether the operator had warning and could escape in time, and whether he or she could warn others to keep clear.

The probabilities of these various outcomes can be estimated with the aid of an *event tree*. See Figure 6-11.

An event tree is a branching structure, each branch representing a possible outcome to a choice. Commonly, but not always, the choices are Yes/No. In some instances multiple outcomes may be possible. On the tree are marked the estimated probability of the various outcomes at each choice. Then the total probability of any final outcome can be calculated by multiplying the probabilities at each choice leading to that final outcome.

Sorting similar outcomes together (noting that the frequency of the heater coils overheating and rupturing was previously assessed as 0.0211 per year) the following results are obtained:

Outcome	Probability per Occasion	Frequency of Occurrence (per year)
Operator burned	0.01188 + 0.00396 = 0.01584	0.01584 × 0.0212 = 0.00034
Operator inconvenienced	0.00035 + 0.00015 = 0.0005	0.0005 × 0.0212 = 0.00001
Major plant damage	0.00396 + 0.39204 + 0.00015 + 0.00285 = 0.399	0.399 × 0.0212 = 0.00846
Minor plant damage	0.01188 + 0.58212 + 0.00035 + 0.00665 = 0.601	0.601 × 0.0212 = 0.01274

For example, if it is estimated that there is a 5% probability of burns to the operator proving fatal, then the fatal accident frequency of the operator would be:

$$0.00034 \times 0.05 = 1.7 \times 10^{-5} \text{ per year}$$

6.6 CALCULATION OF RELIABILITY OF UNITS WITH INSTALLED SPARES

A special case, which is important in a process plant, is the reliability of equipment which has installed spares. A typical case is a pair of pumps, installed in parallel, with one normally operating and the other acting as a standby spare.

Example 6.6.1:

Consider the case of two such pumps, which are identical (see Figure 6-12).

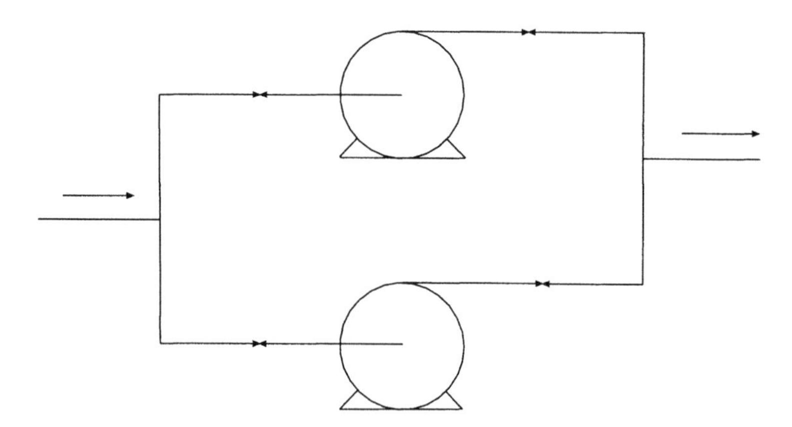

Figure 6-12. Pump station.

Examination of records provides the following information:

- There have been 20 failures of one or other of the pumps in 10 years, and there is no reason to believe that one pump has had significantly more failures than the other.
- On average, when a pump breakdown occurs, and the other pump is put on-line, the pump which has failed is out of service for 5 days for repair.

How often will there be no pump available, because the operating pump has failed before the other pump has been repaired?

Solution:

The frequency of failures of the operating pump is 2 p.a.

The average duration per year for which a pump is being repaired is thus

$2 \times 5 = 10$ days per year

Therefore the probability that there is only pump in operable condition is

$10/365 = 0.0274$

The operating pump fails with a frequency of 2 per year. The probability that it will fail when there is no operable spare is 0.0274. Therefore the frequency of the operating pump failing when there is no operable spare is:

$2 \times 0.0274 = 0.0547$ per year

that is, around once per 18 years on average.

6.7 AVAILABILITY AND MODELING THE PRODUCTION CAPABILITY OF A PLANT

Where a plant comprises numerous items of equipment, it can be very important to determine the production of which it will be capable. This is not as straightforward as it may appear at first sight.

For example, consider a hypothetical plant comprising five sections, all in series, all with the same production rate of 100 units per hour. It may be planned that the plant operate for the whole year except for a planned maintenance shutdown of all sections at the same time. While the plant is producing, it is expected that the production rate will be 100 units per hour.

If each section is expected to be off-line for 1% of the time because of random breakdown and repair time then, assuming that there is no buffer storage between sections, a breakdown of any section will result in the other sections being shut down while the failed section is repaired. The production time, for the period when production is planned (i.e., not the

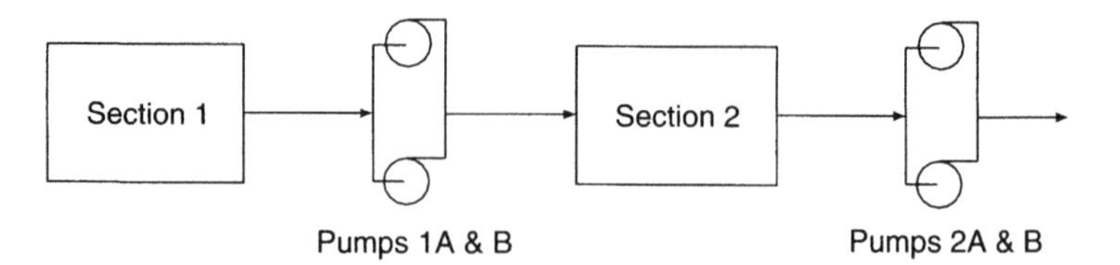

Figure 6-13. Simplified production configuration.

planned maintenance shutdown), will be $(0.99)^5 = 0.951$, that is, around 95%, not 100%.

In undertaking such an assessment, it is necessary to have an estimate of the availability of the sections.

$$\text{Availability} = \frac{\text{up time}}{\text{uptime} + \text{downtime}}$$

$$= \frac{\text{MTBF}}{\text{MTBF} + \text{MTTR}}$$

where MTBF is mean time between failures and MTTR is mean time to repair.

(In the preceding example, each section had an availability, between planned maintenance shutdowns, of 99%.)

Where it is desired to determine the likely production performance of a complex facility, comprising many items of equipment in series and parallel configurations, each with its own availability and with variable production rate, it is very helpful to develop a mathematical model.

One approach to such a model can be to use a computer package such as the @RISK package, which is an "add-on" to the better known common computer spreadsheet packages. This package enables one to insert into any cell in the spreadsheet, not just a single number, but the specification of a distribution.

For example, consider the simple plant configuration in Figure 6-13. This plant can be described in a spreadsheet as shown in Table 6-2.

The @RISK package then undertakes a "Monte Carlo" simulation (with multiple runs using random numbers) to determine the probability distribution of the output of the plant. The results are shown in Table 6-3 and plotted in Figure 6-14.

Table 6-2
Computer Spreadsheet Showing Simulation Coding

	A	B	Comment
1	Equipment	Production	
2	SECTION 1 CAPACITY	=RiskTriang(95,100,105)	Min rate 95, most likely 100, max 105
3	SECTION 1 ON LINE TIME	=RiskDiscrete({0,1},{0.01, 0.99})	1% off-line time
4	SECTION 1 OUTPUT	=B2*B3	
5	PUMP 1A	=RiskDiscrete({0,100}, {0.02,0.98})	2% downtime, normal rate 100
6	PUMP 1B	=RiskDiscrete({0,100}, {0.02,0.98})	2% downtime, normal rate 100
7	PUMP 1 CAPACITY	=B3+B4	Total capacity available (on paper) if needed
8	PUMPING MAXIMUM	100	Only one pump on-line at a time
9	PUMP SET 1 OUTPUT	=MIN(B7,B8)	
10	SECTION 2 CAPACITY	=RiskTriang(90,100,110)	Min rate 90, most likely 100, max 110
11	SECTION 2 ON LINE TIME	=RiskDiscrete({0,1},{0.01, 0.99})	1% off-line time
12	SECTION 2 OUTPUT	=B10*B11	
13	PUMP 2A	=RiskDiscrete({0,100}, {0.05,0.95})	5% downtime, normal rate 100
14	PUMP 2B	=RiskDiscrete({0,100}, {0.05,0.95})	5% downtime, normal rate 100
15	PUMP 2 CAPACITY	=B7+B8	Total capacity available if needed
16	PUMPING MAXIMUM	100	Normally only one pump on-line at a time
17	PUMP SET 2 OUTPUT	=MIN(B15,B16)	
18	OUTPUT FROM PLANT	=MIN(B4,B9,B12,B17)	Max rate is determined by the bottleneck at the time

The graph displays the probability of the output being equal to or less than the nominated value. For example:

- there is a 45% probability (approx.) of the output rate being 98 or less;
- some of the time there is no output; and
- the time-weighted average output rate is 95.8.

The full output from the analysis enables analysis of the contribution of each component to the overall loss of production.

Table 6-3

5 percentile	92.5
10 percentile	94.0
20 percentile	95.8
30 percentile	96.8
40 percentile	97.6
50 percentile	98.3
60 percentile	99.0
70 percentile	99.7
80 percentile	100.0
90 percentile	100.0
100 percentile	100.0
Minimum rate:	0
Maximum rate:	100
Mean rate:	95.8

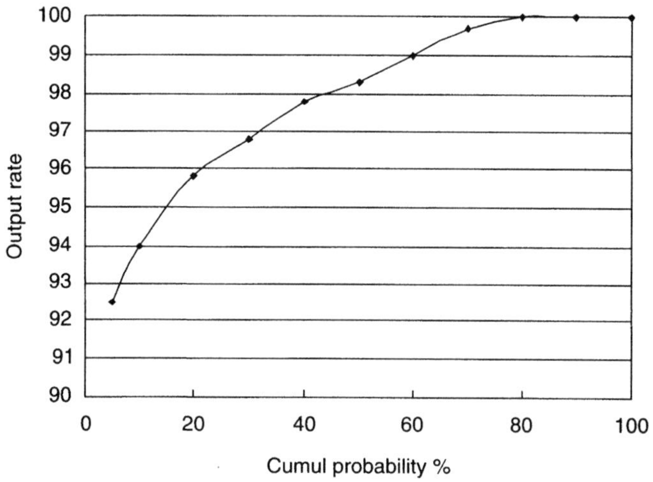

Figure 6-14. Output rate versus probability.

Where there are buffer storages which cushion loss of production by upstream and downstream equipment, the simulation becomes more complex, and special-purpose simulation models need to be developed. Typically, such models represent the production pattern for a series of time periods (e.g., each day per quarter), with planned shutdowns of equipment or plant sections defined for specific days, and unplanned breakdowns, etc., of equipment or plant sections generated randomly by the model. Thus the production by each unit over the quarter, and the rise and fall of levels in intermediate storages, can be seen. By running the model for many iterations, with the random numbers generating different patterns of breakdowns, etc., the interaction of the various items of equipment and their effect on production can be seen.

Such a model is a very useful guide when investigating options for debottlenecking—for example, investigating the case for increasing buffer storage capacity at various parts of the process, increasing capacities of particular equipment or plant sections, or installing additional standby equipment.

6.8 METHODS OF IMPROVING RELIABILITY OF CONTROL AND PROTECTIVE SYSTEMS

Approaches to improving reliability of control and protective systems are discussed in more detail in Chapter 8.

In outline, where reliability of hazardous systems is to be improved, it is preferable in principle to attempt to reduce the inherent hazard first; then to attempt to reduce the demand frequency by improving the containment and control components (including the "human factors"); then to attempt to improve the fractional dead time of the protective systems; and finally to attempt to improve the separation and strengthening or resistance of exposed "targets." These, too, are discussed in Chapter 8.

Where a protective system has an inadequate reliability, that is, too high a fractional dead time, there are several possible approaches:

- reduce the failure frequency by selecting different protective equipment or by diagnosing the causes of failure and engineering them out;
- increase the testing frequency; or
- install "redundant" systems.

If two independent protective systems are installed to perform the same duty, the overall reliability of protection is usually improved, because if one of the systems fails, there is a probability that the other will still be operational. There is said to be *redundancy* in the system; normally the two protective systems would not be needed.

On paper, if each of two redundant protective systems had a fractional dead time of 0.01, then at first sight it would be expected that the FDT of the two would be:

$$0.01 \times 0.01 = 0.0001$$

However, there are several reasons why this very attractive FDT is normally not achieved in practice:

- the two systems are probably tested on the same schedule, so they are not fully independent;
- there are usually environmental and other effects which, if one of the systems has failed, lead to a higher probability that the other will fail soon also (common-mode failure);
- as each of the two systems has a probability of operating spuriously (*spurious trip*), with two such systems the frequency of spurious trips will be double, and there will be a tendency for operators to attempt to deactivate the trip system altogether.

The fractional dead time of simple redundant systems which are tested on the same schedule is as follows:

"One out of two" system: $FDT = (4/3)\ (FDT1 \times FDT2)$

"One out of three" system: $FDT = 2(FDT1 \times FDT2 \times FDT3)$

"One out of four" system: $FDT = (16/5)\ (FDT1 \times FDT2 \times FDT3 \times FDT4)$

("One out of two" means any one of the two systems can actuate the protective function.)

Common-mode failures can be minimized by using different types of system for the two protective systems. This is termed diversity. Where high reliability is sought, diversity should be incorporated to a high degree, with different types of system, based on different principles, with instrument cabling following different routes, possibly with different design teams and different maintenance tradesmen used for the systems. This effort is needed because of the difficulty of positively identifying the many forms in which common-mode failures can occur.

Where redundant systems are installed because high reliability is needed, and where the consequently increased frequency of spurious trips may be a problem, an approach to minimize spurious trips is to use a *voting system*, in which three separate, diverse protective systems send a signal to a simple and reliable logic processor such that:

- any one protective system operating will raise an alarm;
- any two protective systems operating will actuate the trip.

Because of the reliance of such a system on the reliability of the voting component and the actuated mechanism (e.g., the shutdown valve), sometimes two parallel voting systems may be installed. The cost of such elaborate systems is very high, both initially and in use, with high maintenance costs. It is therefore preferable to attempt to avoid the need for very high-integrity protective systems, for example, by designing for minimal inherent hazards.

Referring to the fault tree in Figure 6-10, there are several options for improving the reliability.

The most important one is to separate the protective system from the control system. If the control system fails because of failure of FE, then the alarm and the low-flow protection system cannot operate. If, however, the low-flow alarm and trip systems are actuated by a low-flow switch actuated independently of FE, the reliability of the system will be greatly improved.

Important Principle: Design protective systems to be able to operate entirely separately from the control systems, and check for the possibility of common-mode failures of the control and the protective systems.

As a general guide, the following levels of FDT are achievable:

FDT = 0.01	A simple system, regularly tested and reasonably maintained.
FDT = 0.001	The practical limit for process plant, unless designed and tested by High Integrity Protective System (HIPS) specialists, and maintained and tested to those standards.
FDT = 0.0001	Only in nuclear installations, or process plant with unusually high standards of operation, maintenance, supervision and management, and a benign operating environment.

6.9 SOURCES OF FAILURE DATA

6.9.1 Introduction

The most appropriate data for a particular installation are data collected from that installation. Generic data from external sources can give a guide, but the relevance of such data to any specific plant is always unclear, especially where there are notable local environmental (e.g., corrosion, high or low temperature) conditions, and dubious local standards of operation, maintenance, supervision, and management.

6.9.2 Some Limitations of Data

The effect on failure frequencies of standards of operation, maintenance, supervision, and management is discussed in Chapter 11.

6.9.3 External Sources of Data

These include:

- *IEEE Std 500 (1984).* "IEEE Guide to the Collection and Presentation of Electrical, Electronic, Sensing Component and Mechanical Equipment Reliability Data for Nuclear Power Generating Stations." (Data, with upper and lower bounds, for very high-quality and expensive systems, maintained to high standards.)
- MIL-HDBK-217F *(2002). Military Handbook—Reliability Prediction of Electronic Equipment.* (For expensive equipment.) U.S. Department of Defense.
- *NPRD-2 (1981).* "Non-electronic Parts Reliability Data." Reliability Analysis Centre at the Rome Air Development Centre.
- OREDA *(1998). Offshore Reliability Data Handbook.* SINTEF: Trondheim, Norway.
- F. P. Lees *(1996). Loss Prevention in the Process Industries.* Butterworth–Heinemann, Oxford. UK.

6.9.4 Preliminary Failure Data

The failure data in Table 6-4 may be used in preliminary risk assessments as indicative of the frequencies which may apply. Note that it is always preferable to use data collected from the particular facility or industry being studied. The applicability of the data given in these tables to any particular facility being studied must be assessed by those undertaking the study.

Some studies suggest that the frequencies tabulated for new pipelines are very pessimistic. (See Corder and Fearnebough, 1987.) For example, it has been determined (in an unpublished study) that the average for Australasia has been 33×10^{-6} per km-year for the total of complete rupture plus major leaks. (See Endersbee, 1992.)

Probability of ignition of gas leaking from major pipeline failure.

Such probabilities are very dependent on the surroundings of the leak and the cause of the leak. For example, if the leak is in a populated area, or if it is caused by mechanical equipment, the probability of ignition is

Table 6-4

Pressure vessels (where these are subject to regular statutory inspection)		
Large leak (50 mm dia)	1×10^{-6} to 10×10^{-6} per year (typically 2×10^{-6} p.a.)	
Small leak (6 mm dia)	10×10^{-6} to 100×10^{-6} per year (typically 20×10^{-6} p.a.)	
Pipelines in plants (not applicable where environment is highly corrosive, or where special care is taken of specific pipework)		
Full rupture (100% cross section)	2×10^{-8} L/D per year	
Large leak (20% cross-sectional area)	5×10^{-8} L/D per year	
Small leak (6 mm dia)	3×10^{-7} L/D per year	
Centrifugal pumps		
Pump (single mechanical seal)—5 mm dia equivalent leak	5×10^{-2} per year	
Pump (double mechanical seal)—5 mm dia equivalent leak	1×10^{-3} per year	
Pump seal (single mechanical)—full annulus leak	1×10^{-3} per year	
Pump seal (double mechanical)—full annulus leak	3×10^{-4} per year	
Shaft pulled out—full shaft dia leak	1×10^{-4} per year	
Major casing leak—pump suction 100% dia leak	1×10^{-5} per year	
Flanged Joint failures (diameter equivalent to one section of gasket between bolt holes)		
Normal homogeneous gasket material[a]	4×10^{-5} per year	
Spiral-wound gasket[b]	2×10^{-5} per year	
Cone roof fuel stock tank fires	Highly flammable	Flammable
Full tank diameter fire	2×10^{-4}	2×10^{-5} per tank year
Bund fire	1×10^{-4}	1×10^{-5}
Floating roof stock tank fires		
Seal fire (peripheral only)	4×10^{-3}	4×10^{-4}
Full-diameter fire	4×10^{-4}	4×10^{-5}
Bund fire	1×10^{-4}	1×10^{-5}

Cross-country oil or gas pipeline (where constructed to a recent code, protected from internal and external corrosion, and regularly patrolled and inspected)

Failure Mode	Complete Rupture of Pipeline	Hole Equivalent to 20% Cross-sectional Area	Hole Equivalent to 5% Cross-sectional Area
External interference			
Thickness component	75 pmpy per km	150 pmpy per km	75 pmpy per km
Diameter component	50	25	—
Construction defect or material failure	25	100	100
Corrosion	1	2	100
Ground movement	—	—	—
Total	151	277	275

[a]Hole area = internal circumference of pipe × thickness of gasket/number of bolts.
[b]Hole area = 4 mm^2.

greater than if the leak is in an isolated region and there is no evident external ignition source. Typical probabilities are as follows:

Description of Leak Source	Probability of Ignition (%)
Pinhole or crack	2
"Hole"	5
Rupture (less than 400-mm dia pipe)	10
Rupture (400-mm dia pipe or larger)	50

Some studies by others suggest that these probabilities may be rather low, that is, that an ignition probability of around 50% may be a reasonable estimate for use in risk studies, if the leak is substantial. But the likely conditions around the leak, in particular the mechanism causing the leak, need to be considered in making the estimate for any particular scenario.

Road bulk tanker—pressurized. The frequency is highly dependent on driver selection, training, vehicle maintenance and traffic conditions, etc.

Significant leak of material (100 kg or more): 0.03 per million km traveled

Road bulk tanker—atmospheric pressure. Also highly dependent on driver selection, training, vehicle maintenance and traffic conditions.

Spills of material	0.1 per million km traveled
Proportions	
<150 kg	50%
150–1500 kg	10%
>1500 kg	40%

Control system (single-loop electromechanical).

Simple temp, pressure, level: fail to danger (without warning)	0.1 per year
Complex temp, pressure, level: fail to danger (without warning)	0.2 per year
Analysis	0.4 per year

Alarm system (electromechanical).

Fail to inactive (Note: failure frequency must be converted to fractional dead time) 0.1 per year

Trip system (electromechanical).

Fail to inactive (may be dominated by actuated equipment, for example, valve; failure frequency must be converted to fractional dead time) 0.2 per year

Boiler safety relief valve.

Fail to lift at or below 150% of set pressure	0.02 per demand
Lift at less than 90% of set pressure	0.1 per demand

REFERENCES

Corder, I., and Fearnebough, G. D., "Prediction of Pipeline Failure Frequencies." British Gas plc Research and Technology Station, Newcastle on Tyne, UK. Proc. Second International Conference on Pipes, Pipelines, and Pipeline Systems, Utrecht, The Netherlands, June 1987.

Endersbee, L., "Risk Assessment of Pipelines." Report to the Minister for Manufacturing and Industry Development, Victoria, Australia, 30 June 1992.

Chapter 7

Quantitative Risk Assessment: Computer Modeling, Uses in Setting Buffer Zones, Strengths and Limitations, Uses and Abuses

When you have studied this chapter, you will understand:

- broadly how computer programs can be used to assess the total risks to people around process plants, and to generate contours of risk as an aid to risk management;
- the use of quantitative risk assessments in determining buffer zones around process plants (to conform to statutory risk criteria as discussed in Chapter 4); and
- the uses and limitations of quantitative assessments of risk.

This chapter addresses the risk management step 8 "Assess Risk" of Figure 1-4.

7.1 MODELING THE RISK

7.1.1 Introduction

Many operating plants have numerous vessels, pipes, and other items of equipment which contain hazardous materials. These materials may present risks of fire, explosion, toxic gas escape, etc.

For any one vessel, pipe, or item of equipment, it is possible to postulate a range of escapes due to various causes, with severities ranging from small to large, and relatively frequent to relatively infrequent.

Further, some types of escape, such as vapor cloud explosions and toxic gas escapes, are markedly affected by the weather at the time of the escape, affecting the direction and extent of drift of the gas cloud, and so affecting the risks to people or plant in various directions.

To calculate the risks at any one point from the full range of possibilities for a range of weather conditions requires a large amount of repetitive calculation. So this type of assessment is commonly done using a computer system. A number of such systems are available as commercial packages.

Note that any such system is only as good as:

- the selection of incident types built into the computer system,
- the base data and models used in it for calculating the severity of the consequences of the postulated incidents, and
- the base data and the models used in it for calculating the frequency of each of the postulated incidents.

All such systems contain highly debatable assumptions. The strengths, limitations, and applications of this type of approach are discussed in detail later.

7.1.2 Description of a Typical Computer System

A typical computer system for calculating risks directly from a process flowsheet does so in two main steps:

- generation of data about possible incidents, and
- calculation of the risks arising from those incidents.

The structure of a very basic computer program for calculation of risks from fire and explosion is outlined below to illustrate those steps. With some adaptation (and inclusion of data about wind and atmospheric stability) the broad approach is applicable to risks from toxic gas escapes.

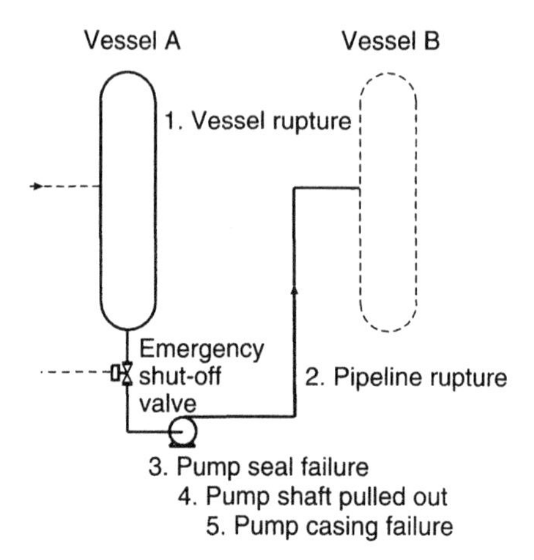

Figure 7-1. Illustration of flowsheet subsection.

7.1.2.1 Generation of Data about Possible Incidents

The flowsheet is subdivided into vessels containing a significant inventory of hazardous material, together with the pipework and equipment connected to each vessel and capable of generating a leak of sufficient magnitude to cause a substantial incident. One such subdivision is illustrated diagrammatically in Figure 7-1.

The following types of incident are considered:

1. Vessel rupture
2. Pipeline leak:
   - 6-mm hole
   - 20% cross-sectional area
   - 100% cross-sectional area
3. Pump seal leak, cross section area equivalent to an annulus of 3 mm width of shaft diameter
4. Pump casing failure, equivalent to 100% rupture of the suction pipeline

The following data are input to the program to enable calculation of the magnitudes and frequencies of the incidents to be used in the risk assessment:

- Vessel location (grid coordinates)
- Vessel inventory (for calculation of size of BLEVE or of vapor cloud explosion)

- Process conditions:

  name of the process fluid

  pressure (for calculation of the escape rate)

  density of the process fluid (for calculation of the escape rate)

  operating temperature (for calculation of the extent of flash vaporization)

  atmospheric-pressure boiling temperature (for calculation of the extent of flash vaporization)

- Pipeline length and diameter (for calculation of the frequency and rate of leakage)
- Whether there is an excess flow valve (yes/no) at the vessel delivery to the pipeline
- Excess flow setting
- Probability that the excess flow valve will fail to operate on excess flow
- Whether there is an emergency shutoff valve at the vessel delivery to the pipeline
- Expected time between start of leak and actuation of the emergency shutoff valve (e.g., 3 minutes)
- Probability the emergency shutoff valve will fail to close when actuated
- Pump suction diameter
- Pump shaft diameter
- Frequency of seal collapse
- Frequency of shaft fracture and pulling out of casing
- Frequency of casing failure

The program incorporates formulae for calculating the frequency of pipeline leaks (based on the pipe diameter and length), the probability of ignition of leaks (based on their magnitude and duration), etc.

The program then generates the following range of possible incident magnitudes and frequencies:

- Vessel rupture and fireball (BLEVE)
- Vessel rupture and vapor cloud explosion
- Pipeline failure (6-mm dia, 20% and 100% cross-sectional area) leak and fire

- Pipeline failure (6-mm dia, 20% and 100% cross-sectional area) leak and cloud accumulation for 1-, 3-, and 6-min durations before ignition and vapor explosion
- Pump failure (seal, shaft, casing) with fire
- Pump failure (seal, shaft, casing) with vapor cloud accumulation for durations of 1, 3, and 6 min before ignition and vapor explosion

The frequencies of all leaks (pipeline or pump) downstream of any excess flow valve are adjusted by the probability of failure of actuation of the valve if the flow exceeds the set point. Similarly, the frequencies of all leaks for durations exceeding the expected actuation time for any emergency shutoff valve are adjusted by the probability that either the valve will not be actuated or it will not close when actuated.

The following incident data are stored in the computer file.

- Vessel name
- Vessel location (grid coordinates)
- Process fluid name
- Leak frequency
- Fire frequency
- Explosion frequency
- Leak rate
- Leak duration
- Fire heat radiated
- Explosion magnitude (equivalent mass of TNT)

7.1.2.2 Calculation of the Risks Arising from the Postulated Incidents

Figures 7-2 and 7-3 illustrate, in simplified form, a typical method of computation of risk contours around a complex plant handling flammable hazards, given data about a number of incidents of various magnitudes and durations at a variety of locations. Those data may be generated by a program such as that outlined above, or by some other means.

The steps are:

1. For the site under study, draw a scaled grid over a site plan.
2. Prepare a data file of the postulated incidents which could arise from the various hazards on the site. The data file would contain the following for each type of scenario, as a minimum:

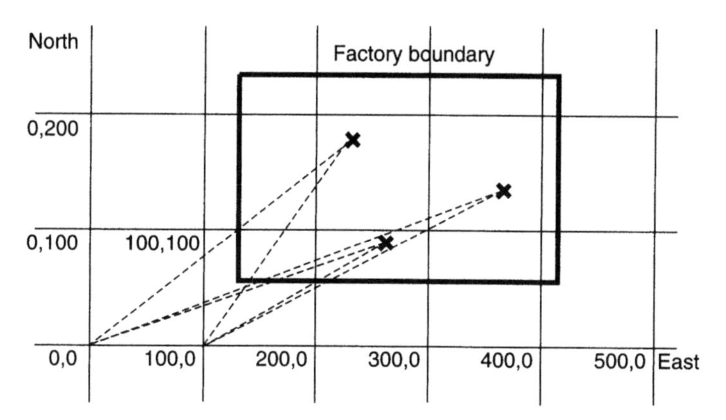

Figure 7-2. Illustration of method of assessing the fatality risk level at grid coordinates. Key: X, Locations of postulated hazardous incidents; 100,0 = coordinates (east, north) of grid points in relation to some datum point.

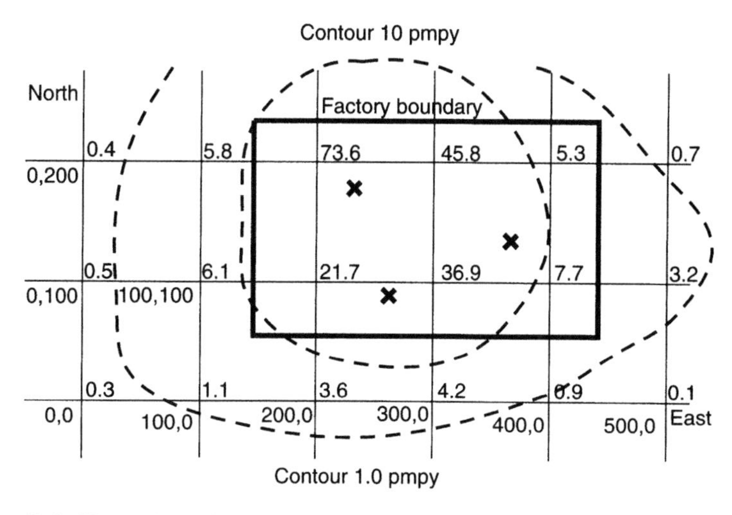

Figure 7-3. Illustration of results of risk assessment showing contour lines drawn manually by inspection, or by computer. Decimal numbers (e.g., 0.3) are accumulated fatality risk from all postulated incidents at that grid point.

- an identifying title or code;
- the coordinates of its location;
- the type of incident (e.g., heat radiation, explosion overpressure);
- the severity of the consequences of the incident (e.g., heat radiated, TNT equivalent mass); and
- the frequency of the incident.

3. Define the limits of the topographical area of interest.

The computer then reads the details of the first incident listed in the data file and undertakes the following processing steps.

1. For each incident in turn:
2. For each grid point in turn:
3. calculate the distance from the source of the incident to the grid point;
4. calculate the physical impact of the incident at that grid point (kW/m^2, kPa);
5. calculate the probability of a person at that location being killed by that impact;
6. using the frequency of the incident and the fatality probability, calculate the fatality risk at that grid point from that incident; and
7. add the risk from that incident to the accumulated total risk at that point.
8. Next grid point....
9. Next incident....
10. Print, either to paper or to a computer file, the tabulated total risks for the various grid points.
11. Either use the printed tabulation of risks at grid points and draw contours manually by inspection, or use the computer file of tabulated risks as input to a computer package for drawing contours.

The same basic approach can be used to assess the frequency of exceeding nominated critical levels of heat radiation or explosion overpressure, by comparing the calculated physical impact at each grid point with the nominated critical level and, if the critical level is exceeded, accumulating the frequency of the incident.

In the case of toxic gas risks, it is necessary to input the toxicity characteristics of each toxic gas (e.g., probit constants), the duration of each incident, and the probability of each combination of wind direction, so as to be able to calculate not only the concentration of the gas at each grid point for each weather combination, but also the probability of such conditions as a step in calculating the frequency of exposure to the toxic gas.

A very valuable addition to such a risk assessment program is the facility to list, for a single nominated location, the incremental component of risk from each postulated incident. This enables the main contributors to any high-risk location to be identified quickly, as an aid to cost-effective risk reduction. This is also useful for subjecting the risk assessment to a "common-sense" check, by reviewing whether the assessed major components would be expected to be so.

7.2 SEPARATION DISTANCES (OR "BUFFER ZONES")

7.2.1 Introduction

An apparently simple question is "How far should proposed Plant 'A' be located from other plants, offices, or housing?" As is often the case, such a simple-sounding question cannot be given a quick answer.

Various codes or regulations specify separation distances for some types of installation. In practice, such distances have often evolved on the basis of "judgment" or compromise between interested parties, with limited technical basis. Sometimes such distances have been based on assessed consequences in the event of the worst possible incident occurring: for example, the distances required around magazines of commercial explosives, which are designed to ensure that housing is not seriously affected if the entire magazine explodes. In the case of a process plant, many of the recommended distances have origins lost in ancient history and have no sound basis at all.

7.2.2 Bases for Setting Separation Distances

There are various bases on which separation distances can be set rationally:

- "worst possible" incident consequences,
- "worst credible" incident consequences, and
- risk.

These are discussed below.

7.2.2.1 Worst Possible Incident Consequences

Following the explosion at Flixborough in the United Kingdom in 1974, some local authorities began to suggest that new installations should be located such that, in the event of the worst possible incident, there would be negligible risk to people (e.g., the only damage to houses would be broken windows). This led to very large separation distances being required. For example, in the case of a pressure storage of 200 tonnes of LPG, if the worst possible incident were postulated (possibly a full release, and the cloud drifting in stable atmospheric conditions), it may be suggested that several kilometers of separation would be needed. This is at least an order of magnitude more than has been required hitherto.

Philosophically there is an inconsistency in the use of this approach to siting of an industrial plant compared with other practices accepted by the community. If the community insists on people being safe in the event of the worst possible incident, then houses and towns should not be permitted downstream of dams, and aircraft should not be permitted to fly over cities.

7.2.2.2 Worst Credible Incident Consequences

To overcome the above inconsistency, a variation was developed: the "worst credible" incident approach. By review of past incidents, views were formed about the types of incident which did and did not occur in practice. The consequences of the worst incidents which were regarded as possible in practice were assessed, and separation distances based on them. Thus the aim was to assure people that, in the event of the worst credible incident, they would be safe.

The difficulty is to agree on the definition of "credible." There has been a wide divergence of opinion.

7.2.2.3 Risk

Philosophically, risk is the best basis for deciding separation distances. All postulated incidents can be taken into account, without any arbitrary cutoff between "credible" and "incredible," and the extent to which the various incidents are taken into account is determined by or weighted by their probabilities. It also becomes possible to define separation distances based on different risk targets for different circumstances.

It is for these reasons that quantitative risk assessment has become extensively used, with the regulatory authorities in Europe and Australia being especially interested.

However, the unquantifiable factors and imprecision inherent in the methods available have led other regulatory authorities (e.g., in the United States—see Appendix E) to prefer evaluation of a selection of quantitatively assessed "worst-case" and more likely "alternative" scenarios for guidance on what separation appears reasonable in the circumstances.

In such an evaluation, the frequency component of risk is covered implicitly by the nature of the worst-case scenario and the selection of the alternative cases.

7.2.3 Use of Risk Assessments to Guide Selection of Separation Distances

An example will be used to illustrate the principles.

It may be decided that the maximum risk to which an individual should be exposed from all risks from a particular industrial facility, including a proposed new plant, is 1 chance of fatality per million per year (pmpy).

It may also be decided that employees on plants adjacent to the proposed new plant should not have their existing risk significantly increased by the new plant. If the existing risks are around an FAR of 2.0 (i.e., 2 fatalities per 100 million worked hours), then this is roughly equivalent to 200 pmpy. It may be decided that people on that plant should not be exposed to risks from any single adjacent plant exceeding 20% of the risks due to their own plant. Thus the maximum risk from any adjacent plant would be set at around 40 pmpy.

To determine the required separation distances, the total risks from the new plant and the existing plants would be assessed (e.g., using a risk assessment system as described above), and the location of the contour for 1 pmpy determined. Noting the philosophical nature of the risk target, the uncertainty of the data, and the approximations inherent in the mathematical representations of the hazardous incidents, it may be concluded that the location of the 1 pmpy contour is an estimate of the location of the nearest housing. In the hypothetical plant shown in Figure 7-4, the required separation from housing would be around 350 meters.

To determine the required separation from the adjacent plant, the risks due to the new plant alone would be assessed, and the 40 pmpy contour plotted. The new plant should, in principle, be located such that the contour does not encroach on the area of the existing plant, provided that allowance may be made for the proportion of time that a person working on the existing plant would be in the area. Similarly, the risks to the proposed new plant from the existing plant would be assessed to ensure that those on the new plant are not exposed to an excessive risk from existing operations. In the example shown in Figure 7-4, the required separation from the adjacent plant would be around 120 meters.

It may be found, when doing an assessment of the risks from an existing plant, that the risks at housing or at the location proposed for the new

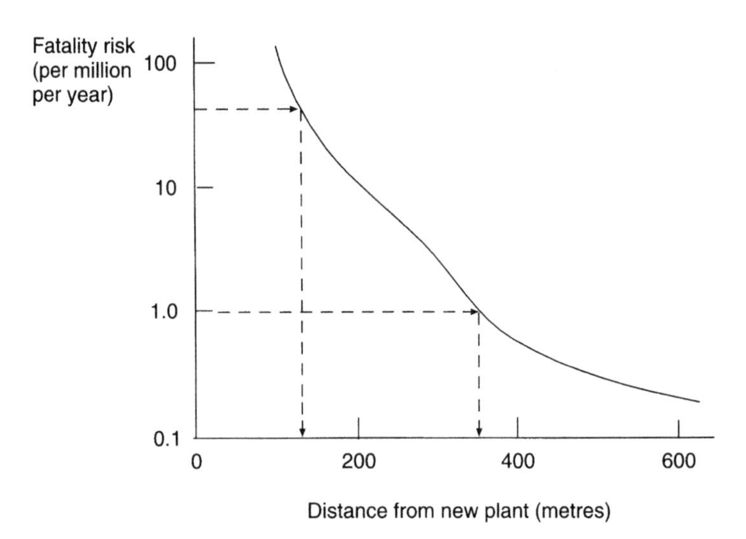

Figure 7-4. Definition of buffer zones by risk assessment.

plant are above the targets, even without the new plant being taken into account. Although the assessments are not precise, it may be decided that construction of the proposed new plant must not result in any net increase of the assessed risks. It may therefore be decided that the proposed new plant could only be permitted if existing risks are reduced by at least an equivalent amount, so that there is no net increase of risk, and preferably a net reduction of risk.

7.3 SOME EXPERIENCES WITH QUANTITATIVE RISK ASSESSMENT

7.3.1 Assessment of Proposed New Plants on an Existing Site

In the late 1970s, a company was planning to build a large petrochemical plant. A crude computer-based risk model was cobbled together by the process safety team and used to assess the relative merits of the competing tenders. The results of the model runs indicated that for two of the tenders the buffer zone between the plant and the nearest housing need be only around 100 meters, whereas in the case of the third it would need to be around 300 meters. (See Figure 7-5 below.) No one believed the absolute magnitude of the figures (100 meters being far too short a distance), but the reason for the third tender needing three times the distance was investigated by the team by exploring back through the calculations performed by the model. The reason was found: a very tangible process feature

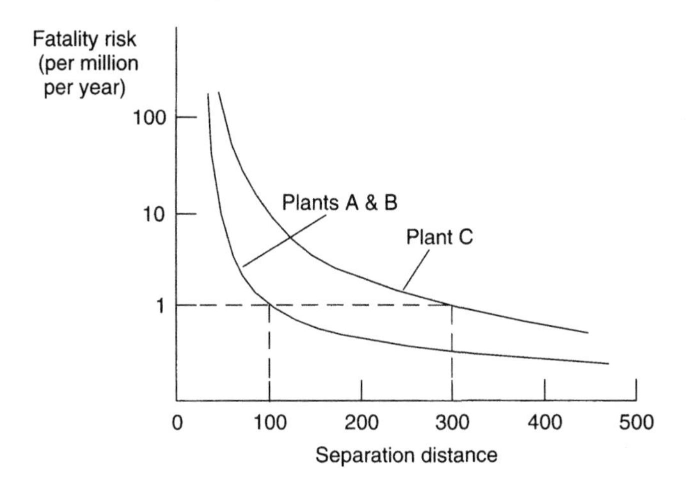

Figure 7-5. Comparison of required buffer zones.

which had not been recognized by anyone else. This was a very valuable discovery.

Following further experience with the crude model, it was completely rewritten. It was tried out on the proposed plant, and a buffer zone of 300 meters calculated, which was substantially less than the 500 meters that was the actual distance. But it was noted that the order of magnitude was reasonable—the model did not say 30 meters, nor did it say 3000 meters, neither of which would have been credible.

Then the new model was run for all the existing hazardous plants on the site. Individually, they appeared to have risks compatible with their location. This gave a sense of comfort to the managers of the plants studied and added to the credibility of the model. Then the model was run to test the cumulative risks from all the plants together (see Figure 7-6).

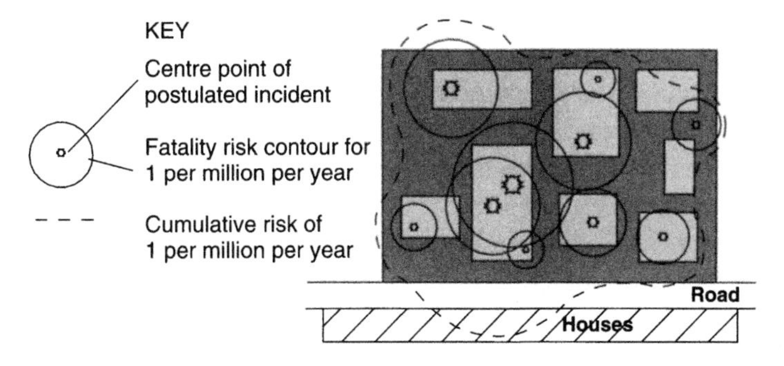

Figure 7-6. Illustration of cumulative risks.

This run found that the assessed risk at the nearest housing was rather above the target (i.e., by a factor of 3), although it was noted that this was probably within the margin of error for the method. At first this raised concern, but then it was remembered that it had been felt by factory staff for some years that housing developments had been allowed to encroach rather too close for safety, but not so close as to be demonstrably unsafe. So the model was expressing quantitatively just what people had concluded intuitively. This further built confidence in the model.

The model was then used to identify the main contributors to the total risk at the housing, enabling attention to be focused on the work that would be most cost-effective in reducing the risk. This was a valuable aid to management.

Then, when another new plant was proposed for a cramped space on the site, the model was run again. With some trepidation the process safety specialist took the results to the project team and said that assessed risks were rather on the wrong side of borderline. The project manager reported that they had been coming to the same view, but for purely qualitative reasons. In view of the commercial importance of the plant, the project team would have had little prospect of insisting on major changes to the design or the proposed location if they had only their subjective and qualitative evidence. But with the quantitative support of the model, which had been widely accepted within the company by that time because of the perceived reasonableness of its earlier findings, these changes were undertaken.

By this time the quantitative model had established a high degree of credibility and acceptance.

7.3.2 Limitations Discovered

A few years later, a risk assessment of a chlorine plant concluded that its level of safety in relation to the nearest housing was satisfactory, apart from some minor improvements to the control and alarm systems. Then a technical manager, while conducting an informal inspection of the plant, noticed that a section of the main liquid chlorine pipeline from the liquefiers to the stock tanks was heavily corroded, with rust scale around 2 centimeters thick on the outside of the 50-mm diameter pipeline (see Figure 7-7). He wondered how the risk assessment could have taken into account the probability of a leak from a badly corroded section of pipeline such as that seen, but then realized that rather than asking and answering that question, the correct action was to replace the pipeline. The risk

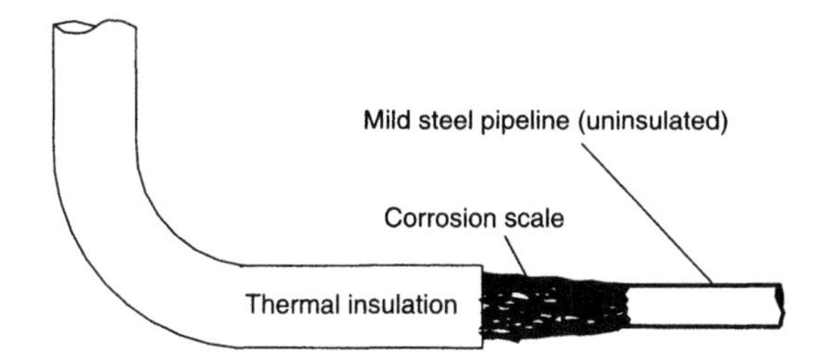

Figure 7-7. Diagram of corrosion on liquid chlorine pipeline.

assessment which had just been completed had assumed that the plant was in good condition, and so was invalidated by the actual condition.

A few months later, the manager of that chlorine plant was visiting the local shopping center around 1 kilometer from the plant. He smelled chlorine and hurried back to the plant. He found that the problem was that the scrubber which removes chlorine from the vent from the chlorine liquefier was not functioning because the scrubber liquor (caustic soda) was fully reacted. During the subsequent inquiry, the following circumstances were discovered:

- The vent was emitting chlorine because the scrubbing liquor was exhausted.
- For some weeks it had been necessary to renew the scrubbing liquor much more frequently than normal, but the operators had not drawn this to the attention of management.
- On this occasion, the operator had forgotten the need to renew the scrubbing liquor so soon.
- The reason that the scrubbing liquor needed renewal so frequently was that the liquefiers were operating poorly, so there was more chlorine gas to be absorbed.
- A recorder chart had been indicating abnormal conditions for some weeks, but this had not been recognized by the operators, supervisors, or management.
- When this was pointed out, the plant staff were unclear what the chart indicated, how to determine whether it was a real or spurious reading, and how to rectify the fault.

The relevance of the quantitative risk assessment to the plant in view of these weaknesses was clearly minimal. It was not appropriate to adjust the assessed risk up (or down) by a factor to take account of the management standard, as the true level of risk is indeterminate, though clearly high.

Rather, it was a situation in which the management standard was not consistent with the standard assumed in the assessment, and hence the risk assessment did not apply to the plant at this time. In the face of such operational ignorance, it is clear that the plant needed, not a quantitative risk assessment, but a program of training at a variety of levels.

7.3.3 The Problem of the "Human Factor"

If one reviews the history of major accidents in the oil and chemical industries over the past 20 years or so, it is evident that the great majority of such accidents have been triggered or enabled not by the types of equipment failures normally predicted and assessed in risk assessments, but from management system weaknesses and human errors such as poor plant isolation before maintenance, poor routine inspection of plant, ill-considered plant modifications, and poor operator training.

Risk assessment relies on probabilistically random and independent events. A poor standard of management systems or management in general results in driven failures, not random failures. It also provides a link and dependence between various contributory factors which may seem, at first sight, to be independent.

Errors like those listed above are not amenable to quantitative risk assessment at present. Some further examples are outlined below.

- In the case of the *Piper Alpha* offshore oil platform disaster in the North Sea, the initial leak and fire did not result from equipment failure, but from a failure in the work permit system.
- An offshore platform in Australasian waters narrowly avoided a similar disaster. It is reported that a pump was isolated for maintenance by closing a power-operated valve, which was tagged "Do Not Operate." There was no positive isolation. While the pump was being worked on, an electrical tradesman worked on the valve actuation system. On completion of his work, the electrical tradesman actuated the controls, the valve opened, and oil escaped and ignited. Fortunately, a manual valve was closed before it was engulfed by the fire.
- A fuel oil tank at a chemical factory exploded during a maintenance shutdown. The vent on the tank had been modified, years before, with an extension almost to ground level intended to act as an overflow. Sparks from flame cutting during nearby maintenance work ignited vapor emerging from the vent pipe, burned back to the tank and exploded, rupturing the tank at the base. The resulting fire killed several people.

• The road transport industry maintains a high standard of training for drivers of hazardous materials. Only experienced and specially trained drivers are said to be given the responsibility of driving LPG tankers. An LPG tanker overturned, fortunately without release of gas, when it was being driven by a trainee driver, who was following an experienced driver. The experienced driver is reported to have commented that he saw the trainee get into difficulties, and that if he, with his experience, had been driving the tanker, he would probably have been able to save the situation.

• A safety analyst visited the site of a major incident involving a storage facility for LP gas a couple of days after it had occurred. What he saw there was, to him, clear evidence of inadequate managerial awareness of the inherent hazards of the operation, and inadequate attention to safeguards. He felt that, if he had been asked to do a risk assessment on the day before the incident, he would have recognized that the risks would be greater than average. But he admitted that he would not have regarded the risks as so high that occurrence of the incident on the next day would have been statistically credible.

These situations illustrate the inadequacy of assuming "good industry standards." In the case of the disasters or incidents listed above, they resulted not from calculated risks occurring, as statistically they always can, but from poor standards of operation, maintenance, supervision, and management.

A quantitative risk assessment can be seen as hanging at the end of a chain of assumptions, each assumption being a link, the failure of any one of which would have the potential for initiating the incident that has been determined, in the assessment, as being of very low probability. It is no good saying that the average strength of the links in the chain is high. It is the strength of the weak link which matters. In hazardous plants, it is no good saying that the average standard of software on a plant is high. It is the standard of the worst example that can determine the full risk of accident or disaster.

Perhaps we ought not to be assessing risks on the assumption of good industry standard of management and operational practices, but rather to be assessing the probability of failure of that assumption.

7.3.4 Development of the "Management Factor"

During the 1980s it became recognized that quantitative risk assessment, as undertaken at the time, did not take account of what was called the "human factor." In an attempt to remedy that weakness, some specialists developed the "management factor." This was a number, between 0.1 and 10, determined by studying the management systems at the facility being studied, by which the generic failure frequencies used in the risk assessment are multiplied. For example, in one case the management systems were assessed as being better than average, and the management factor was assessed at 0.52 (!), so all the generic data-bank failure frequencies used in the assessment (for pipelines, pumps, valves, instruments, etc.) were multiplied by 0.52. There was no empirical validation of the selection of a range of two orders of magnitude for the management factor, nor of the method of scoring the management systems in deriving the magnitude of the management factor. Further, the approach was philosophically badly flawed. This is discussed more fully in Chapter 11.

7.3.5 The Purposes of Risk Assessment

A risk assessment report has no intrinsic value greater than that of the recycle value of the paper on which it is written. It has value only if it is acted on.

The purpose of risk assessment is to facilitate risk management. It is important to note the management component of the aim because management, like engineering, is not a science but an art which uses sciences. Therefore in risk management the dictum of Lord Kelvin is not an absolute.

The main uses of risk assessment in risk management are to assist in providing answers to the following types of questions.

- What sorts of risk are there in this situation?
- Is this risk too high to be acceptable?
- What are the main components of this risk?
- What should we do about this risk?
- How can we reduce the risk most rapidly?
- How can we reduce the risk most cost-effectively?
- How can we keep this risk low?
- Are managing this risk effectively at present?
- What do the exposed people need to know about the risk?
- What should those exposed people do in the event of the risk being realized?

Only in the case of the second question is there a strong case for knowing the magnitude of the risk in absolute terms; all the other questions can be answered by comparative assessments of the risk, or by purely qualitative analysis. Even in the case of the second question, for the absolute value of the risk to be useful it is necessary to have an agreed standard for "acceptable risk": a contentious matter. (See Section 4.2.)

7.3.6 The Limitations of Historically Determined Risk

Although it is possible to determine what a particular risk has been in the past by study of statistics, the practical use made of such information needs careful consideration.

For example, the risk of fatality in industrial accidents in a country over the past 5 years can be determined by dividing the number of such fatalities in that period by the average industrial population. But whether that risk is relevant for the future, or for any particular factory, is unclear. If there is a change in the industrial economy, such that costs are drastically cut and maintenance and training, etc., are neglected, the risks are likely to rise dramatically and in a way that defies prediction. Similarly, whether average industrial risks apply to a particular factory is also unclear.

So a historical risk, which has some justification for being regarded as a "real" risk, may not represent reality in the future.

7.3.7 A Trap for Managers

A phrase in common use is, "It is the bottom line that counts!"

This presumably originated in business and accounting, where the profit or loss revealed by the bottom line of the statement of accounts is ultimately much more important than the components detailed above it.

In quantitative assessments of risk there are usually so many assumptions of dubious reliability, and so many uncertainties about data, that the absolute value of the total risk is very imprecise, quite apart from questions about relevance, as in the case of the chlorine plant discussed earlier. (Risk assessment practitioners often assert that their assessments are within an order of magnitude, but they offer no verifiable bases for such assertions.)

If that is so, why do such assessments? The answer is that the value of the assessments lies not in the bottom line, which is usually of too dubious precision or applicability to be relied on, but in the insights gained in undertaking the analysis, and in the relative magnitudes of the components of the assessed risk. The insights enable consensus to be reached quickly about the relative priorities of different components of risk for action to reduce them, and facilitate recognition of the most cost-effective and rapid means of achieving the desired reduction.

Lesson: Direct the focus of managerial attention away from the bottom line of a risk assessment, toward the components of the risk and the potential for action to reduce risks.

It is normal management practice, when reviewing a list of costs as in a budget or in a routine accounting report, to focus attention on the largest unattractive figures, penetrate the basis of them, and seek improvement. In the case of budgets, this leads to tighter estimates, and resulting pressure on performance to achieve those levels. When reviewing a budget, it is more likely that a manager will seek to tighten the unattractive figures than invite a more relaxed estimate of the attractive ones.

Managers often review risk assessments in the same way. If the total risk as displayed in the "bottom line" is unattractively high, they will query the basis for the major components of that total. For example, in Table 7-1, it is likely that managers would probe the assumptions used in calculating the risks for components B and F. When it is found, as is inevitable in risk assessment, that some of the assumptions are subjective and open to debate, they press for greater optimism. The analyst then adjusts the values as far as he or she believes is credible. The problem is that the manager does not invite review of the components which were assessed as having low risk, with a view to increasing them. For example, in Table 7-1, it is unlikely that the assumptions behind the assessments

Table 7-1
Illustration of Importance of Risk Components, Compared with Cost Components

Component	Cost	Risk
Component A	23	**23**
Component B	1075	**1075**
Component C	240	**240**
Component D	11	**11**
Component E	312	**312**
Component F	4361	**4361**
Total	**6022**	6022

for components A and D would be checked for excessive optimism. Thus the managerial review leads to an optimistic bias.

Lesson: Where the components of a risk assessment are reviewed by management, ensure that both the high-risk and the low-risk components are reviewed, with equal attention being given in each case to the potential for the components to need upward and downward revision.

If care is not taken here, the assessed risk will approximate more closely to managerial dreams than to reality.

7.4 SUMMARY OF THE STRENGTHS AND LIMITATIONS OF QUANTITATIVE RISK ASSESSMENT

7.4.1 Strengths and Benefits

The uses of hazard analysis and risk assessment, and benefits obtained from those uses, include the following:

- The systematic approach is very helpful in approaching the task of safety improvement in a systematic manner.

- The effort spent debating risk criteria, including definition and agreement of the types of criteria to be used, clarifies key safety issues and the responsibilities of the organization and helps build consensus about the degree and general direction of efforts for risk reduction.

- Similarly, the effort spent following one of the available systematic methods of identifying inherent hazards, and the scenarios by which they could lead to hazardous incidents, leads to increased understanding of the hazards, and to consensus that the resulting list is a satisfactory basis for taking further action.

- Preparation of the fault trees and event trees necessary to determine the frequency or likelihood of postulated hazardous incidents leads to improved understanding of the scenarios that could lead to the hazardous incidents, and to development of consensus about the most likely causes. Even though such assessments of frequency or likelihood are not precise, they are usually sufficient to guide judgments about the relative likelihood of the incidents, and about whether the incidents are likely, possible, unlikely, or not really credible. This facilitates agreement about the necessary actions to reduce risk.

• Provided that the assessment of the severity of the consequences of the postulated hazardous incidents is undertaken by people who are aware of the limitations of the available methods, sufficient precision can usually be obtained for the incidents to be classified with reasonable confidence as serious, moderate, or minor. This narrows the field of uncertainty and is usually sufficient for planning of cost-effective risk reduction.

• Although the assessments of risk derived from the assessed frequencies and consequences may not be precise, in many cases the results are either so much lower or higher than the criteria that it is clear whether the risk is satisfactory, or unsatisfactory and needing reduction.

• Although assessments of the required buffer zones between hazardous plants or operations and potentially exposed areas, such as housing, are often very approximate and dependent on continued good management, there is usually no better method available for defining the zone.

• Although assessments of the required buffer zones (as above) are very approximate, and may be based on unprovable assumptions, use of such a method provides a means of undertaking consistent assessments provided that the same assumptions are used throughout. In other words, although it cannot be proved that the buffer zones are sufficient in any particular case, at least they are derived on a basis that is consistent with earlier work which was judged acceptable.

7.4.2 Weaknesses and Difficulties

The weaknesses and difficulties in undertaking hazard analysis and risk assessment, and in making use of the results, include the following:

• Risk criteria are unavoidably philosophical, subjective, and arguable.

• In many regulatory jurisdictions, risk criteria have no formal standing.

• Where risk criteria have a formal legal standing, it will almost certainly be found that any accident that does occur will show that the actual risks did not, in fact, meet the criteria, regardless of the risk levels calculated in advance.

• Risks as perceived by members of the public may be very different from actual risks as determined from study of history. As major planning decisions are made in the political arena, the decisions are based more on the public perceptions of the risk levels than on the calculated or estimated levels.

• It is never possible to be sure that all hazards have been identified, and that all hazardous scenarios have been postulated.[1]

• In the case of the process industry, most of the major disasters in recent years have resulted primarily from failures of management systems, which would not have been included in the quantitative assessment of risk, and not from random equipment failures such as are statistically assessable using data from data banks. This is a most serious limitation, with important implications for the use of quantitative risk assessment in the regulatory process for that industry.

• The mathematical methods used for assessing the severity of the consequences of hazardous incidents are of varying precision, and many types of hazardous incidents can only be assessed by making a variety of simplifying assumptions of dubious relevance.

• The data used for assessing the frequency or likelihood of occurrence of the postulated hazardous incidents is usually from a variety of sources, often not defined, with doubtful applicability to the particular situation being assessed.

• In many assessments, it is difficult to identify potential common-mode failures, where a hazardous incident could occur only in the event of a number of apparently independent and unrelated conditions coinciding, but where in fact the events are not truly independent, as they are all influenced by some other single event.

• Whereas equipment usually has a finite and definable variety of ways of failing to operate correctly, the creativity and initiative of people leads to an often unimaginable variety of ways of making mistakes or of doing unexpected things, or of not doing what they should. This adds difficulty to identification of scenarios leading to hazardous incidents, and adds uncertainty to estimates of the frequency of the incidents occurring.

• As a result of the above uncertainties, the assessed level of risk is correspondingly uncertain and imprecise.

---

[1] Leonardo da Vinci knew this long ago. He wrote "Il mondo è pieno d'infinite ragioni che non furono mai in isperienza," which means: "The world is full of causes that were never experienced before." (Quoted by Robert Hughes in *A Jerk at One End.*) Although it is true that most process plant incidents occur from general types of cause that have occurred before, and therefore these types can be identified if one has made a study of past incidents on one's own facility and those elsewhere, there are numerous combinations of possible weaknesses, that is, scenarios, that cannot all be identified. The task is to recognize and then to eliminate or minimize as many as possible of such weaknesses, so as to minimize the probability of unexpected combinations occurring.

• It is not possible to verify risk assessments by measuring the actual risk on the ground. So poor assessments may not be discovered until too late.

• There are many difficult decisions related to risk and safety that need to be made where quantitative risk assessment is still unable to provide real guidance, and where practical experience is the best guide. An example is the separation between buildings for fire safety. While, in theory, the rate of combustion of one building and its contents could be postulated, the heat radiated could be calculated, and a separation could be determined which aims to keep the heat radiation level at the adjacent building at a safe level, the approximations inherent in the calculations render the absolute value of the calculated separation distance too uncertain for use in decision making. The required separation distances for prevention of fire spread by flame impingement and spread of burning fragments is best judged by experience, and will depend on the construction of the buildings and the effectiveness of the response to the fire.

• The sheer volume of work needed to undertake some types of risk assessment necessitates careful and rational short-listing and selection of applications. There is a large and unavoidable component of experience and judgment required for this to be done well, with a correspondingly large scope for error.

7.5 APPLICATIONS OF HAZARD ANALYSIS AND RISK ASSESSMENT

7.5.1 Incorporation of Safety into the Design of a New Plant

Systematic approaches have been developed for managing the incorporation of safety into the design of new plants from the outset.

Numerous examples could be given of projects which have been designed using this approach, and which are demonstrably safer as a result. For example:

• A petrochemical plant was proposed for a site around 500 meters from housing. In this first study, it was decided that the worst credible vapor cloud explosion should not be capable of doing more damage to the housing than to break windows. So a requirement was specified that the tendered designs should be accompanied by a demonstration that the largest credible leak, determined in a defined way, should not result in a vapor cloud in excess of a defined number of tonnes. The result was a plant with remarkably low inventories of flammable materials.

- A large refrigerated storage of LPG was proposed. There are two approaches to refrigeration of such facilities in common use; one with large inventories of LPG outside the high-integrity bulk tank, one with very small inventories. Because of the need to keep the chance of vapor cloud explosion extremely low, the small-inventory approach was specified. Thus the inherent safety of the plant was greatly improved, and at little if any extra cost.

7.5.2 Review of the Safety of Existing Plants

If existing plants handle or store significant quantities of hazardous materials, or if they operate potentially hazardous processes, management is responsible for ensuring that the facilities are operated safely, that is, with a low risk.

It may be appropriate to survey the plants and operations to determine whether the safety precautions are adequate for the inherent hazards.

Such a survey will need to include the following:

(a) a review of the design, to identify the nature and magnitude of the inherent hazards, and the nature and extent of safety precautions incorporated in the design;

(b) an inspection of the plant, to check its condition and to ensure that hazardous materials are adequately contained, operations are adequately controlled, and protective systems (active or passive) are in operable condition; and

(c) an inspection of the "software" (the management systems and human factors), both as they are intended to be and as they actually are, to check that they match the hazards and the physical features of the plant and operations.

Although such a survey may be expected to reveal any major weaknesses and deficiencies, it will not necessarily reveal the priority with which improvements should be made, nor will it indicate the necessity, in absolute terms, to make the improvements. To do this, an extra component would need to be added to the survey:

(d) a hazard analysis and quantified risk assessment of the main identified hazards, for comparison with criteria agreed by management (preferably also with some statutory basis).

7.5.3 Determination of the Buffer Zones Required around Hazardous Facilities

In the case of the new petrochemical plant (Section 7.5.1 of this chapter) which was planned for a location where it would have around 500 meters separation from existing housing, a preliminary risk assessment was carried out of the three tenders received for design and construction of the plant.

Two of the plants were assessed as requiring around 100 meters separation, while the third required 300 meters. Initial attention was then directed at identifying the reasons for the third plant needing so much more separation, to check that the reasons were real, and not a computational eccentricity (e.g., where the assessment method was inapplicable). On examination, a major deficiency (from the hazard viewpoint) was identified.

Although no great confidence was placed in the assessed separation distances, it was noted that the results obtained were of the right order of magnitude: the results were not 10 meters nor 1000 meters, neither of which would be sensible, regardless of the separation distance actually available. This appeared to confirm that the assessment was reasonably appropriate. It was also noted that the assessed risk at the housing (at 500 meters) would be very low; less than 5% of the target. Thus there was confidence that there was room for error in the assessment, in that if more detailed assessment later in the design process found an increased risk, it was unlikely that it would be 20 times more than the preliminary assessment. It was therefore decided that the preferred location was feasible from the safety viewpoint.

The same approach can be taken with assessments of the safety of existing plants and their existing buffer zones.

For worked examples illustrating the principles behind estimation of the required buffer zone, see Sections 14.13 and 14.14 of Chapter 14.

7.5.4 Focusing Attention on Critical Issues

At the design stage of a large project, there was debate about the location of a flare, on the grounds that in the event of a release of gas from a relief valve it would perhaps be possible, in unusual atmospheric conditions, for the escaping gas to be ignited. A risk assessment showed that the frequency of any escapes from the relief valve would be very low, as would the frequency of the abnormal atmospheric conditions. On the other hand, there was significant potential for an escape of flammable liquid from another section of the plant, for which no leak detection or isolation facilities had been provided. So the attention of the design team was redirected from the relatively unimportant issue of which of two locations was to be preferred for the flare, and directed toward the more important matter of leak detection and isolation.

7.6 FAULTS IN THE APPLICATION OF HAZARD ANALYSIS AND RISK ASSESSMENT

7.6.1 Divorcing Risk Assessment from Design

It is common, when seeking statutory approval for a proposed new plant, for a risk assessment to be required. A consultant may be sought and, in effect, given the brief of undertaking a study to demonstrate the safety of the proposed plant, the design of which is effectively fixed. This is a difficult position for any consultant and puts a strain on his or her objectivity.

Alternatively, although the design may not be finalized, the risk consultant may be expected to undertake the assessment in isolation, and may not be given the opportunity to influence the development of the design.

When the risk assessment is divorced from the design process, the risk assessment becomes a sterile and cynical exercise, and little benefit is obtained from the effort.

7.6.2 Assessing Risks without Auditing the Equipment and People

A risk assessment was undertaken of a plant, and a satisfactory result was obtained. During a detailed inspection of the plant a few months later, it was found that a pipeline carrying a liquefied toxic gas was heavily corroded at the edge of thermally insulated sections. The condition of the rest of the pipeline under the insulation could not be determined without removing the insulation. So the assessment was based on inapplicable data. Further, the correct approach would not be to attempt to estimate the likelihood of a leak from the corroded pipe, but to replace it.

About the same time, there was a "near miss" at the plant. During the investigation that followed, it was found that there had been an instrument on the panel showing an abnormal reading for 3 weeks, but that the operators, the supervisors, and the plant professional staff had not noticed the reading, nor did they understand the significance of the reading when it was pointed out to them in the investigation, nor did they know how to check whether the reading was real or due to the instrument being defective, nor could they say how to correct the fault if it were real.

The risk assessment was thus almost worthless, as the assessment of the likelihood of a major leak was unrelated to the physical condition of the plant and the technical competence of the people involved.

Further, the real need of the plant was not a risk assessment, but a detailed physical inspection, restoration of any dubious items of equipment, and a rigorous program of improving procedures, training, supervision, etc.

This shows the priority of good engineering and management.

7.6.3 Incorporating Management Weaknesses by Adjusting Equipment Failure Data

Because of the recognition of importance of "software," that is, the management-determined factors of:

- organization,
- procedures and methods,
- knowledge, skills, and training,
- documented standards and records, and
- attitudes,

and because of the difficulty of quantifying the frequency of hazardous events due to human error arising from software faults, some consultants have taken to conducting a management audit, then multiplying equipment failure data by a factor (greater or less than 1) to take account of management, which is worse or better than some "average" level. Although a poorly managed plant is much more likely to have a major incident than a well-managed plant, any estimate of the factor by which equipment failure frequency would be changed by the management standard is a pure guess. Further, as a large proportion of incidents due to human error are unrelated to incidents due to equipment failure, to base the frequency of incidents due to human error on the frequency of equipment failure is fallacious. See Chapter 11 and Tweeddale (1992).

As set out earlier, a risk assessment is invalid where there are demonstrable management weaknesses. The correct action in such circumstances is to correct the weaknesses, not to attempt to quantify their effect on the assessed risks.

7.6.4 Biased Review of Data

Although it is good practice to review the data used in a risk assessment once the results have been obtained, to identify which of the data are critical, it is important that favorable and unfavorable data be reviewed equally critically.

There is sometimes a tendency for the designers, or the management of the client organization, to attempt to renegotiate data that have been used in a risk assessment that produced inconvenient initial results—challenging the unfavorable data, and accepting without question the favorable data. Because most of the data are "flexible" and unable to be determined with precision and confidence, this leads to an optimistic bias.

7.6.5 Assuming Unachievable Precision

An assessment may show the contour of the defined maximum acceptable risk level as almost, but not quite, reaching a potentially exposed facility such as housing. As the contour does not reach the facility, it may be decided that the risk level is satisfactory. Yet, a change to the data well within the range of its precision may produce a contrary result.

The preferred actions, in the event of an assessment showing a nominally acceptable but possibly marginal result, are:

- to identify the data which could make the proposal unacceptable, and to attempt to refine their precision; and
- to identify the major components of the nominally acceptable risk, and to adopt a systematic approach to their reduction, so that the risk is reduced to a level substantially below the limit.

7.6.6 Use of Data of Unclear Relevance for the Particular Application

The failure data used in risk assessment can be derived from various sources. The difficulty is to know the applicability of that data to the particular plant or operation being studied. For example, the particular application being studied may be different from the sample used in collecting the data in relation to:

- material;
- process conditions (temperature, pressure, etc.);
- environment (degree of external corrosion, etc.);
- standard of operation, maintenance, inspection, supervision, etc.

7.6.7 Reliance on Risk Assessment in the Absence of Practical Experience

Because of the unavoidable need to adjust data to suit each specific situation, and to interpret the results in the light of the approximations inherent in the data and methods, it is important that risk assessment be undertaken by those able to apply experienced judgment based on knowledge of the technology involved. This is often not the case.

7.7 CONCLUSION

Risk is an important concept for judging the safety of a plant with the potential for serious but infrequent hazardous incidents.

The discipline of hazard identification, analysis, risk assessment, and systematic risk reduction is very helpful in gaining a general understanding of the opportunities and priorities for risk reduction and safety improvement by such means as engineering and management. As an analytical tool, it is very valuable.

The data and methods available for quantifying the levels of risk in many fields (especially human error) are currently very imprecise, such that assessed absolute levels of risk are less precise than many proponents of risk assessment would assert. As a means of determining the actual level of risk in absolute terms, it is very unreliable. Thus no decisions related to process plant safety should be made principally on the absolute value of a quantitative assessment of risk.

Many critics of risk assessment argue that the data are too subjective and the methods too imprecise for risk assessment to be a valid technique. But this is also true of the only methods available for assessing the economic viability of projects (e.g., market forecasting), or much of engineering, which is, after all, an art which combines science with experienced judgment, rather than an entirely scientific discipline. If engineers were limited to methods which could be demonstrated to be unarguable and precise, nothing would ever be built.

Although risk assessment has been proven to be a very valuable tool in improving plant safety, it is open to misunderstanding and abuse. It therefore needs to be used with care, with critical review, and with maximum use of nonquantitative information, as an aid to engineering and management, and not in isolation from them.

REFERENCE

Tweeddale, H. M., "Balancing Quantitative and Nonquantitative Risk Assessment." *Trans. IChemE* **70 (Part B)** 70–74 (1992).

Chapter 8

A Systematic Approach to Risk Reduction

When you have studied this chapter, you will have an understanding of some of the engineering and managerial options for reducing risks of process incidents of various types.

This chapter addresses risk management step 10, "Reduce the Risk," in Figure 1-4.

8.1 PRINCIPLES

8.1.1 Introduction

The factors which determine the safety of a hazardous plant, machine, or operation, or a transport operation, are:

1. inherent hazard,
2. adequacy of containment and control of the hazardous facility or operation (both hardware and software),
3. adequacy of the protective systems (both hardware and software), and
4. damage limitation.

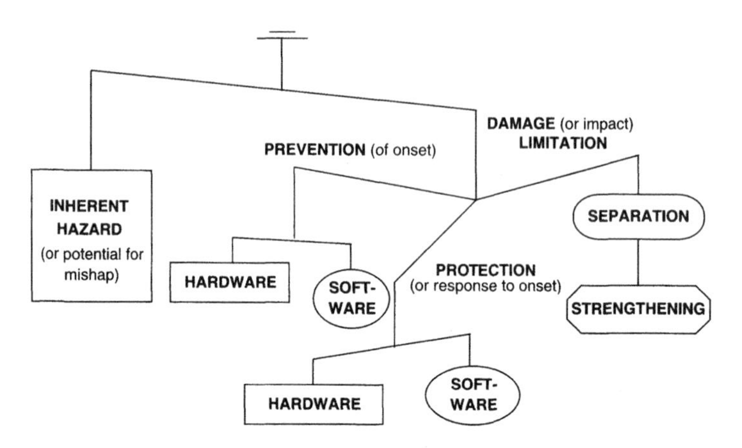

Figure 8-1. The Risk Balance.

Conceptually, the inherent hazard needs to be balanced by the total safety package. This is displayed in the Risk Balance[11] of Figure 8-1 (Tweeddale, 1985).

In principle, the preferred approach to risk reduction is to work from the top of the list, downward, that is, from the left-hand side of the diagram toward the right-hand side.

First, the inherent hazard should be eliminated or reduced as far as reasonably practicable. This can be done most easily at the outset of a proposed design, but it is often possible to make significant improvements to an existing plant.

8.1.2 Elimination or Reduction of Inherent Hazard (i.e., Strategies for Inherently Safer Design)

The dimensions of the inherent hazard are:

- the quantity of the hazardous materials present (the more material present, the greater the potential consequences of a mishap);
- the physical and chemical properties of the materials (some materials are more reactive or toxic than others);
- the process conditions under which the material is being handled or stored (e.g., extremes of pressure and temperature increase the difficulty

[11] Adapted from Tweeddale, H. M., "*Improving the process safety and loss control software in existing process plants.*" Institution of Chemical Engineers, London, Symposium Series 94, Multistream 85, April 1985 with permission from IChemE.

of containing the materials and controlling the operations and hence increase the inherent likelihood of a mishap);

- the complexity of the plant and equipment (e.g., the more complex the plant and equipment, the greater the inherent likelihood of failure).

A matching strategy for eliminating or reducing the inherent hazard is set out by Hendershott (1999a,b):

1. *Minimize*: Use small quantities of hazardous materials.
2. *Substitute*: Replace a hazardous material with one that is less hazardous.
3. *Moderate*: Use:
   - less hazardous conditions;
   - a less hazardous form of material; and
   - facilities that minimize the impact of a release of hazardous material or energy.
4. *Simplify*: Design facilities that:
   - minimize or eliminate unnecessary complexity;
   - make operating errors less likely; and
   - are forgiving of errors which are made.

These strategies are discussed in detail in the references, for example, IChemE (1995) and Bollinger et al. (1996).

For example, a company proposed an expansion of its production facilities for a particular product range. This involved increased production capacity for a hazardous intermediate product, and increased capacity for those plants which would use the intermediate product to manufacture a variety of consumer products. It was initially proposed to increase the stock tank capacity for the hazardous and potentially unstable intermediate material to provide the necessary production flexibility to cope with periods when either the upstream plant or the downstream plants were shutdown or limited in production rates. The increase of inherent hazard presented by the large increase in on-site storage was to be balanced by numerous complex and expensive containment and protective systems. Finally it was decided to manage with the existing storages, without any increase of capacity, and to increase slightly the designed capacity of the extensions to the upstream and downstream plants, to enable them to catch up in the event of lost production. The net extra cost was not regarded as significant in the total project cost.

There are at least two processes available for manufacture of the end product manufactured at Bhopal. Both were widely used by the chemical industry at the time. Only one of the processes involves the manufacture of methyl isocyanate (MIC) as an intermediate. With hindsight, if the Bhopal plant had used the other process, there would have been no MIC to escape, and the incident could not have happened. The Bhopal plant relied on good containment, control, and protective systems to prevent incidents from happening. With so little separation from the shanty town just outside the factory, with the dwellings there having no "strengthening" (or ability to be sealed from the outside) to safeguard the occupants from the effects of any gas escape, the standard of the facilities and software for containment, control, and protection would have to be very good for the risk to be low. In the event, again with hindsight, the standard was not sufficient. (See Section 13.3 of Chapter 13.)

8.1.3 Improved Preventive Measures

In the process industries, most accidents arise because of loss of containment of hazardous materials. Some arise because of loss of control of hazardous processes.

A large proportion of the fires and explosions at oil refineries and petrochemical plants are caused by failures of pipelines. Many of these failures have been of small-diameter branches which are less robust physically than larger diameter pipework. Management of some refineries and petrochemical plants has undertaken programs of removing such small-diameter branches and other relatively fragile components such as bellows and sight glasses where possible, and physically strengthening the remainder.

A common cause of large leaks and fires on plants handling flammable materials, and leaks of toxic materials on plants handling such materials, is failure of mechanical seals on pumps. A major cause of sudden failure of mechanical seals is the severe vibration or shaft movement resulting from collapse or other failure of the ball or roller bearings used on the pumps. One component of software aimed at improving containment is to monitor the vibrations generated by bearings, using the nature and extent of the vibrations to determine the condition of the bearings and the likelihood of imminent failure.

In some processes, it is very important that some process variable (temperature, pressure, concentration, etc.) be closely controlled, as any significant deviation from designed conditions can result in a serious incident. In such cases, extra attention may be given to the control systems, such as relying on one control system normally, but providing another control system, set to take over routine control if the variable deviates slightly beyond its normal control range. In some processes, this may be preferable to relying on a single control system, supported by a protective response such as an alarm and a plant automatic shutdown system.

8.1.4 Improved Protective Systems

A *protective system* is whatever is provided for use only when the normal facilities for containment or control fail. Protective systems may be active, such as an alarm or trip system, or passive, such as a bund.

(An installed spare piece of equipment for backing up the normal equipment, such as a second control system or a spare generator alongside the operating generator, may sometimes be treated as a protective system, depending on the circumstances.)

A typical active protective system may be a high-temperature alarm on a reactor, to prevent mishap in the event of a failure of the cooling system, or some other fault condition. The alarm is only part of the protective system; the operator is also a part of the system, as he or she will be needed to respond to the alarm and to take the appropriate action. Because one such alarm may not operate, or may be ignored, it may be backed up with an automatic quenching or shutdown system.

Some equipment must, for safety, be fitted with very reliable shutdown systems, designed to operate if any one of a number of conditions arises. For extra reliability, critical components of the shutdown system may be duplicated to increase the probability that at least one of the components will operate if necessary.

Every protective system needs testing if one is to have confidence that it is in an operable condition. Because such systems remain for long periods without being called upon to act, their condition can easily deteriorate to inoperability long before the emergency arises. It is then too late to discover a need to repair the protective system.

The testing of protective systems is so critical to their reliability that many organizations prepare detailed schedules for protective system testing at the time a new plant is being designed, and the protective systems are designed with the need for testing taken into account.

The testing of passive protective systems, such as bunds, may be in the form of an inspection.

Major reasons for unreliability of protective systems include:

- failure to test;
- failure to test *all* the components in the protective system; and
- failure to analyze the results of the tests, and to investigate and rectify the causes of any recurrent faults.

8.1.5 Damage Limitation

It is usually difficult to improve separation distances once a plant is built. The location of a hazardous plant or operation is critically important in determining the effort that must be given to improving the inherent safety, containment, and control, and the protective systems. These have a great impact on design. Therefore it is important for the location of proposed hazardous equipment or operations, and the resultant separation distances from potential "targets," to be considered at the outset of the design in relation to the potential consequences of hazardous incidents.

It is sometimes possible to strengthen potential critical "targets" on existing plants. For example, an organization examined the ability of major storages to resist explosive overpressure. It found that horizontally mounted pressure storage tanks were very resistant, but that spherical pressure storages would be moved sideways on their supports sufficiently for the supports to collapse under the weight. The diagonal cross bracing between the supports was designed for normal wind forces, not for explosion overpressures. It was a relatively simple matter to design and install improved cross bracing that could absorb the energy from an explosive blast without the spherical pressure vessel collapsing.

In all of the above fields for risk reduction, it is necessary to be thorough in investigating and identifying the options for improvement. In almost every case where a risk reduction is needed, careful examination of the above options reveals a range of options, from which those that are most practicable and cost- effective can be chosen.

8.1.6 Cost- and Time-Effectiveness

Although it may be preferable, in principle, to approach risk reduction in the sequence listed above, for cost effectiveness it is necessary to consider other factors.

A good approach is to list the various options for risk reduction, then for each to estimate (usually this can be done only very roughly) the following:

- the expected extent of risk reduction that would be achieved;
- the cost of achieving that reduction, where the cost is expressed in the appropriate units, such as money, staff time, or other resources; and
- the expected elapsed time from start of work on that option before the benefit would be achieved.

Then, by selection of the appropriate criterion (or a combination of criteria) such as:

- risk reduction per unit "cost" and
- risk reduction per elapsed unit of time,

it is possible to sort the options into the priority that is expected to produce the most risk reduction with the available resources, or within some target time period.

This may be done very easily on a computer spreadsheet, trying different criteria for priority setting.

8.2 TRANSFERRING THE RISK

There are several ways of transferring the risk, but they all amount to a form of insurance.

Where the impact potential loss can reasonably be expressed in monetary terms (e.g., property damage or production interruption, but not injury to people or damage to the environment), then it is theoretically possible for insurance to cover the loss entirely. However, in practice this is rarely the case.

Faced with the risk of a loss that could be made good by monetary compensation, an organization may transfer the risk to an insurance company in return for the certainty of paying a regular premium.

There are several "players" in insurance, including:

- the insured, that is, the person or organization facing the risk of a loss, and who does not wish to bear the cost of the loss if it occurs;
- the insurance broker, who acts as an adviser to clarify the type of insurance coverage the insured needs, and an agent to identify an insurance underwriter (see below) who can provide that coverage at an appropriate premium;
- the insurance underwriter, who accepts the risk of having to pay a claim if the loss occurs, in return for a regular premium payment by the insured; and
- re-insurers, who provide backup to underwriters, typically in the event of large claims, in return for a regular premium from the underwriter.

The insurance broker is paid by a commission on the premiums paid by the insured.

Insurance companies are businesses, not charities. Out of the premiums they receive, they must pay their own administrative costs and hope to make a profit. So, on average, the claims paid across the field are less than the premiums paid.

Therefore it is not economically sensible to insure against a potential loss that would not be a significant problem to the organization if it were to occur. Similarly, where losses are relatively frequent, the insurance premium will exceed the claims paid.

In contracts, the risk to the client of unforeseen costs may be transferred to the contractor by means of a fixed-price contract. But the contractor, depending on the degree of competitiveness of the field, will normally add an amount into the price to act as a premium to cover the risk of being caught with unforeseen costs.

8.3 REDUCING FIRE RISKS IN PROCESS PLANTS

8.3.1 General Design Features

Where there are flammable or combustible materials in a section of a process, the general features of design for fire safety to be considered for adoption are:

- minimized inventories (to maximize inherent safety);
- high standard of containment, or leak avoidance (most fires require a leak);
- layout (to separate main inventories from areas vulnerable to heat radiation or combustion products;
- ventilation (so any leak of flammable vapor is able to disperse without igniting);
- curbing, bunding, and drainage (to prevent flammable liquid spills from spreading; the size of a flammable liquid pool fire being dependent on the size of the pool. The more likely leak sources should be located near a drain opening to minimize the area of the leak path over the paving);
- prompt leak detection (to enable a prompt response);
- means of isolating any leak (thus starving a fire, or potential fire of fuel);
- control of ignition sources (to reduce the chance of a leak igniting); and
- appropriate firefighting facilities.

Typical specific safety measures include the following:

8.3.2 Pipelines

- Minimal joints would be used, consistent with access for maintenance.
- Flanged joints would be used where necessary, with the gasket type being carefully selected for reliability and leak-tightness in contact with the material being handled. Screwed joints would not be used in hazardous service.
- Relatively fragile small-bore pipework (less than 40 mm nominal size) is to be minimized. Where it is necessary, it would be mechanically supported to prevent damage and leakage.
- No sight glasses would be used on flammable or toxic duty, as these have a history of failure.
- Where a very volatile liquid or flammable gas is being handled through complex pipework, flammable gas detectors would be installed in the vicinity.
- Remotely operable emergency isolation valves would be fitted to the discharge nozzles of all large vessels handling flammable materials, to enable any leak in a downstream pipeline to be isolated.
- No glass pipework would be used.
- Appropriate attention would be given to choice of materials of construction and methods of fabrication to avoid potential problems from corrosion and other metallurgical attack by process materials (e.g., hydrogen) and environmental materials (e.g., traces of chloride from salt spray in rainwater).
- Where it is especially important for reverse flow of process materials to be avoided, appropriate design measures such as double block and bleed valving would be used rather than nonreturn valves.

8.3.3 Tanks Holding Flammable Liquids

- Storage tanks would be the minimum practicable size.
- Tanks would be installed within bunds to comply with the relevant codes.
- Tanks would be protected as appropriate for the materials contained by such means as nitrogen blanketing, venting and vacuum breaking systems, flame arresters on vents, grounding cables, etc.

8.3.4 Pumps Handling Flammable Liquids

- Pumps would be located outdoors where practicable.
- Close attention would be paid to the materials of construction.
- Mechanical seals would be used, and where appropriate for special security of sealing, double mechanical seals would be selected with the interseal flushing systems fitted with leak detection alarms.
- Where double mechanical seals are not used, throttle bushes would be fitted to minimize the rate of leakage in the event of a seal failure.

8.3.5 Reactors Handling Flammable Liquids

- Discharge nozzles would be fitted with remotely operable emergency isolation valves.
- Means would be provided for opening reactors, or charging solid raw materials, without discharge of flammable vapor into the working area.
- Relief valves would be sized to handle mixed liquid and vapor discharge where appropriate, with any discharge being appropriately directed to a blowdown drum, flare, etc.
- Special attention would be given to design of seals on reactor stirrers. Top-entry stirrers would be preferred to minimize the chance of a large spill if a seal fails.

8.3.6 Buildings

- Process buildings would be of noncombustible construction.
- Drains from elevated levels would be constructed of noncombustible materials.
- Fire detection and firefighting equipment would be designed and installed to match the fire hazards. Where smoke detectors or flammable gas detectors are to be installed indoors, their final location would be confirmed after startup by using smoke tests to verify actual air currents resulting from both air conditioning and convection.
- Any office requirements related to warehousing operations would be located outside the warehouse, to minimize the chance of fire resulting from office equipment.

- Any battery charger needed for forklifts would be located where malfunction could not ignite stored materials.
- The storage pattern would be designed to avoid close contact between incompatible materials, and to segregate any flammable materials.

8.3.7 Catalyst Handling

- Where catalysts can self-ignite if exposed to air for a time, they would be stored under a blanket of an inert gas such as nitrogen, and plants for handling it would be designed to exclude air.

8.4 STEPS IN DESIGN OF A NEW PLANT TO MAXIMIZE FIRE SAFETY

A logical approach can be taken to design of the layout, drainage, and fire protection systems of a new plant, which will result in inherently greater fire safety. They are:

1. Divide the plant into discrete blocks, or strips, with good separation between them.

The layout on the left in Figure 8-2 is typical of older oil, gas, and petrochemical plants. The layout on the right is common now. The sections of the plant are separated typically by 15 meters to provide adequate fire separation in most instances.

Blocks with the greatest chance of flammable leaks should be located as far as possible from blocks with permanent or unavoidable ignition sources such as furnaces. Where adequate separation cannot be provided,

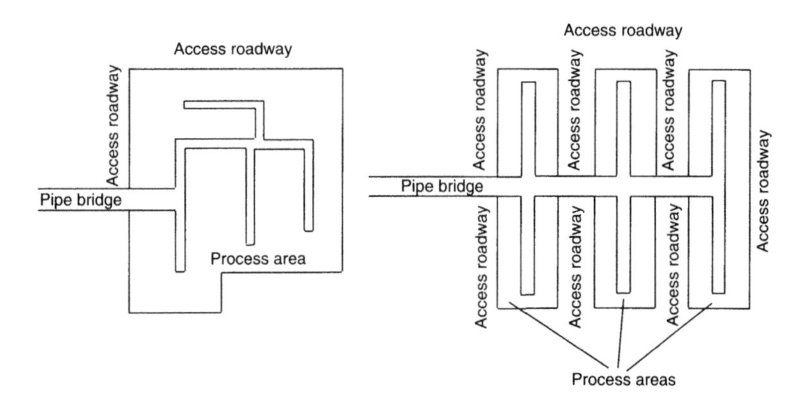

Figure 8-2. Layout of a plant in blocks.

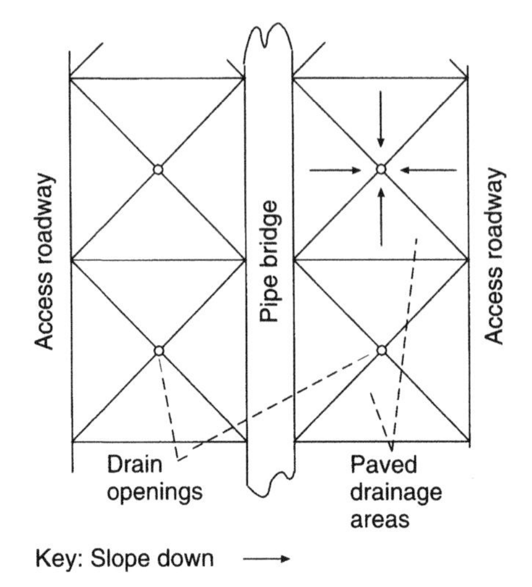

Figure 8-3. Plan view of a plant block and pipe bridge, with drainage zones and drain openings.

consider installation of steam curtains or water curtains to reduce the chance of flammable vapors drifting to the ignition sources.

Check that pipe-bridge routes enable access to all parts of the plant by vehicles (such as mobile cranes) without passing under pipe bridges (to reduce the chance of a pipe bridge being damaged by such vehicles).

2. Divide the paving of the process area into distinct drainage catchment zones of limited size to limit the surface area of any spill, and so to limit the size of any pool fire.

Typically, wherever there is a risk of spillage of flammable liquids, all ground under process plants should be hard paved and divided into catchment zones of around 100 square meters (i.e., around 10×10 meters), sloped at around 1 in 50 to the drain opening.

By limiting the area of any spill, the height and heat radiation of any spilled burning liquid is limited, thus limiting the need for fire protection. See Figure 8-3.

3. Locate plant equipment such that vulnerable plant (large inventories, columns, etc.) is on high points and ridges of the paving (so that flammable spills flow away from the vulnerable plant), and locate the main potential leak sources (pumps, etc.) as near as possible to the drain openings to minimize the leak path from the leak source to the drain, thus limiting the area of any pool fire and the damage done. See Figure 8-4.

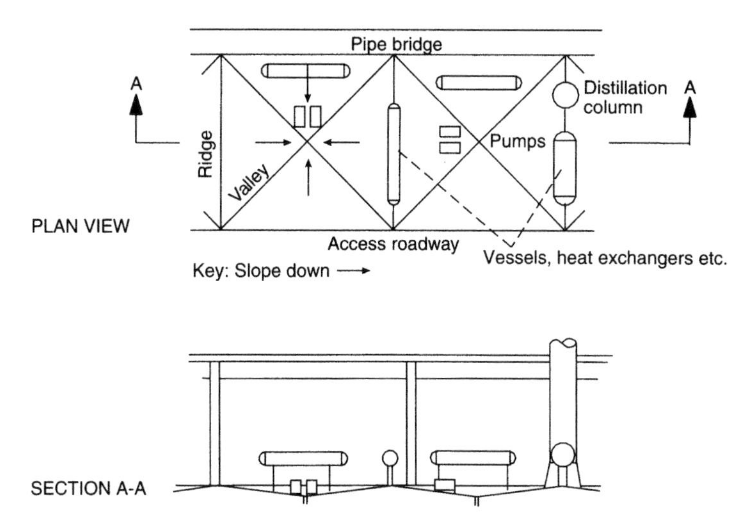

Figure 8-4. Equipment location.

A common fault of older plants is the location of pumps under pipe bridges, as pumps are particularly prone to seal failure and fire. Such fires can be very damaging under pipe bridges, cutting the cable runs very quickly and exposing the pipe-bridge supports to flame impingement, necessitating passive fire protection. This has been only partly rectified in many recent designs, where the pumps are located only just outside the pipe bridge, such that a large seal fire (a relatively common event) would be expected to cause rapid damage to cable runs under the pipe bridge. Further, location of pumps between vessels, columns, etc., and the pipe bridge means that, in the event of fire, access is difficult to cool the exposed plant. If, on the other hand, pumps are located toward the outside of plant blocks, less plant is exposed in the event of fire, and access for firefighting is easier (as is access for maintenance). See Figure 8-5.

4. Design the drainage system such that each drain opening can take a flow of firewater of 2000 liters per minute, connecting with a main capable of taking 8000 liters per minute. Design the drain system such that each drain opening is on its own spur. For petrochemical plants, design for a fully flooded drain system to prevent flammable vapors collecting in the drains, and spreading via the drains to other areas. See Figure 8-6.

5. Review the layout, identifying the main potential leak sources (e.g., pumps, complex pipework manifolds, vessels with many flanged pipe connections).

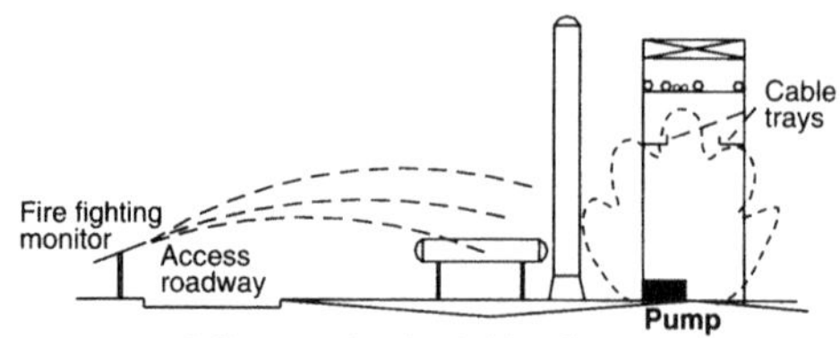

1. Pump under pipe bridge; fire threatening cable trays, and heated plant and structures difficult to reach from the monitor.

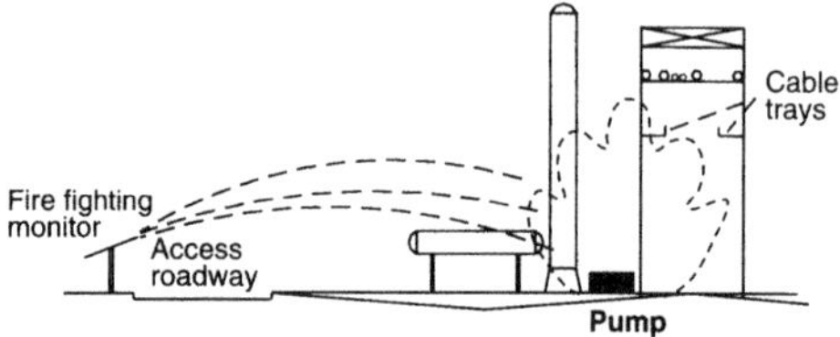

2. Pump beside pipe bridge, fire still threatening cable trays and heated plant and structures difficult to reach from the monitor

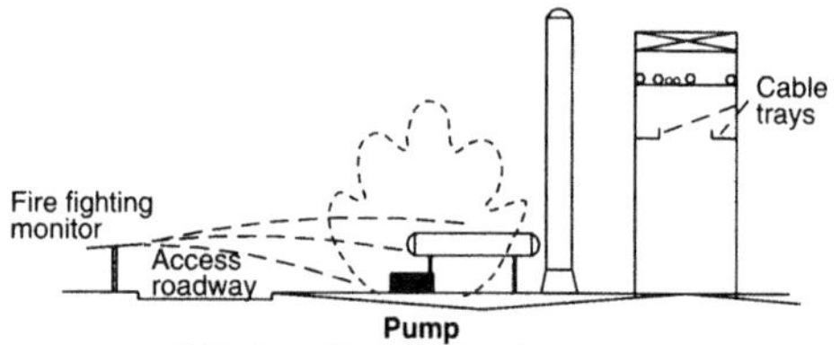

3. Preferred location for the pump, with the only threatened plant easily reached from the monitor

Figure 8-5. Pump location.

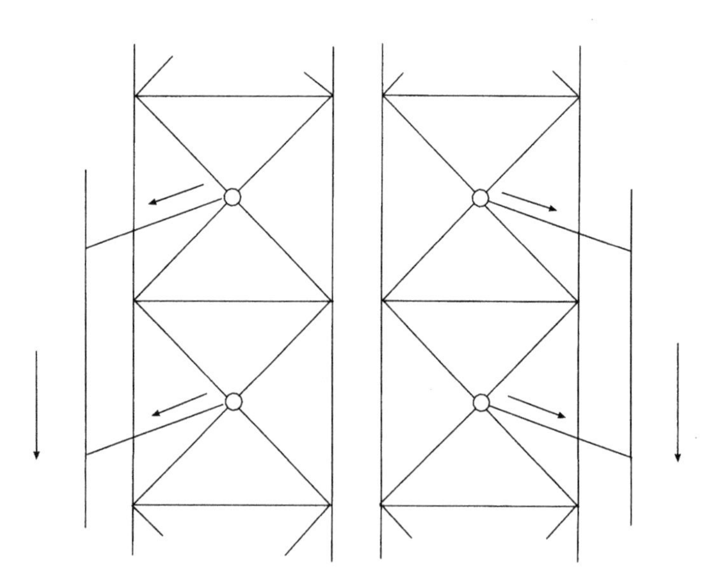

Figure 8-6. Drainage design.

- Where such leak sources may be supplied with a substantial inventory of flammable material (e.g., 5 tonnes of LPG, 10 tonnes of flammable liquid), provide a remotely actuated isolation valve (with 10–15 min fire insulation) upstream of the potential leak source (or two manual valves well separated so that if one is inaccessible because of fire, the other should be useable), and a nonreturn valve at some convenient point downstream. For example, a pump fed from a large inventory may be fitted with a suction remotely actuated valve, and a delivery nonreturn valve.
- Specify passive fire protection (e.g., fireproof cladding) of structural steelwork within a radius of 6 meters of such leak points to a height of 9 meters.
- Similarly plan for fire shielding or other protection of large cable runs within a radius of 12 meters of the potential leak sources. (Concrete fire cladding is preferable to intumescent or subliming paints, which lose effectiveness suddenly at the end of their rated exposure and that may be dislodged by firefighting water.)

6. Locate flammable gas detectors and fire detectors at suitable locations. For example, a pump bay handling LPG materials should be provided with flammable gas detectors, possibly at four points around the bay. Fire detection systems require careful selection depending on the application, but may be based on UV, IR, fire wire (in which the insulation between two strands breaks down with heat), or pressurized plastic tube (which ruptures and releases the pressure if exposed to heat).

7. Define requirements for fixed active fire protection (deluges, sprinklers, etc.). For example, deluges with provision for "light water" additive induction may be located at large pumps handling flammable liquids to help smother any fire resulting from such causes as seal collapse. The actuation point should be at a safe distance.

Pumps handling flashing flammable liquids such as LPG should not be fitted with deluges, as the water would increase the vaporization of the LPG. However, a plant in the vicinity of such pumps may need deluging to provide cooling.

Decisions about provision of fixed active fire protection will be influenced by:

- access for firefighters in an emergency;
- the speed of response possible; and
- manning available to respond to an emergency.

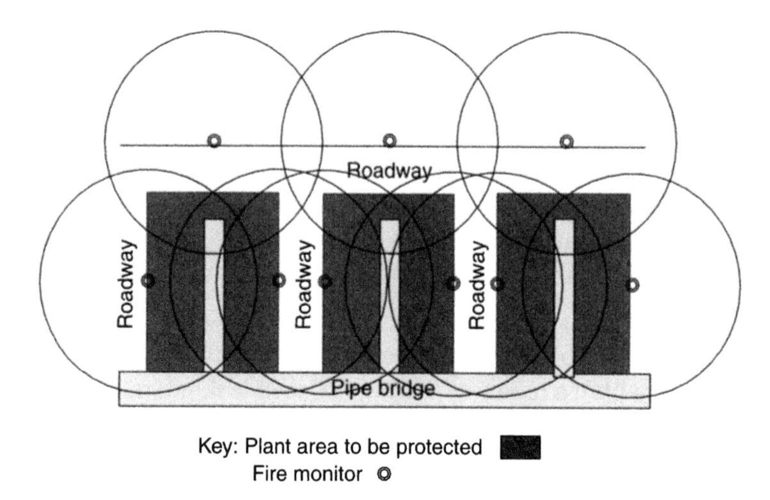

Figure 8-7. Fire monitor location.

Where a good level of protection can be provided manually, it may be judged that there is less need for fixed active protection.

8. Define the requirements for fixed fire monitors and hydrant points. See Figure 8-7.

Monitors should be located such that all vessels containing significant inventories of flammable materials can be reached by two monitors from substantially different directions. Note that a common error is to locate monitors too close to the plant being protected, or too far away. Guidance should be sought from suppliers at the time, taking account of the designed pressure of the firewater system at maximum design flow.

(Sometimes the monitors are located such that they appear, from the plan, to be able to provide the required cover, whereas if viewed in three dimensions, e.g., on a plant model, it is clear that the coverage will be obstructed by structures and equipment. This should be carefully considered.)

Then, assuming that two monitors are being applied (for plant cooling) to any single fire, calculate the number of fire hydrant outlets needed to supply the total firewater requirements. Typically, a monitor will supply around 2000 liters per minute, and a 38-mm fire hose (with a spray nozzle) around 400 liters per minute. If the worst fire for design purposes is defined as covering four catchment zones (i.e., around 400 square meters plan area), and if the plant equipment is reasonably congested on that area, then around 20,000 liters per minute total firewater may be required. Therefore, with two monitors in use, there would be a need for around 16,000 liters per minute from hoses, or portable monitors. This would necessitate around 40 hydrant outlets. These should be located within reasonable reach of the fire, for example, within a distance from the edge of the fire equal to around the throw of a hose (e.g., 15 meters) plus two hose lengths. This may best be arranged by use of hydrants with four valved outlets each.

Note the manning requirements for application of the required water via hoses, and if that manning could not be made available promptly, plan on provision and use of portable monitors that can be set and left, with periodic redirection.

If a plant model is available, it should be examined to ensure that the important plant needing protection will not be shielded from firewater coverage by other plants. (It is difficult to determine this from a plan.) When the plant has been constructed, it should be inspected to ensure that such shielding does not exist.

8.5 CASE STUDY: UPGRADING A FIREFIGHTING WATER SYSTEM

It is common for the changed requirements of the firefighting water system to be inadequately considered as an oil, gas, or chemical site expands. Quite often all that is done is installation of additional water mains, monitors, and hydrants to provide cover for the new plants or facilities. Thought may be given to additional pumping capacity. But often the result is that the system grows in an unengineered manner, such that it would fail if called on.

The following case study illustrates both the possible problems, and a systematic approach to upgrading the system.

A petrochemical site had a number of plants. (See Figure 8-8.) Firewater was supplied to each plant via a ring main system, fed from a set of three pumps (A), supplied from water storage (B), which was in turn supplied from a water main external to the site. At various times, additional plants were built and additions made to the firewater system supplying the ring main.

After one expansion of production facilities, it had been decided to construct a separate tank farm (C) to hold intermediate materials and products from the various plants. A separate firewater system was constructed to supply deluge and firefighting water to the tank farm.

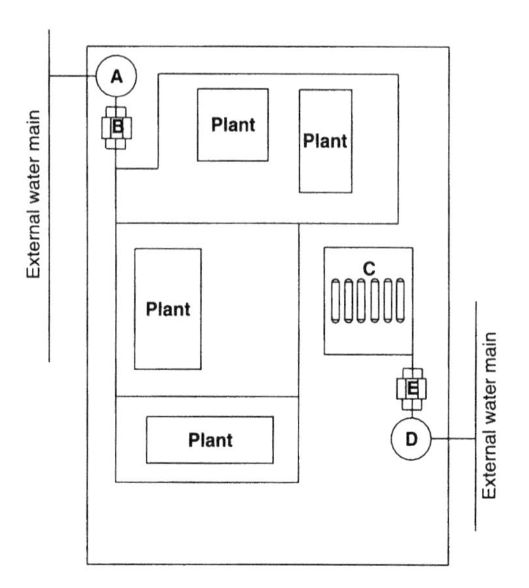

Figure 8-8. Outline of original firefighting water system(s).

It was supplied from an external main, via a storage tank (D) and a set of three pumps (E). At various times, more tanks were added, and modifications made to the firewater system supplying the tank farm.

In due course, it was decided to replace one of the plants with a much larger one. As part of this project, the firewater system for the site was to be reviewed.

As part of this review, a series of tests were undertaken of the firewater system, measuring flows and pressures in the system.

A number of important weaknesses were discovered.

- As plants had been added to the site, and tanks added to the tank farm, the pumping capacity of the two systems had been increased by increasing the impeller sizes of the electric pumps (one electric pump and two diesel pumps per pump set) and increasing the governed speed of the diesel pumps. This led to several serious problems:

  the hydraulic friction in the suction pipework from the water storage tanks to the pumps would cause the pumps to cavitate (if all pumps were operating) by the time the water level in the tanks was at 50%, thus reducing the effective water storage capacity;

  the dead-head pressure, if the diesel pumps were operating and supplying the mains, was extremely close to the rated pressure of the ring main, potentially leading to rupture; and because the pumps had been modified to supply the required pressure at the hydrants at high flow, the pressure at the hydrants at low flow was too high to generate foam successfully with the standard equipment.

- As a result of the pump modifications, their pressure/flow characteristics no longer matched, and if all pumps were operated together, some of them would operate at full flow, and some would contribute little.
- There was potential for a number of single failures to lead to total failure of each of the two firewater systems (the plant firewater system or the tank farm firewater system). Examples include failure of the main supplying the particular system, rupture of the suction line to, or delivery pipeline from, the pump set.

The following work was undertaken:

- The structure of the firewater system was considered in detail, and it was decided to combine the two systems, so as to have two, largely independent, sources of water, and supply from up to six pumps, to any fire in a plant or in the tank farm (see Figure 8-9).
- A series of fire scenarios was defined for each plant, and the firewater requirements (flow, duration) estimated for each.
- A hydraulic analysis of the network was undertaken for the various scenarios to determine the pumping delivery pressure needed for each, given the required pressure at the monitors, hydrants, and deluges.

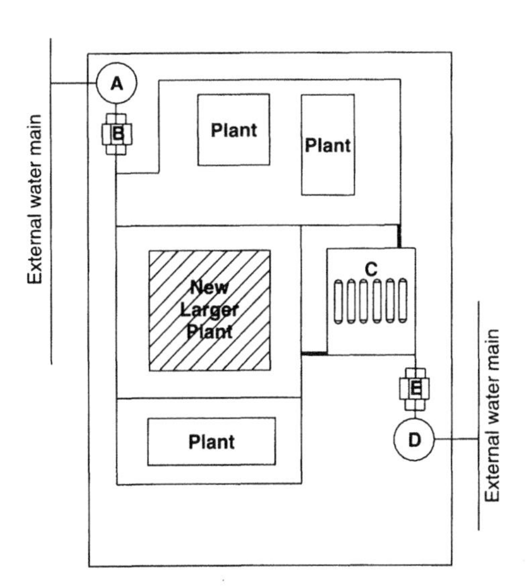

Figure 8-9. Modified firefighting water system.

- The ring main system was checked that there would not be excessive pressure drop in any of the fire scenarios (it was found that there was no problem with the integrated system and water supplied from two pumping stations at different points on the ring main system).
- The necessary water storage requirement was determined from the worst-case flow/duration, taking into account the estimated rate of replenishment from the external mains (it was found that the combined storage of the two existing tanks was adequate, whereas with the separate systems this was not the case).
- The pumps were restored to their original configuration with matching characteristics.

The starting sequence for the pumps was re-engineered. It was decided that, as there was an increasing requirement for firewater, pumps would start alternately at the two pumping stations. Further, the original design for each pumping station provided for the electric pump to start as soon as there was a demand, and for the diesel pumps to start automatically as the pressure in the ring main dropped progressively lower. It was realized that, with six pumps linked, each to start as the pressure dropped progressively lower, the sixth pump would start with the ring main at a very low pressure, such that the pressure at the hydrants would be unacceptably low.

It was therefore decided that, when there was a demand for water, the first electric pump would start at once, and a start sequence initiated for all the other pumps, but with increasing delays. If the first pump restored the ring main pressure to its working level, then the start sequence for the other pumps would be cancelled. As the firefighting effort became more geared up, and more water was demanded, the start sequences would be initiated again, and more pumps would start until the ring main pressure was restored. By this means, all the pumps could come into service without the ring main pressure falling to an unacceptably low level.

8.6 PRINCIPLES OF FIREFIGHTING

8.6.1 Introduction

Firefighting is a very large subject and is only briefly outlined here. In many fires in industrial plants, three phases can be identified:

- first-aid firefighting,
- works team response, and
- external brigade response.

Where the works has no fire team of its own, the second phase may not exist, or it may be undertaken by plant employees. These phases are discussed below in turn.

8.6.2 "First-Aid" Firefighting

This starts as soon as a fire is detected. The aims of first-aid firefighting are:

- to summon assistance, and
- to extinguish the fire while it is small, or to limit its growth.

These aims are to be met without injury to people, and without anyone being "courageous" or foolhardy.

If the fire is detected visually by someone in the vicinity, the rule should be established that the first action is to sound the alarm, so as to get assistance on the way. For this to be possible, alarm points should be located throughout the plant area so that no one needs to go more than about 30 meters to reach one. On a multilevel plant with significant fire hazard, it may be appropriate for an alarm point to be located at the stairway at each level, so that the alarm can be initiated by the person while escaping from the fire. Such alarm points should be well marked, so as to be visible from a distance.

There is often conflict between sounding the alarm, and the wish to use an extinguisher or hose reel. This is best resolved by locating the extinguishers and hose reels (if provided) at the alarm points, so the person seeing the fire can go to the one location, sound the alarm, and collect the firefighting equipment at the same time.

Equipment suitable for first-aid use should be of a type which needs no preparation before use. Such equipment includes portable fire extinguishers and hose reels (solid hose, permanently connected to the water supply, coiled on a drum ready for immediate use). Conventional fabric fire hoses are *not* suitable for first-aid use, as they take too long in practice to deploy, especially by inexperienced works employees.

8.6.3 Works Team Response

When the alarm is sounded, either manually or by an automatic system, the works fire team can respond.

Typically, works teams may comprise a nucleus of one or more full-time firefighters, normally employed on training and equipment maintenance, supplemented by plant employees who can be spared from their normal duties in an emergency. It is important not to count on use of employees needed to control plant and other operations. The works team may include the other operators on the plant on which the fire has started.

The aims of the works team response are:

- to make the plant safe for firefighting;
- to take action that will limit supply of fuel to the fire;
- to take action to safeguard people; and
- to provide extra firefighting effort either to extinguish the fire or to minimize its spread while waiting for arrival of the external brigade.

These actions may involve:

- shutting down the plant if that can be done quickly,
- isolating electric power from the part of the plant in which the fire has started,
- isolating supply of flammable liquid or gas from the leak, by closing valves either remotely or on the plant,
- depressuring the vessels and systems supplying fuel to the fire,
- depressuring vessels, etc., which could become exposed to the fire,
- deploying monitors to cool exposed plant, actuating any deluge systems, etc., and
- running out fire hoses and starting to cool exposed plant.

Note that in petrochemical and oil processing plants, it is often not possible or even safe to extinguish fires while the fuel is still leaking or present. In particular, a leak of a flammable gas should generally not be extinguished except by isolating the leak, because of the possibility of the unburned gas accumulating and exploding.

The aim of much firefighting in such circumstances is to limit damage to adjacent plants, in particular to prevent initiation of additional leaks that would add fuel to the fire.

Further, as firefighting water can spread burning oil, etc., and so make the fire much worse, the use of firewater should be minimized. Water should not be applied for cooling unless the cooling is necessary. One test is to see if "steam" clouds are evolved as the water is applied. If not, then cooling is not needed. Note that cooling water should be applied as a spray: jets from the old fixed nozzles often supplied in fire-hose boxes can thermally shock hot vessels and plant and are not as effective as a spray from an adjustable nozzle.

Where there are chemicals present that can damage the environment, care needs to be taken to minimize the opportunity for firewater contaminated with such chemicals to reach the drainage system and to escape from the site to the environment. Although the better chemical sites will have catchment basins, etc., to hold contaminated firewater, it may be necessary for the works team to arrange to build dykes to catch the firewater. Except in a prolonged fire, this will be possible if emergency plans have been prepared in advance.

8.6.4 External Brigade Response

The external brigade provides extra resources of:

- people,
- equipment, and
- fire experience.

However, in the case of chemical or other process plants, the expertise in the chemical hazards would be expected to be with the plant employees. Therefore it is most desirable that sufficient liaison be established on a routine basis, before any fire, for the external brigade officers to have confidence in the specialist chemical hazard guidance that the plant employees may be able to offer during the fire.

8.6.5 Fire Plans

Great value can be obtained from preparation of *fire plans*, which are a special form of emergency plan. The typical content of a fire plan includes:

- the inventories of flammable materials present,
- typical postulated types of fire,
- typical firefighting approaches, and the firefighting media to be used in each case,
- sources of ignition on the plant,
- environmental protection measures to be considered,

- organization and control of the firefighting effort, before and after arrival of the external brigade, and
- communication channels, radios, etc.

8.6.6 Firefighting Media

There are several types of firefighting media. They include the following:

- water;
- foam (different types for oil-type fires, i.e., liquids that do not mix with water; for alcohol-type fires, i.e., liquids that mix with water; and for special materials, e.g., LPG, where a high expansion foam may reduce the surface area available to the fire without increasing the vaporization);
- dry chemical; and
- inert or fire-extinguishing gases.

It is beyond the scope of this text to elaborate on the applicability of the various media. Fire equipment suppliers are a good source of such information.

8.7 REDUCING FIRE RISKS IN WAREHOUSE OPERATIONS

8.7.1 Introduction

Good practice in warehousing of chemicals has been defined over the years by consideration of the types of inherent hazard that can be presented by various chemicals, and by analysis of historical incidents at warehouses to determine their causes and the factors that contributed to the severity of the consequences.

By far the most likely serious hazardous incident at a chemical warehouse is fire, which can lead to serious consequences of several kinds:

- injury to people from burns caused by the heat,
- impairment of health of people from exposure to smoke containing unburned (and possibly vaporized) toxic or irritant materials, or to toxic or irritant combustion products,
- environmental damage from runoff of contaminated firefighting water,

- property damage to both the warehouse itself and nearby facilities, and
- fire residues, leaving the site contaminated.

Spillages at warehouses have also led to environmental damage, but the chance of this, and the scale of the damage, has been less than in the case of fire.

In summary, the risks from chemical warehouses arise because of the combination of a number of factors:

- flammable, unstable, explosive, toxic materials may be stored in large quantities;
- the packaging is often combustible;
- the storage pattern may provide excellent ventilation around the packages, allowing rapid flame spread; and
- vertical stacking, with air gaps between layers, allows rapid vertical flame spread;
- enclosed space, retaining the heat from a local fire, results in heating of other materials at a distance within the building, resulting in rapid horizontal flame spread.

Thus hazardous materials are stored, and in a configuration that facilitates spread of fire. All that is needed is ignition.

With the large inherent risk of fire in chemical warehouses, it would be expected that ignition sources would be minimized or rigidly controlled. In practice, this is commonly not the case. Ignition sources that have caused warehouse fires include:

- smoking,
- naked flames such as are used for shrink wrapping or flame cutting,
- welding,
- arson,
- office equipment, igniting papers, etc., and the fire spreading to the warehouse,
- electric radiators and other heaters,
- electric battery chargers,
- electric equipment,
- defective electric lighting (e.g., a fluorescent light tube that fell from its fitting is believed to have started one fire),

- electric lighting located above storage areas where the stacking height has allowed the stored material to be heated by the lights,
- steam lines, electric cables, etc., where stored materials are too close allowing the temperatures to rise, and
- sparks from the exhausts of forklifts and other vehicles.

The principles of good location, layout, design, and operation, set out below, are derived from experience with warehouse incidents and have been assembled from the practice of large organizations that have experience with such incidents and have adopted the practice to reduce the chance of recurrence.

8.7.2 Location

Because of the chance of fire, and the resulting problems from toxic smoke, it is desirable that chemical warehouses be located remote from housing, schools, hospitals, and shopping centers. It is difficult to be specific about the distance required in any particular case, and not possible to generalize about chemical warehouses because of the range of toxicity and flammability of materials that may be stored. However, several hundred meters from housing may be appropriate in the case of large warehouses.

It is also desirable that chemical warehouses be separated from food processing for the same reasons. In the one installation, food processing and warehousing of toxic chemicals would present a hazard that would be better avoided altogether (i.e., should both be on the one site) or the most strenuous measures taken to prevent contact between the two classes of material in both normal and abnormal conditions.

The separation needed between a chemical warehouse and adjacent facilities or buildings can best be decided on the basis of experience, as the means by which fire has been spread in the past (flame impingement, burning fragments carried in the updraft) are mostly not really amenable to formal mathematical analysis.

There should be road access for the fire brigade from at least two sides, with at least 10 meters separation from other buildings or installations.

8.7.3 Layout

In order to limit the spread of any fire, large warehouses should be subdivided into smaller compartments with fire-resistant walls.

Similarly, where a warehousing facility shares a building with other activities such as production or office and other auxiliary rooms, the warehousing area should be separated from those activities by fire walls. This includes such rooms as:

- offices,
- toilets and locker rooms,
- recreational rooms,
- rooms for electrical supply equipment,
- heating plant rooms, and
- battery charging rooms.

...wherever there is a potential for fire to be started by such means as smoking, electrical equipment, radiant electric radiators, or office machinery. These auxiliary rooms must not be used for product storage.

At least two escape routes must be available from each area within a warehouse.

8.7.4 Design

8.7.4.1 Construction

The structure should be fire resistant, to prevent collapse. Reinforced concrete is preferred to unprotected steel. If unprotected steel is used, consideration should be given to coating it with vermiculite concrete on vertical load-bearing members, or with sprayed mineral fiber.

It is most important that fire walls be designed and constructed to a recognized code, incorporating insights from recent warehouse fires. Common faults with fire walls include:

- inadequate sealing around the edges, enabling fire to spread to the adjacent area,
- not projecting through the roof, so that flame emerging through the roof can damage the roof of the adjacent section, and allow fire spread,
- being used to support the roof, or being tied to the roof members or the roof columns, such that structural collapse due to the fire results in the firewall being pulled over, and
- having penetrations for doorways or windows that are not fitted with fire doors, etc., designed to a recognized code.

Floors should be impervious and with a smooth finish to facilitate cleaning. This reduces the chance of incompatible chemicals coming into contact as a result of spillages during handling. (In one warehouse operated by a major international chemical company were found stored bags of potassium nitrate, sulfur, and activated carbon, without any physical separation other than not being stored side by side. The Chinese knew of the properties of such materials, when mixed, several thousand years ago!)

Roof cladding should be lightweight to minimize the loading on structures threatened by fire. It is desirable to install means of venting smoke and heated air, to reduce the potential for fire spread in the early stages and to facilitate access by firefighters. The use of plastic panels in roofs may not be satisfactory, as these often do not rupture until subject to direct flame, when the fire is well advanced.

8.7.4.2 Heating, Lighting, Electrics

The means of heating adopted should be such that ignition of stored products, paper, etc., in the warehouse is not possible. Direct electric heating and open oil or gas burners should not be used.

Electric lighting should be installed above aisles or traffic routes, not above the storage racks.

Battery chargers should not be installed in warehouses unless the battery charging area is fire separated from the storage areas. Any battery chargers should be located in a well-ventilated area to enable dispersion of hydrogen evolved from the batteries. Electrical equipment should be minimized in the battery charging area to minimize the chance of ignition of the hydrogen, and should preferably be of flameproof design.

8.7.4.3 Drainage

Where a warehouse may be used for storage of environmentally active materials, either the floor should not be fitted with drains, or such drains should discharge to a pit with sufficient capacity to hold any credible spillage, and the entire building should be bunded or curbed to prevent escape of contaminated firewater.

If stormwater downpipes lead from the roof via the interior of a warehouse to be used for environmentally active materials, for example, along columns down to the floor to subfloor stormwater drains, the pipe should be surrounded at floor level by a curb of at least the same height as the curb around the building (which should be at least 150 mm).

Where flammable liquids are to be stored, there is a good case for arranging for the floors to drain to a sump outside the building, so that any spill is quickly removed from the vicinity of other stored flammable materials. Such a sump should not be able to be drained directly to the sewer system, without deliberate manual intervention such as by being pumped out.

8.7.5 Firefighting Equipment

The main elements of a fire response system are:

- detection of the fire (especially important as a high proportion of fires start after hours), and
- firefighting:

  automatic (e.g., sprinklers)

  manual

In working hours, if the warehouse is normally occupied, there is a good chance that a fire would be detected manually, and that an initial response with fire extinguishers or hose reels would extinguish it.

When someone sees a fire, there is often a conflict of priorities: to attempt to extinguish it, or to sound the alarm. This may be minimized by locating the alarm buttons at the same points as extinguishers and hose reels. The principle should be to sound the alarm at once for any fire. Even if a fire appears small, and well within the capability of employees to extinguish, it is important that the brigade be called.

Fire hoses, as distinct from hose reels, are not first-aid firefighting equipment, as they take too long to deploy.

As a high proportion of fires start after hours, when warehouses are unattended, it is important that fire detection systems be installed and connected to both a local alarm and to the external brigade. Early response is necessary to increase the probability of stopping the fire before it grows beyond the capability of the response effort. There is no doubt that sprinklers (designed to appropriate codes and properly maintained) are very effective in detecting fires and extinguishing them while very small. Further, the frequency of spurious initiation of sprinklers is extremely low.

In the early stages of a fire, the smoke and heat from the localized fire rise to just under the roof and spread sideways through the building. This both impedes access and effective response by firefighters and raises the temperature of the rest of the warehouse, increasing the ease of fire spread. For this reason, smoke vents in the roof are valuable. In allowing smoke to escape, they also allow fresh air and oxygen to be drawn in, but the net effect in the early stages of a fire is reported to minimize fire spread. Sometimes plastic roof panels are used in place of manually operated or automatic vents, but such panels often do not open until heated by flame impingement, which may be too late.

8.7.6 Good Operating Practice for Chemical Warehouses

8.7.6.1 Using Warehouses Only for Warehousing

Warehouses are designed with the prime purpose of storing large quantities of material. If any of the stored material has inherent hazards of flammability, toxicity, etc., abnormal occurrences involving the materials have the potential to have severe effects.

As hazardous incidents do not just happen, but are caused, it is good practice for any activity not essential to the warehousing activity to be kept out of warehouses. Activities that are often found in warehouses, which have a history of causing or contributing to hazardous incidents, include:

- production or processing operations, as these have the potential for fire, or for needing maintenance involving welding, flame cutting, or grinding, with the potential for a fire being started in, or spreading to, the stored materials;
- workshop and maintenance activities, with the potential for ignition from machine tools, sparks from grinders, or flame from welding;
- offices, including the store's office, with the potential for cigarette butts, electric radiators, and office machinery igniting the unavoidable papers in the office, leading to a fire that spreads to the rest of the warehouse;
- any avoidable activity that causes people to be in the warehouse, as people are the cause of most incidents (note that some activities, such as packing containers, which are seen as warehousing activities, should preferably be located where they can be fire-separated from the

storage area itself. Wherever there are people, there is an increased chance of a fire starting); and

- shrink wrapping with a naked flame or an oven, with inadvertent ignition of the film and spread of the fire to the stored materials.

8.7.6.2 Operation

Good operating practice involves meticulous attention to what may appear to be mundane details, such as:

- security to prevent unauthorized access or presence of people who do not need to be there,
- rigid enforcement of no-smoking rules and other measures to prevent ignition (as set out above),
- orderly stacking in defined storage areas with no "temporary" storage in aisles,
- not storing materials too high, which can exceed the design load for any sprinklers, or can impede distribution of water from sprinklers,
- general housekeeping, including cleaning up and disposing of spilled materials,
- storing flammable, toxic, and other materials separately,
- having fire safety equipment checked regularly by people competent to do so,
- having, and practicing, emergency procedures and firefighting procedures, and
- rotation of stock, to ensure that old stock is not accumulated where it can deteriorate.

8.7.7 Toxic Gas Effects from Warehouse Fires

Burning materials generate combustion gases. Common materials, such as wood, paper, natural and man-made fibers, plastics, and chemicals generate a range of combustion gases that are asphyxiant or toxic to varying degrees. The gases include:

- carbon monoxide and dioxide (from a wide range of materials),
- hydrogen chloride (from burning plastics such as PVC),
- hydrogen cyanide (from burning plastics such as polyurethane),

- water vapor,
- phosgene (from involvement of chlorinated solvents in the fire),
- sulfur dioxide,
- oxides of nitrogen, and
- vapors or dusts of unburned material.

Completeness of combustion is enhanced by:

- high combustion temperatures,
- oxygen surplus, and
- long residence time of the combustion gases in the high-temperature zone of the fire.

These vary depending on the position of the burning materials in relation to the rest of the fire. At the edge of the fire, the temperatures may be less, and toxic materials may be evaporated unburned, or only partly burned, in spite of a surplus of oxygen. Nearer the center of the fire, the temperatures and residence times may be higher, but oxygen may be in short supply. Thus it is difficult to predict the quantitative composition of the combustion gases from particular fires.

Odors may be a nuisance at a substantial distance. If toxic chemicals are involved in the fire, and if the combustion gases have a marked "chemical" odor, members of the public may be anxious about the long-term effect of exposure.

8.8 REDUCTION OF RISKS IN TRANSPORT OF HAZARDOUS MATERIALS

8.8.1 Introduction

There is extensive media coverage of hazardous incidents that occur during transport of hazardous materials. This is understandable because of:

- the serious incidents that have occurred, and
- the fact that such incidents usually take place in public areas such as roadways and near housing or schools.

There is a great deal that can be done to minimize the risks arising from transport of hazardous materials.

8.8.2 Minimization of Inherent Hazard

Options include:

- minimization of the transport of hazardous materials, for example, by manufacturing the hazardous materials on the same site as the plants that will use them;
- use of safer means of conveying hazardous materials from place to place, such as suitably designed pipelines; and
- transport of substitute materials, such as shipping ethylene dichloride (which is not highly flammable, is a liquid at normal temperatures, and is difficult to ignite when cool) instead of vinyl chloride (which is flammable, a compressed liquefied flammable gas, and easily ignited).

8.8.3 Prevention of Hazardous Incidents

Options include:

- design of tankers to have robust barrels, with recessed valves, etc., such that, in the event of collision or overturning, the containment of the hazardous materials is not lost;
- design of the chassis and prime movers for structural integrity and roadworthiness, to minimize the chance of traffic accident that could expose the barrel to mechanical damage;
- selection of drivers for:

  the appropriate temperament for careful patient driving (noting that studies have shown a strong correlation between a safe driving record and a steady personality),

  suitability for long-haul and long-duration trips, if that is required, and

  ability to cope with the specialist training on hazards and emergency response;
- preparation of a suitable syllabus for training drivers, and conducting such training to a high standard including testing, covering hazards, good practice, and emergency response;
- encouraging and requiring drivers to inspect their vehicles before starting any trip and to report any unusual or unsafe condition of the vehicle, or any unusual or unsafe event while transporting hazardous materials;

- ensuring that only trained and authorized drivers (without exception) undertake transport of hazardous materials;
- establishing a routine of auditing the condition of the vehicles and the tanks, and the procedures used; and
- selecting routes that minimize traffic hazards.

8.8.4 Protective Response

Options include:

- preparation of emergency plans, defining the required equipment, responsibilities, and communication channels, and ensuring that the equipment is provided and ready, that all involved know their responsibilities and how to carry them out, and that the communication routes are established;
- in particular, training drivers to take the necessary emergency actions; and
- conducting emergency practices as realistically as possible.

8.8.5 Damage Limitation (Separation and Strengthening)

Because the vehicles travel on public roads, it is difficult to arrange for separation of the public or property from an accident, other than by the police establishing barriers or arranging evacuation of nearby areas.

8.9 REDUCTION OF BLEVE RISKS

Following a series of BLEVEs in various parts of the world during the 1970s, the causes and means of preventing them were studied in detail and are well understood.

The cause is flame contact to the steel of the pressurized storage vessel above the level of the liquid inside. (Below the liquid level, the heat from the flame outside is removed from the steel by convection and the liquid inside boiling.) The boiling of the liquid increases the pressure until the relief valve operates to limit further pressure rise. However, the steel above the liquid level may be heated to a level where it is no longer capable of withstanding the design stress, and it ruptures, releasing the contents, which ignite.

BLEVEs are entirely preventable by adoption of simple design and operating principles. The design principles include the following:

- Minimize the number and size of vessels containing liquefied flammable gases or flammable liquids above their normal atmospheric-pressure boiling points.
- Minimize the chance of fire exposure of such vessels, and also of closed vessels of flammable liquids with a relatively low boiling point. (Ordinary 200-liter drums of flammable solvents can BLEVE if exposed to fire for a sufficient time for the contents to boil and to build up pressure inside.)
- Because many historical BLEVEs have resulted from leaks from pipework, the principles of fire safety listed earlier should be adopted.
- Ensure that any liquid leak from the vessel or pipework flows away from the vessel and does not burn where the flame can impinge on the vessel. This necessitates hard paving of the ground in the vicinity of the vessel, with a slope of between 1 in 50 and 1 in 100 away from the vessel.
- Further, pipework in the vicinity of the vessel should be designed to minimize the chance of leakage. Typically this is achieved by minimization of nozzles on the vessel, especially below the liquid level, avoidance of sight glasses, minimization of small-diameter pipework and bracing of any that is unavoidable, minimization of pipe joints, fitting of an emergency isolation valve as the first valve in the liquid discharge line, and locating pumps and other equipment at a safe distance.
- Protect the vessel from fire by either fireproof insulation, or a deluge system able to be actuated from a distance.
- Install flammable gas detectors to initiate an alarm in the event of a leak of the flammable material in the vessel.
- Install fire detectors, if the vessel is in an area where it is not regularly observed, with actuation of the detectors automatically initiating the deluge system, if fitted, and actuating the alarm.

8.10 REDUCTION OF VAPOR CLOUD EXPLOSION RISKS

A vapor cloud explosion can occur if a leak of flammable vapor mixes with air and is ignited.

In the open air, a large quantity of flammable vapor (typically more than 5 tonnes) is needed for an explosion to occur, which necessitates a very rapid rate of release. In practice, such a rate would be possible only from rupture of a very large and high-pressure gas pipeline, or from a leak of a liquefied flammable gas or a flammable liquid stored at a temperature above its normal atmospheric-pressure boiling point.

Indoors, a much smaller quantity of vapor can cause a vapor explosion, because of the confinement of the ignited cloud. As a result, even small leaks of flammable gas, or evaporation of flammable liquid solvent, can produce a sufficiently large cloud for a confined vapor explosion. Such an explosion can be very damaging to the building and its contents, but because of the small amount of heat energy available to the explosion, the range of the effects is limited to the immediate surroundings of the building.

Accumulated vapor from leaks of solvents, such as acetone or isopropanol, can also cause a vapor explosion if the vapor is confined by a building. In practice, such an explosion would not have a significant effect beyond the boundary of a factory, but could seriously injure employees and the factory property itself.

The general principles of design adopted to minimize the risks from vapor cloud explosion are:

- Minimize the potential size of the cloud. (The larger the cloud, the more likely it is to reach an ignition source, the more likely the ignited cloud is to explode, and the greater the power of the explosion.)
- Minimize the confinement of the cloud. (The greater the confinement, the greater the chance that the cloud would remain sufficiently concentrated to be ignited, the greater the chance that an ignited cloud would explode, and the greater the power of the explosion.)
- Minimize ignition sources.

The design details to be considered for adoption include:

- Minimized inventories of high-pressure flammable gas, liquefied flammable gases, and flammable liquids stored at temperatures above their atmospheric-pressure boiling points. (Such liquids, when depressurized on leaking, flash into vapor either entirely or in part, thus generating a large amount of vapor.)
- Minimized pressures and temperatures, so as to minimize the leak rate in the event of a leak, and the amount of vapor generated.
- Minimized opportunities for leakage, with minimized pipe joints, robust pipework, no sight glasses, etc.
- Leak detection, such as by use of flammable gas detectors.

- Installation of remotely operated emergency isolation valves, so as to isolate any leak promptly.
- Locating the major potential sources of leaks of such materials outdoors, and with minimal confinement by adjacent equipment. There have been numerous very damaging explosions of very small clouds (e.g., 100 kg or less) in enclosed spaces, such as gas compressor buildings.
- Control of ignition sources.

In the case of offshore platforms, where there is the likelihood of escape of condensate at some time during the life of the platform, yet confinement or congestion is unavoidable, there are a number of design features that can be adopted to minimize the likelihood of explosion (rather than a less damaging flash fire) and to minimize the damage in the event of an explosion. These include the following:

- Attempting to locate the more likely sources of condensate leak (e.g., pumps) in locations where there is the least confinement, for example, near the edges of the platform, and attempting to avoid the use of solid walls in the vicinity.
- Where walls are necessary for protection from the external environment, using blast-relief design on the outer walls, and blast-resistant design for the inner walls.
- Paying close attention to avoidance of ignition sources.
- Avoiding or minimizing congestion between the potential source of condensate leakage and the outer edge of the platform or the blast-relief walls, so as to minimize turbulence in the event of ignition and formation of an expanding burning cloud.
- Taking special care to minimize small-bore pipework and to strengthen or protect whatever small-bore pipework is unavoidable.
- Installing the usual gas detection, remote isolation, and excess flow protection facilities.
- Paying special attention to permit-to-work procedures and modification control procedures in the area.

8.11 REDUCTION OF DUST EXPLOSION RISKS

The effects of a dust explosion are mostly limited to the immediate vicinity of the building in which it occurs.

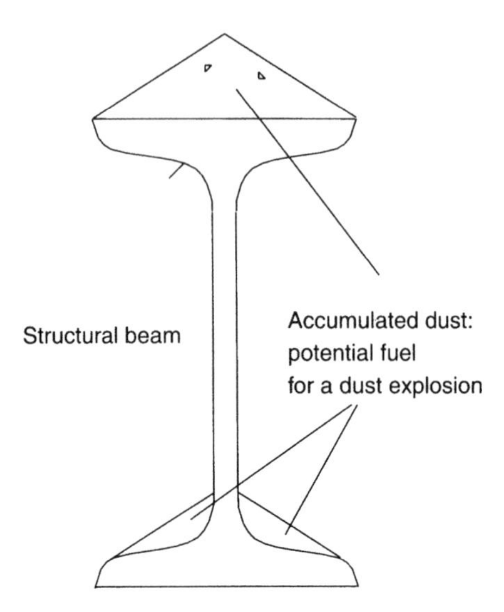

Figure 8-10. Dust collecting on a beam.

Typically, a dust explosion occurs when a combustible dust is ignited within a confined space such as a machine or a building. Very commonly, a small and relatively harmless initial dust explosion shakes loose accumulated dust on horizontal surfaces within the building, such as roof beams and ledges, leading to a second and more severe explosion a few seconds after the first. This cycle may repeat itself several times. See Figure 8-10.

The design principles that would be adopted to avoid dust explosions include:

- avoidance of dust-generating operations, for example, by use of wet processes,
- containment of any dust,
- exhaust ducting to remove any dust unavoidably released,
- avoidance of elevated ledges and horizontal surfaces (e.g., on roof structures) on which dust can accumulate with time)—the proposed plant would have a false ceiling to cover the roof trusses and prevent any dust reaching them,
- control of ignition sources, especially within machines handling dust,
- inerting of atmospheres within machinery with nitrogen within which dust clouds could be generated (but taking careful note of the serious dangers of asphyxiation by nitrogen),

- installation of suppression systems to process containers in which a dust explosion could occur, to flood the container with explosion suppressant on the onset of a pressure rise due to the start of an explosion,
- avoidance of propagation paths between vessels, etc., in a process to prevent propagation of an explosion throughout the process (e.g., rotary valves, flooded feed screws), and
- explosion relief to outside the building for equipment with the potential for dust explosion (e.g., bag filters).

A most important control measure is housekeeping, with regular removal of any accumulated dust. This would be necessary in the proposed plant to meet pharmaceutical standards, quite apart from explosion safety standards.

8.12 REDUCTION OF TOXIC GAS RISKS

The appropriate measures required for minimization of toxic gas risks are very similar to those required for fire and explosion safety: inherent safety, good containment and control hardware and software, good protective hardware and software, and facilities for damage limitation. (The main difference is that ignition sources are not an issue.)

The following principal design features should be adopted:

- Integrity of containment of the toxic materials, so as to avoid leaks.
- Minimally sized pipework, consistent with robustness, to minimizethe possible escape rates from possible leaks.
- Gas detection systems to give prompt warning of commencement of a leak.
- Fitting of excess flow valves to stop flow at once in the event of a major leak.
- Fitting of remotely operable emergency isolation valves to inventories of toxic materials so that any leak, large or small, can be promptly isolated. In cases where actuation of such valves will not cause a serious plant upset (with attendant hazards), the valves could be automatically actuated by the gas detection system, so as to minimize the duration of any leak, to minimize the physiological effects.
- In the case of toxic materials that do not pose a flammability problem if enclosed and that are water soluble, storage of the gas cylinders within a building fitted with a water spray system automatically

operated by gas detectors, as well as manually operable. Examples of such materials are hydrogen chloride and ammonia, which are highly soluble in water and would be rapidly scrubbed out of the air in the building and thus prevented from escaping.

8.13 REDUCTION OF ENVIRONMENTAL RISKS

8.13.1 Introduction

The term *environmental risks* will be limited here to sudden and unexpected events, not the chance of environmental damage from continuing emissions such as from stacks, or from continuing fugitive emissions from such sources as valve stems and pump seals.

Generally, environmental damage is caused by loss of containment of environmentally active materials. This can result in escape of such materials to the environment via:

- vapor leaks to the atmosphere,
- liquid leaks to the ground, or to the drainage system and then to the ground or rivers, lakes, or the ocean, and
- solid materials being disposed of to waste depots, etc., which are unsuitable for the environmentally active materials.

Among the most damaging environmental incidents have been incidents where a fire has damaged containers of chemicals in storage, and firefighting water has carried the materials to rivers or the ocean.

The options are considered here in outline only.

8.13.2 Reduction of Inherent Hazard

The options include:

- avoidance or minimization of processing, storing, or transporting of environmentally active materials,
- substitution of less environmentally active materials, and
- use of processes that are less likely to generate abnormal circumstances that would result in escape of environmentally active materials.

8.13.3 Prevention

The options include:

- storage and handling of environmentally active materials apart from other materials or operations, and in locations where the containers are unlikely to be exposed to fire or mechanical damage,
- use of robust containers,
- inspection and monitoring of plant equipment and operating procedures to minimize the chance of escape to the environment,
- paying close attention to fire avoidance, and
- training all employees on the importance of avoiding environmental damage, the routes by which environmental damage could occur, and the means of avoiding such incidents.

8.13.4 Protective Response

The options include:

- storage of environmentally active materials in bunded or curbed areas where any leakage will not soak into the ground or flow to water courses or drains, either on its own or with rainwater, etc.;
- provision of means of diverting normal drainage to catchment basins so that any material accidentally discharged to the drainage system can be held pending suitable treatment;
- providing those sections of plants handling environmentally active materials with curbing and separate drainage systems to prevent spills of those materials entering the normal drainage system;
- providing warehouses containing environmentally active materials with bunds sufficient to hold the likely maximum quantity of firefighting water, or curbing the warehouse and providing a suitable drainage system and catchment basin to hold contaminated firewater; and
- training all employees on the emergency steps to be taken in the event of a spill of environmentally active material, and in the steps to be taken in the event of a fire necessitating the use of firewater in the vicinity of such materials.

8.14 REDUCTION OF THE RISK OF LOSS OF RELIABILITY

8.14.1 Reliability-Centered Maintenance (RCM)

8.14.1.1 Introduction

Reliability-Centered Maintenance (RCM) is a special case of the general approach described above for hazard identification and ranking. Where scenarios with apparently high risks, or scenarios with apparently severe consequences, are identified as arising from possible equipment failure or breakdown, then the options for preventive maintenance are considered. It is wise to link a reliability-centered maintenance program with Risk-Based Inspection (RBI—see later discussion), so that the observations in the inspection program are used, as appropriate, to trigger preventive maintenance.

It is helpful, in developing a reliability-centered maintenance program, to focus attention on critical *capabilities*, rather than critical equipment. Although this may seem purely a matter of words, it can lead to a sharper focus on what needs to be maintained, and it also helps nontechnical people (such as accountants and senior managers) to appreciate the need for expenditure before equipment has failed.

To develop an effective reliability-centered maintenance program, it is important to involve an appropriate range of experience and expertise in the team, including the various branches of engineering and related trades, and with people at various levels in the organization, so as to gather the full range of information about the equipment and aid in building commitment.

Much reliability-centered maintenance has, in the past, looked only at the repair costs and the cost of lost production due to the downtime. It is important in high-hazard process industries (which are not necessarily high-risk industries), such as smelting, chemicals, oil and gas, to consider all the consequences of equipment failure, including injury to people, environmental damage, damage to public image possibly inhibiting statutory approval of future developments and causing imposition of punitive controls on existing operations, prosecution, and civil legal claims with the threat of custodial sentences.

8.14.1.2 History of the Development of RCM

Years ago, maintenance was mainly undertaken when equipment failed. Then it was realized that some breakdowns were very damaging, and that there was a case for maintaining equipment to prevent breakdown. This led to the rise of *planned maintenance*.

It was difficult to strike a balance between—on one hand—doing too much planned maintenance (with the cost of doing work that may not have been necessary) and—on the other hand—too little (with the excessive cost of breakdowns). Many organizations monitored the ratio of planned to breakdown maintenance and aimed for some arbitrary ratio such as 50:50.

But it was found that planned maintenance of many types of equipment did not necessarily reduce the rate of breakdown maintenance, and in some cases actually increased it.

Further, it was realized in those industries where breakdown could be very costly or dangerous (or both), initially in air transport and later in many other industries, that a better way of managing maintenance was needed.

An aviation industry task force established in the United States, involving the industry and the regulators, concluded that:

(a) scheduled overhaul had little effect on the reliability of complex equipment unless there is a dominant mode of failure; and

(b) there are no effective forms of scheduled maintenance for many types of equipment.

The task force produced a report on recommended principles for maintenance, which went through several revisions. The third version became unofficially known in the industry as Reliability-Centered Maintenance.

United Airlines was commissioned by the US Department of Defense to prepare a guidance document on the relationships between maintenance, reliability, and safety. The document was called "Reliability-Centered Maintenance."

8.14.2 The Foundation of RCM

The foundation of RCM rests on four principles:

1. The purpose of preventive maintenance is to minimize or prevent the consequences of breakdown, not breakdowns themselves. Where equipment has many modes of breakdown, some will have more serious consequences than others. The emphasis should be on

prevention of these breakdowns, or minimization of their frequency or their impact.

2. The consequences of failure depend on where the equipment is installed and operated, so a preventive maintenance program for a particular type of equipment in one organization may be entirely inappropriate for the same equipment in another organization.
3. It is not assumed that all failures can be prevented by preventive maintenance, or that it is desirable to do so.
4. The fundamental focus is on the capability required from the equipment, that is, its function, not the equipment itself.

8.14.2.1 The Steps in the RCM Process

The RCM process involves the following steps. (These are identical, in principle, to the steps in risk management.)

1. Define the objectives of maintenance management. In doing this, it is important to define the priority of the various types of performance, for example, safety, uninterrupted production, environmental performance, and asset preservation. Performance standards should be defined for each of these.
2. Define the required capabilities (or functions) of the equipment in relation to these types of performance. Capabilities or functions are defined by verb–noun pairs. For example, in the hot oil heating plant illustrated in Chapter 6, the required capabilities of the pump include:
   - generate pressure
   - contain oil
3. Define modes of failure that will have an impact on the required types of performance.
4. Identify the types of effect of those modes of failure.
5. Assess the magnitude of the effect on each of the required types of performance.
6. Analyze the priority capabilities or functions, failure modes, and effects, to identify opportunities for improving, eliminating, or reducing the failures or the impacts. The priority assessment process should take account of both the potential consequences of failure and the estimated relative likelihood of occurrence. This

assessment requires active involvement of experienced maintenance staff; experienced both in general maintenance and in the maintenance of the specific plant being studied.

7. Define the preventive tasks to be undertaken. These will be of the following kinds:
   - scheduled inspection (to detect in advance any condition of the equipment that indicates impending failure, to enable rectification at low cost, rather than experiencing later failure at greater total cost);
   - scheduled restoration tasks (to bring the condition back to a newer state, such as is routinely undertaken in the case of some components of aircraft); and
   - discard and replacement.

These can be summarized as follows:

- Focus on preserving the capabilities of the system (rather than preserving the condition of the equipment for its own sake).
- Identify the specific failure modes that will result in loss of capabilities.
- Prioritize the failure modes in order of importance of their impact on the capabilities, preferably taking into account the relative likelihood of occurrence.
- Identify effective and workable preventive maintenance work that will minimize or prevent the high-priority failure modes.

It is important that the above analysis be undertaken preferably by, but at least with, the operational staff responsible for the plant achieving its function, rather than by a specialist "back-room" team in isolation.

8.14.2.2 Suitable Techniques

Both Hazop studies (see Appendix A) and Failure Mode and Effect Analysis (FMEA—see Appendix B) are very suitable techniques for identifying failure modes and (normally qualitatively) assessing the magnitude of their impacts on the required capabilities. Each must be undertaken as a team study involving the operational staff together with specialists in the particular types of technology involved.

(FMEA is commonly extended to Failure Mode and Effect Criticality Analysis—FMECA—by quantification of the severity of the consequences and of the likelihood. See Appendix B. Thus FMECA is a special form of rapid ranking.)

FMEA or FMECA were developed and are widely used in mechanical plants, whereas Hazop was developed and is widely used in process plants. Each can be adapted to suit other types of industry.

The rapid ranking technique, as described in Chapter 3, uses the same principles as Hazop and FMECA and can be structured to suit a wide range of types of technology by the team considering and defining the types of failure mode to be considered in the analysis to match the particular technology. It is a particularly convenient method for considering the different types of impact (safety, environmental, production continuity, asset damage, etc.) at one time, taking account of the estimated relative likelihood of occurrence.

8.14.3 Risk-Based Inspection

8.14.3.1 Introduction

One of the uses of the rapid ranking approach is in development of inspection programs that focus on the equipment or systems where such inspection will be most cost-effective overall.

Where scenarios with apparently high risks, or scenarios with apparently severe consequences, are identified as arising from possible equipment failure or breakdown or from human error, of such types that detection by inspection could be sufficiently early for prompt and effective preventive action to be taken, then an inspection schedule is established. It is necessary, of course, to consider the full costs of the postulated failures and their likely frequencies so as to derive the extent of the annualized cost saving that would be expected, for comparison with the cost of the inspections and the preventive actions.

Thus, some of the factors to be considered when setting up the risk-based inspection program, and when adding any new inspection to it, are:

- the full costs of failure (including tangible costs of damaged equipment and loss of production, etc., plus intangible costs such as injury, environmental damage, loss of market, and loss of reputation);
- the speed with which the incident develops, as a basis for determining how often inspection would be needed, and whether it needs to be undertaken on shift by the shift operators or tradespeople, etc.;

- the nature of the inspection and the test equipment (if any) needed, as an indication of the level of training and experience needed by those doing the test;
- the likelihood that inspection would detect the onset in sufficient time for effective action to be taken;
- the experience and expertise needed to determine the nature of the corrective action, and to authorize it; and
- the costs of inspection, and of the action that would need to be taken, so as to determine whether the proposed inspection is actually an economic option.

In some cases it may be decided that the cost of inspection is clearly low compared with the consequences of undetected failure. This is particularly likely with protective systems, where a failure may not be detected until a situation arises in which the protective system is expected to operate, but it does not. By then it is no help to know that the failure has occurred.

One approach is to define a maximum acceptable probability of the postulated situation in which both a situation arises needing protection, and the protection system fails—for example, the frequency of a pressure controller failing at a time when the high-pressure alarm and trip systems are in a failed state.

Then, using mathematics set out in Chapter 6, and using data for the annual probability of failure of the control system and the alarm/trip systems, it is possible to calculate the desired minimum frequency to test the instruments. This can be done quickly for a large number of protective systems by setting up an appropriate computer spreadsheet.

8.14.4 Risk-Based Inspection Frequency (See Figure 8-11)

Given that:

T years is the time for a fault condition to develop from barely detectable to failure;

p (range 0–1) is the probability of an inspection successfully detecting the developing fault condition within the period T; and

c (range 0–1) is the required probability (confidence) of detection of the developing fault before failure;

then the probability of an inspection *not* detecting the developing fault condition is:

$$(1-p)$$

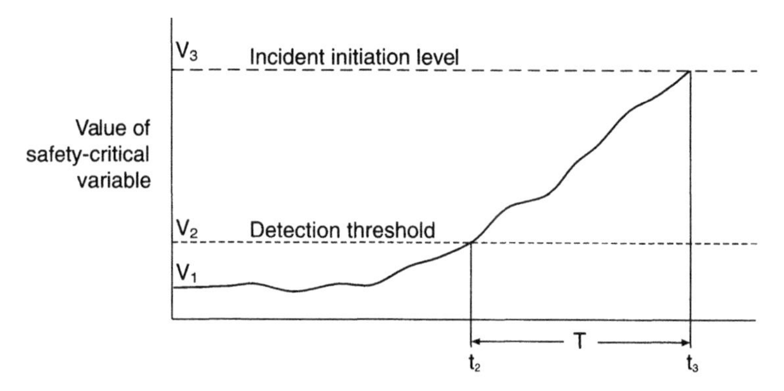

Figure 8-11. Illustration of fault development.

The maximum acceptable probability of the developing fault condition not being detected before failure is:

$(1-c)$

Then $(1-c) = (1-p)^n$, where n is the number of tests needed within time T to achieve the required confidence of detection. Thus:

$n = \log(1-c)/\log(1-p)$

so

Required frequency of inspections per year $F = \log(1-c)/T \log(1-p)$

Example 8.14.1:

In the event of changes in process conditions, it is possible that corrosion could develop in a vessel. The corrosion is expected to take 3 months to develop from being capable of detection to a vessel failure. Any single inspection is expected to have a 75% probability of detecting any fault. Because of the severity of the consequences if a failure occurs, the required probability of detecting the developing fault before failure is set at 95%. What inspection frequency is required?

$T = 0.25$ years

$p = 0.75$

$c = 0.95$

$F = \log(1-0.95)/0.25 \log(1-0.75)$

$= (\log 0.05)/(0.25 \log 0.25)$

$$= (-1.30)/0.25 \times (-0.6)$$

$$= 8.6 \text{ per year}$$

that is, around every 6 weeks. (It may be more practicable to inspect the equipment monthly or quarterly.)

Example 8.14.2:

A large pump handling liquid butane has a double mechanical seal. The seal could be wrecked in the event of bearing failure. It is expected that most types of bearing failure on this particular pump would take around 2 days to develop from being detectable by observation of the noise to being sufficient to cause collapse of the seal. It is believed that any specific inspection by an operator listening to the pump has a 50% chance of detecting the problem. Because of the potential for a large fire or even a vapor cloud explosion, the probability of detecting bearing deterioration before seal collapse is required to be at least 90%. How often should the operator visit the pump to listen for bearing noise?

$T = 2/365; \quad p = 0.5; \quad c = 0.9$

$$F = (365 \times \log 0.1)/(2 \times \log 0.5)$$

$$= 365 \times (-1)/2 \times (-0.301)$$

$= 601$ per year, that is, in practice, around twice per day, or once per shift, whichever is administratively neater.

8.14.5 Economic Viability of Inspection

Given that the impact of a failure is able to be expressed purely in monetary cost (not always possible with any precision), and that:

- C is the total cost of conducting a test per occasion ($);
- F is the frequency of testing;
- D is the total cost of a failure ($);
- H is the frequency of the failure (p.a.); and
- c is the confidence that the testing will detect the fault condition before failure,

then

$$\text{Annual cost of inspection} = C \times F$$

and

Annualized cost of failures detected by the inspections $= D \times H \times c$

Therefore the inspection regime required to prevent failure is economically viable only if

$$D \times H \times c > C \times F$$

No account is taken here of the time value of money. The approach can be varied for NPV concepts. Note also that this takes no account of the safety, environmental, or other possible impacts that cannot be realistically expressed in monetary terms.

8.15 DESIGN FOR RELIABILITY OF CONTROL AND PROTECTIVE SYSTEMS

The mathematical bases for designing and operating reliable control and protective systems are set out in Chapter 6 and are summarized and assembled here.

The principles are as follows:

- Control and protective systems should be designed by experienced control and instrumentation engineers, with their judgment and experience augmented by reliability theory, and not by use of reliability theory alone. The aim is to ensure that the control and protective systems are *capable* of performing the required duties in the applicable process conditions.
- The quality of the components selected for the systems should be of a high order. It is futile to assemble reliable systems, even if these systems are redundant, with unreliable components. The reliability depends on both the general type of the component and the manufacturer.
- The systems will need to be maintained to the same standards as the original specification. In other words, a system that is to be very reliable will need very high standards of maintenance. This necessitates a team of skilled and trained tradespeople, with high standards of professional management and technical guidance, and a high standard of workshop facilities and resources.
- Where it is important that control not be lost (i.e., where a protective response is little help), redundant control systems can be installed, with the control ranges of the systems set progressively wider, such that only one system is normally operating but the other systems will come into operation if the normally operating system fails for some reason. (But note: if the second control system range is too wide,

control may be effectively lost before the second system comes into action; but if the second control range is too narrow, there is potential for conflict between the two controllers.)

- Control systems need to be recalibrated and retuned periodically.
- Especially in the case of more advanced or elaborate control systems, there is a need for the following:

regular evaluation of the performance of the control system (i.e., has it met the performance targets?), and

regular formal retraining for operators in the function of the control system and the role of the operators.

- Protective systems should be entirely independent of the control systems for which they aim to provide protection.
- Protective systems need to be tested. An untested protective system cannot be relied on, as by the time it is called upon to act there is a substantial probability that it will have lapsed into a failed state.
- A test of a protective system should be as close a simulation of fault conditions as possible and should test the system as fully as possible from the sensor of the fault conditions to the active component that protects the plant (e.g., from a thermocouple sensing high temperature, right through to the fuel shutoff valve).
- Where additional confidence is needed that a control failure will result in a protective response, then the choice is either a more reliable single protective system or installation of redundant protective systems.
- A single protective system can be made more reliable, in principle, by use of more reliable components for the particular duty required, by more frequent testing, or by improved testing including checks that the system is left "armed" following the test. (However, often when a single protective system is found to provide inadequate protection, the reason is associated with a set of conditions that was overlooked when the original process or control system, or both, were first designed. Therefore an apparently excessive unreliability in practice should be investigated first from the viewpoint of the process, and then from that of the control and protection system design.)
- Redundant protective systems increase the probability that at least one protective response will operate, but also increase the probability of a spurious protective response due to an erroneous "fail safe" response arising from component failure. (One problem with spurious

alarms or trips is that operators become tired of them and disconnect the protective systems altogether.)

- The probability of a spurious protective system response can be reduced by use of "voting" systems.
- The probability of a common-mode failure of redundant systems, whether they are control or protective systems, can be reduced by use of diversity in the total design process and in maintenance of the system when in operation. Examples include use of:

  different process variables as indicators of loss of control,

  different principles of operation in the equipment chosen,

  different routes for instrument cables,

  different tradespeople for calibrating, testing and maintaining equipment,

  different teams and contractors to design and provide backup systems, etc.

In summary, the essential principles are:

- capability,
- redundancy,
- diversity, and
- routine calibration and testing.

8.16 EQUIPMENT DESIGN FOR RELIABILITY AND SAFETY IN THE OIL AND GAS INDUSTRY IN PARTICULAR

There are numerous design details that experience has shown to be good practice for safety and reliability in the oil and gas industry. Numerous examples of these are described in a practical manner by Lieberman (1988).

REFERENCES

Bollinger, R. E., Clark, D. G., Dowell, A. M., Ewbank, R. M., Hendershott, D. C., Lutz, W. K., Meszaros, S. I., Park, D. E., and Wixom, E. D., "Inherently Safer Chemical Processes: A Life Cycle Approach." (D. A. Crowl, Ed.). American Institute of Chemical Engineers, New York, 1996.

Center for Chemical Process Safety (CPPS), "Guidelines for Engineering Design for Process Safety." American Institute of Chemical Engineers, New York, 1993.

Hendershot, D. C., "Inherently Safer Process Design Philosophy." Proc. Conf. Process Safety, The Israel Institute of Petroleum and Energy, Herzliya, Israel, May 5–6, 1999a.

Hendershot, D. C., "Inherently Safer Design Strategies." Proc. Conf. Process Safety, The Israel Institute of Petroleum and Energy, Herzliya, Israel, May 5–6, 1999b.

The Institution of Chemical Engineers (IChemE) and the International Process Safety Group (IPSG), "Inherently Safer Process Design." The Institution of Chemical Engineers, Rugby, UK, 1995.

Kletz, T. A., "Cheaper, Safer Plants, or Wealth and Safety at Work." The Institution of Chemical Engineers, Rugby, UK, 1984.

Kletz, T. A., *Plant Design for Safety.* Taylor and Francis, Bristol, PA, 1991.

Kletz, T. A., *Process Plants: A Handbook for Inherently Safer Design.* Taylor and Francis, Bristol, PA, 1998.

Lieberman, N. P., *Process Design for Reliable Operations*, 2nd ed. Gulf, Houston, 1988.

Tweeddale, H. M., "Improving the Process Safety and Loss Control Software in Existing Process Plants." Institution of Chemical Engineers, London. Symposium Series 94, Multistream 85, April 1985.

Chapter 9

Management of Risk and Reliability of New Plants

When you have studied this chapter, you will understand:

- how safety and reliability must be incorporated into a new design from the outset; and
- the methods by which that can be done.

This chapter sets out an integrated approach to the process of risk management for a new project.

9.1 INTRODUCTION

It is much easier to incorporate safety into the design of a new project at the outset, when the concept is being developed and defined, than to attempt to add safety on once the design is finalized. For example, if the proposed location of the plant is inappropriately close to housing, then a great deal of safety equipment may need to be added to compensate. This will add to the capital and maintenance costs. On the other hand, if safety studies are started at the outset, an inappropriate location will be avoided, or the inherent safety of the plant may be improved by such means as selection of a different process, or by designing for minimal inventories, such that no incident could be serious enough to place housing at risk, in spite of a limited buffer zone.

One satisfactory approach to management of the safety of a new project during the design and construction phases, which has been used repeatedly with success, comprises a number of studies conducted at appropriate stages of design and construction. These match the information available and the activities undertaken at each stage of the project and are described below. They are summarized in Figure 9-1, which is consistent with the phases of a project as defined by the Project Management Institute (PMI, 1996).

(A variation on this approach is described by IChemeE, 1994.)

9.2 SAFETY, RELIABILITY, AND ENVIRONMENTAL SPECIFICATION

This study is undertaken at the conceptual stage of a project (i.e., during Phase 1 in Figure 9-1), possibly when the design is at the block diagram stage.

The objectives of the study are:

1. to identify the inherent hazards of the materials and processes that it is proposed be used,
2. to define the criteria that the plant will need to meet in the safety, reliability and environmental fields, and
3. to define guidelines for the design of the plant, for incorporation in the technical specification sent to tenderers, which would be expected to result in a design that will meet those criteria.

This study may be conducted as two meetings, separated by several weeks of study and investigation.

The first meeting comprises the senior people responsible for various aspects of the plant. Typically, those present may include:

- the project manager,
- the project design manager,
- the project construction manager,
- the works manager of the factory in which it is to be built, and
- the process safety manager.

The conceptual design of the plant is outlined to the meeting. Then a list is prepared of all the process materials to be present in the plant, either as raw materials, intermediates, finished products, etc. A Process Materials worksheet (see Chapter 2) is prepared on a flipchart, listing the materials across the top, with types of hazard or environmental problems listed down the side, including headings such as fire, explosion, BLEVE, toxic escape, chronic toxicity, unacceptability in liquid effluent, etc. The

PHASE	Phase 1 Conception and Definition	Phase 2 Flowsheet Design	Phase 3 Detailed Design	Phase 4 Construction and Commissioning	Phase 5 Operation
Project Activity	Review of Concept to give direction to design Technical Specification Finalized →	Tender Selection Contract signed →	Detailed Design Design frozen →	Construction/Fabrication/ Commissioning Handover to customer/user →	Use/Maintenance
Reliability and Risk Activity	• Hazard Identification • Development of guidelines for design	• Check compliance with specification • Preliminary risk analysis and assessment • Compare tenders • Guide detailed design	• Contribute to detailed design • Check "final" design • Guide Phases 4 and 5	Ensure construction/ fabrication meets specification, contract requirements and design intent, i.e., ensure "fitness for purpose"	Review any modifications made during construction • Ensure future risks remain low
Reliability and Risk Methods	• Hazard Identification Check sheets	• Risk Management Process • Risk Balance for risk reduction • Tangible/Intangible risk elements identified	• Hazop • FMEA/FMECA* • Other detailed identification methods	• Quality Assurance methods	• Post-startup review of modifications • Ongoing risk monitoring, auditing and review

Figure 9-1.
* See Appendix C.

team then discusses each material in turn to assess whether it has a theoretical potential to present each listed hazard or environmental problem.

Then the team defines the plant sections, such as raw material storage, various processing sections, finished product storage, tanker loading, transport etc., and review each against the list of possible problems, using the two worksheets for plant sections and process operations (see Chapter 2).

Having identified the hazards, potential environmental problems, and main potential causes of lost production, the next step is to discuss and start to understand typical ways in which they could be realized.

The next step is to start to define the engineering and management initiatives that are appropriate to eliminate or minimize the hazards and risks. This is considered more fully in Chapter 8, but would typically involve consideration of the following:

- Elimination or reduction of the inherent hazard, such as by avoidance of use of hazardous materials or by use of lower temperatures, pressures, flows, etc. As Kletz says, "What you don't have, can't leak."
- Improved containment of hazardous materials to prevent leaks, or control of process operations to prevent reaction runaway or other forms of loss of control. This will involve a combination of equipment ("hardware") measures and management and operational ("software") measures.
- Improved protective systems to respond in the event of loss of containment or loss of control. Again, this will involve a combination of "hardware" and "software" measures.
- Damage limitation, to limit the impact of any incident that occurs and that the protective response does not cover. This will comprise either increased separation between the location of the hazardous incident and the exposed people or property, or strengthening of the exposed facilities to resist the impact of the hazardous incident.

By the time these steps are complete, all those present have a clear view of the main types of problem that should be considered. These are then listed—for example:

- Fire
- Vapor explosion
- Auto-ignition
- Acute toxic escape

- Unacceptable atmospheric emissions
- Unplanned downtime
- Unplanned restriction of production rates

The convenor of the meeting then allocates responsibility to individuals present to arrange for the preparation of a briefing document about each hazard. Each briefing document is to address three topics:

1. the nature and sources of the threat to safety, the environment, or to reliability;
2. appropriate acceptance criteria; and
3. design measures to be adopted to enable those criteria to be met.

The meeting then disperses. For a very large plant, this first meeting will probably last up to a full day.

Each person with the responsibility for preparing a briefing document is expected to consult and use whatever expertise can be obtained. In practice, preparation of many of the briefing documents may be delegated to other staff members. When completed, draft copies of the documents are distributed to all other members of the team. Each member of the team reviews each document.

When all documents have been reviewed, the second meeting is called. The documents are reviewed together, page by page, and all comments are discussed and an agreed position adopted. The agreed documents are then passed to the engineering contracts staff for incorporation of the engineering guidelines and requirements in the technical specification to be sent to tenderers (or, in the case of in-house design, to the design team).

Where the study identifies requirements more within the scope of the responsibilities of production management, they are noted and kept for handing on to the production team when appointed.

Examples of the value of this approach include:

1. A petrochemical plant was proposed for a site around 500 meters from housing. In this first study, it was decided that the worst credible vapor cloud explosion should not be capable of doing more damage to the housing than to break windows. So a requirement was specified that the tendered designs should be accompanied by a demonstration that the largest credible leak, determined in a defined way, should not result in a vapor cloud in excess of a defined number of tonnes. The result was

a plant with remarkably low inventories of flammable materials, and thus a clearly lower inherent hazard.

2. A large refrigerated storage of LPG was proposed. There are two approaches to refrigeration of such facilities in common use: one with large inventories of LPG outside the high-integrity bulk tank, one with very small inventories. Because of the need to keep the chance of vapor cloud explosion effectively noncredible, the small-inventory approach was specified. Thus the inherent safety of the plant was greatly improved, and at little if any extra cost.

9.3 SAFETY, RELIABILITY, AND ENVIRONMENTAL REVIEW

During Phase 2 in Figure 9-1, that is, when the tenders have been received (or, in a nontendering situation such as an in-house design, when the flowsheet is complete in draft), the broad design as far as it has developed is reviewed, with the following objectives:

- to check that the design guidelines specified as a result of the Safety, Reliability, and Environmental Specification have all been included to this stage of design;
- to assess as far as practicable whether the design will meet the risk and environmental criteria; and
- to identify design features, or sections, which will require special attention during the following detailed design.

Where several tenders have been received, it is common to undertake a relatively quick SRE review of each of them as part of the tender-selection process. Then, before negotiations about the contract have been finalized and while the preferred tenderer is still receptive to requests for changes and reluctant to increase the tendered price, a more detailed SRE review is undertaken to identify desired improvements.

In any event, the review usually comprises three distinct programs of work, to match the above three objectives: (a) a quick review of all tenders before selection of the preferred tenderer; (b) a more detailed review before finalizing the contract with the preferred tenderer; and (c) an ongoing scrutiny of the design as it develops to ensure that the safety and environmental requirements are met. This is often one of the roles of the client's team located in the design offices of the successful tenderer.

The first of these comprises the following steps:

1. A careful review of the tender is undertaken against each item in the SRE Specification to ensure that all physical requirements have been addressed.
2. A preliminary risk assessment is undertaken as far as is practicable, with the objectives of checking the overall safety, and of identifying the main contributors to the risk as a basis for specifying detailed design (e.g., very high integrity of control and protective systems in some specific sections of the process, special design requirements for the pumps, pipework).

This preliminary risk assessment may be structured as a "rapid ranking" study, as there may not be sufficient design detail for a fuller assessment to be undertaken (see Chapter 3). However, if regulatory authorities require a preliminary quantitative study, it may be appropriate to go further, and to attempt a more thorough analysis and quantification of the short-listed risks. In any event, if a preliminary assessment is needed for review by regulatory authorities, a rapid ranking study, demonstrating a wide-ranging review of the possible risks, and a short-listing of those risks needing careful attention in design and subsequent management, is an excellent foundation for dialogue with those authorities and creation of mutual understanding.

This preliminary assessment, including the rapid ranking study, also provides much information about topics to be explored with the preferred tenderer before the contract is finalized, with a view to getting agreement to design changes to minimize risks at minimal or zero additional cost.

Examples of the use of this approach include the following:

1. As discussed in Chapter 7, in the case of a large petrochemical plant, all three tenders were assessed using a rudimentary quantitative risk assessment computer system to determine the indicative separation distance required from housing. Before starting the assessment, it was expected that they would all be about the same. It was found that two were about the same, but that the third would require around three times the distance. The assessment was examined, and the reason for the difference was found to be in a particular approach to the design of a section of the process, which required a much larger inventory of hazardous material, with large-diameter pipework and pumps. Once this was noticed, the greater inherent hazard was recognized and agreed to be real. This enabled

consideration to be given to the possibility of redesigning that section of the plant (not feasible) and to the desirability of not choosing that tender (the actual outcome—for a variety of reasons including safety).

2. A plant would incorporate a very large gas compressor driven by a steam turbine. It was recognized that it would be important for the emergency shutdown system for the compressor and turbine to be very reliable. This led, at this stage, to specification of a higher reliability for that system than was common. This led to discussions with the manufacturers of the equipment, and provision of unusual features in the design to enable regular testing of the shutdown system including the steam valves, without shutting down the whole plant.

9.4 HAZARD AND OPERABILITY STUDY (HAZOP)

9.4.1 Introduction

During Phase 3 in Figure 9-1, that is, when the detailed design is completed to the "Process and Instrumentation Diagram" stage (equivalent to the final flowsheet, showing schematically every equipment item, every pipeline, valve, instrument, and control system), it is possible and very valuable to conduct a Hazop study.

In such a study, the design is subjected to a systematic and very detailed study, by a team of people with a range of backgrounds and expertise, looking for ways in which upsets could occur in the process with serious safety and operational results for the plant.

9.4.2 Objectives

The objective of Hazard and Operability Studies (Hazop) is *to facilitate smooth, safe and prompt commissioning of new plant, without extensive last-minute modifications, followed by trouble-free continuing operation.*

The track record of Hazop studies is impressive. Wherever the technique has been applied in accordance with the principles set out below, the results have been:

- smooth, trouble-free commissioning and startup;
- a great reduction in expensive, last-minute modifications;
- well-briefed staff; and
- smooth subsequent operation (except where Hazop recognized possible problems that were not subsequently resolved).

9.4.3 Essential Features of a Hazop Study

Hazop studies can take a variety of different forms, which can lead the casual observer to wonder what it is that makes a Hazop study different from some other form of meeting or review.

The essential features of a Hazop study are:

- It is a systematic, detailed study following a preset agenda.
- It must be conducted by a team composed of members with a variety of backgrounds and responsibilities, representing all the groups with a responsibility for the operation (e.g., a Hazop of a new project in design would have representatives from design, construction, and ultimate operation).
- It concentrates on exploring the possibility and consequences of deviations from normal or acceptable conditions.
- It is an audit of a completed design.

In outline, a study takes the form of a discussion, examining each element of a design or operation in turn, considering a checklist of possible deviations for each element. For each postulated deviation, an attempt is made to envisage ways in which the deviation could occur, and for each such way a judgmental estimate is made both of the severity of the possible consequences and of the likelihood of the deviation. If the meeting comes to the view that the combination of the severity and the likelihood together is sufficient, the deviation is noted as a problem to be resolved. If resolution is likely to require little discussion, it may be tackled in the meeting. Deviations apparently requiring significant effort for resolution are listed for attention outside the meeting.

There are many variations of the approach to conducting Hazop studies that incorporate the above principles. One such approach is described in Appendix A.

An approach that is particularly appropriate to machinery and individual items of equipment is FMEA (or FMECA)—see Appendix B.

9.5 CONSTRUCTION QUALITY ASSURANCE AND AUDIT

In addition to the normal extensive quality assurance procedures during construction and commissioning (i.e., during Phase 4 of Figure 9-1), it is very helpful to check the physical plant as it is completed, specifically to ensure that the requirements identified during earlier safety and environmental studies have actually been incorporated in the plant as built.

This inspection can be made part of the training and familiarization program of new employees.

9.6 PRECOMMISSIONING SAFETY INSPECTION

Also during Phase 4 in Figure 9-1, before the plant is commissioned, preferably before any hazardous materials are fed to it in the startup preparations, it is valuable to have the plant inspected by a senior safety manager not otherwise closely associated with the plant.

Although it may not be expected that the senior manager will identify any significant fault in a brief inspection at this stage, the knowledge that the inspection will take place, and that the manager will ask to see the minutes of the Hazop studies, the pressure vessel register, the list of "reservations" or the list of jobs yet to be completed, the approvals from the various statutory bodies, etc., is an encouragement to undertake the work correctly in the period leading up to the time of the inspection.

9.7 POST-STARTUP HAZOP STUDIES

It is common, during the later stages of construction and commissioning of a plant, for minor changes to be made to the plant without submitting the changes to Hazop study as would normally be required. (If the earlier studies have been done carefully, the number of last-minute changes will be only a small fraction of what would otherwise be required.)

When the plant has started up, that is, in the early stages of Phase 5 in Figure 9-1, and the pressure of the startup period has relaxed somewhat, those changes should be submitted to Hazop study to check that they will be allowed to remain as installed.

9.8 SPECIAL CASES

The principles embodied in the six stages of study outlined above can be adapted for special cases.

For example, it was proposed that the capacity of a plant be increased by installing a larger reactor. This reactor would have the potential, in theory at least, to explode and injure operators and damage property in the vicinity. At the conceptual stage of the project, the engineering project manager wished to know whether the instrumentation for control and protection of the plant would need to be of "high integrity" (i.e., especially complex to design, install, and maintain) to ensure an adequate safety with the larger reactor, or whether a high standard of conventional instrumentation would suffice. The difference in cost of the two approaches would amount to about 10% of the total project cost.

Working back from defined safety criteria for operators and the public, a maximum frequency of explosion was determined. (It was, as would be expected, very low, a probability of around 1 in 1000 per year.) A fault tree was constructed, showing the structure of events (including instrument failure) that could lead to an explosion. Staff experienced with the existing small reactor estimated the frequencies and probabilities of the various events, and the required integrity of the control and protective instrumentation systems was determined, for the maximum frequency of explosion not to be exceeded. It was found that a high standard of conventional instrumentation would suffice in all cases, provided that a few additional systems were installed to cover hazards identified in the investigation that had not been covered in the existing small reactor, and that in a few cases additional protection was provided for backup. Note that, as should be expected in such studies, the outcome of the study was not principally a statement that a particular approach would be sufficiently safe, but a recognition and specification of how to make the design safer.

The work involved in this study included all that normally included in the Safety, Reliability, and Environment Specification, and part of that normally in the Safety, Reliability, and Environment Review, which was correspondingly restructured when the design had progressed to the flowsheet stage.

REFERENCES

Bollinger, R. E., Clark, D. G., Dowell, A. M., Ewbank, R. M., Hendershot, D. C., Lutz, W. K., Meszaros, S. I., Park, D. E., and Wixom, E. D., "Inherently Safer Chemical Processes: A Life Cycle Approach." (D. A. Crowl, Ed.) American Institute of Chemical Engineers, New York, 1996.

Center for Chemical Process Safety (CPPS), "Guidelines for Engineering Design for Process Safety." American Institute of Chemical Engineers, New York, 1993.

Hendershot, D. C., "Inherently Safer Process Design Philosophy." Proc. Conf. Process Safety, The Israel Institute of Petroleum and Energy, Herzliya, Israel, May 5–6, 1999a.

Hendershot, D. C., "Inherently Safer Design Strategies." Proc. Conf. Process Safety, The Israel Institute of Petroleum and Energy, Herzliya, Israel, May 5–6, 1999b.

IChemE, *Safety Management Systems*. European Process Safety Center, published by The Institution of Chemical Engineers, Rugby, UK, 1994.

IChemE (The Institution of Chemical Engineers) and IPSG (The International Process Safety Group), "Inherently Safer Process Design." The Institution of Chemical Engineers, Rugby, UK, 1995.

Kletz, T. A., "Cheaper, Safer Plants, or Wealth and Safety at Work." The Institution of Chemical Engineers, Rugby, UK, 1984.

Kletz, T. A., *Plant Design for Safety*. Taylor and Francis, Bristol, PA, 1991.

PMI, "Project Management Body of Knowledge." Project Management Institute, 130 South State Road, Upper Darby, PA 19082, USA, 1996.

Chapter 10

Management of Risk and Reliability of Existing Plants and Operations

When you have studied this chapter, you will understand:

- how the essential elements of process safety and reliability (and risks of any type) can be classified under six headings;
- how to structure the approach to managing the risks of operating a plant;
- the importance of both monitoring and auditing and the difference between them;
- how to upgrade safety software, including the degree of commitment by those involved; and
- the reasons underlying the OSHA requirements in 29 CFR 1910.119 and the EPA requirements in 20 CFR 68.

This chapter addresses the risk management steps 11, "Set Up Quality Operations," 12, "Monitor," and 13, "Audit" of Figure 1-4.

10.1 SOME PRINCIPLES FOR GOOD MANAGEMENT OF PROCESS SAFETY AND RELIABILITY

The fundamental requirements for safe operation of a hazardous plant or when handling hazardous materials can be classified as follows (after Hawksley, 1987):

1. All those involved must understand the hazards to safety and reliability, the potential consequences of incidents involving those hazards, typical scenarios leading to incidents, and the safeguards in place.
2. Appropriate equipment and facilities must be provided which is "fit for the purpose" of reducing the risk from the hazards as far as is reasonably practicable, by containing the hazardous materials and controlling the operations.
3. Systems and procedures must exist to operate that equipment in a satisfactory manner within the design intent and to maintain its integrity. These include systems for:
   - operating and maintaining the plant to quality standards;
   - monitoring performance in relation to risk and reliability, as part of routine control;
   - auditing the risk and the reliability, against all six of these principles; and
   - progressing risk and reliability improvements.
4. An appropriate organization must be established with appropriate staffing, communication, and training to operate and maintain the facilities, equipment, systems, and procedures.
5. Adequate arrangements must exist to detect the onset of, and to handle, foreseeable emergencies.
6. Effective arrangements must exist for promoting safety, and building a strong safety and reliability climate and culture.

A suitable outline as an aide memoire is:

- Understanding
- Hardware
- Software
- People
- Emergencies
- Culture

Other parallel sets of requirements have been defined by other organizations such as CCPS (1989) and API (1990), or are included in various statutory regulations (e.g., see Appendix E).

Several corporate approaches to design of management systems for achieving a high standard are set out and discussed by IChemE (1994). Broadly, what is required is a "quality" approach, entailing:

- definition of policy by consideration of the requirements of the organization, the community, employees, and government;
- defining standards of performance and guidelines for systems and procedures to be used by each operation;
- preparation of site-specific or activity-specific procedures;
- nomination of those responsible for performance;
- monitoring performance;
- auditing the systems and the adequacy and accuracy of the monitoring;
- formally assuring senior or corporate management of compliance with the standards;
- reviewing the policies;
- then restarting the cycle by reviewing and updating the standards, etc.

Selected important tasks are discussed in more detail below.

10.2 ONGOING MONITORING AND AUDITING OF PROCESS SAFETY AND RELIABILITY

10.2.1 Responsibilities of Production Management

In the operation of a process plant, the following features of its performance must be managed:

- the quantity of product made, in relation to the production plans prepared in the light of the market requirements,
- the quality of product made, in relation to the quality standards defined to meet the market requirements,
- the costs of production, in relation to the budget prepared in the light of what the market will pay, and
- the safety and environmental performance, in relation to standards defined by both statutory authorities and the industry in the light of what the community expects.

"Management" can be defined in various ways, of which the following classifications are typical:

- planning,
- motivating,
- organizing,
- controlling.

The "controlling" element (which is common to all definitions of the activities of management) comprises:

- defining a standard,
- measuring the actual performance,
- analyzing the reasons for any difference between the actual and standard performance, and
- taking action to bring the actual performance back to the standard.

In management of production volume, the rate of production is monitored either continuously or frequently, logging perhaps hourly, and calculating the quantity made daily, weekly, monthly, etc. This production is then compared with the production plan, and arrangements are made to change the production rate to correct any significant deviation.

In management of product quality, materials are sampled frequently and put through the QC testing program; the causes of any quality problems are identified; and arrangements are made to correct the quality.

In management of cost, cost information is collected continually and reported regularly, the causes of high (or low) costs compared with budgets are identified, and arrangements are made to improve costs further (consistent with meeting other market requirements).

In management of occupational safety, information is collected regularly about lost time and other accidents, the causes of these are identified, and arrangements are made to prevent recurrence.

It is common for the level of performance in all of these four responsibilities of production management to be reported regularly (e.g., monthly) through the management structure, often together in the one monthly report, and for them to be reviewed at management meetings.

But process safety, although at least as important as any of the above, presents some problems for routine management.

10.2.2 Difficulties in Managing Process Safety and Reliability

The field of occupational safety, mentioned above, is characterized by a relatively high frequency of relatively minor accidents, most of which are related to the physical work being undertaken by the injured person, and the physical hazards of the workplace.

On the other hand, the field of *process safety* is characterized by a relatively low frequency of relatively serious incidents that are related to the inherent hazards of the process.

The frequency of occupational accidents in any large organization is usually sufficient to enable management to detect whether performance is improving or deteriorating. Further, the absence of accidents is usually good evidence of a high level of occupational safety.

On the other hand, very few production facilities are sufficiently large for the frequency of major fires, explosions, toxic releases, and catastrophic breakdowns to be used as the basis for control. The absence of such incidents is not evidence of a high level of risk management. A plant that has operated without major incident for 20 years may be on the brink of having one.

(Consider the following. If the frequency of major incidents on large process plants averages around 1 in 3000 per year—some estimates have it as around this level or lower—and a particularly dangerous plant has a risk that is 33 times the average, then the expected frequency of major incidents of the dangerous plant would be around 1 in 100 per year. Clearly 20 years of operation without a major incident is too short a time for any valid conclusion to be drawn.)

Further, the level of occupational safety cannot be used as an indication of the level of management of the risk of major incidents. For example, the years after World War II saw a dramatic improvement in the occupational safety of the process industries, but at a time when the occupational safety of the industry and individual companies in it was at a record high (late 1960s and early 1970s), a rise began in the number of major incidents (see Figure 10-1, which illustrates the results for one major company; it is typical of many others). Many examples can be cited of major incidents that occurred in plants with excellent occupational safety. The reason for the lack of correlation between occupational safety and process safety lies in the different fields they cover, as outlined above.

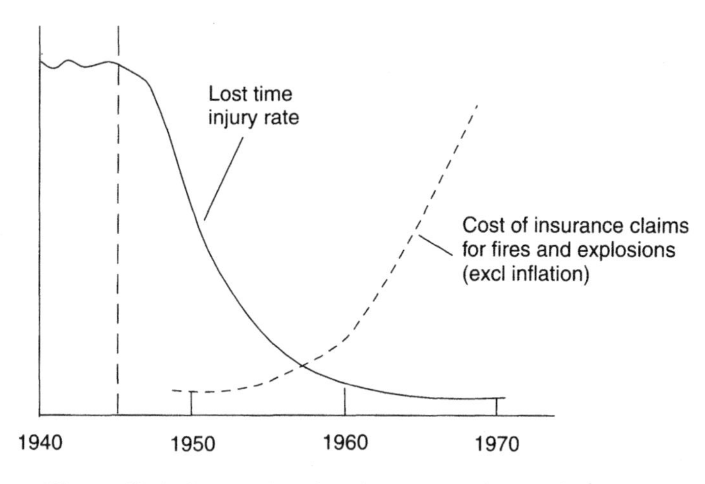

Figure 10-1. Occupational and process safety performance.

(A reliable warning sign that senior management do not understand process safety and major hazards generally, and hence are probably not managing them effectively, is if they cite good occupational safety as proof that their operations are safe. This type of remark is encountered remarkably often.)

As it is (fortunately) not possible to use the frequency of fires, explosions, and toxic releases as a basis for managing the risks of major incidents, another approach is needed.

Any such approach must be structured around the requirements for process safety, such as are displayed on the Safety Balance (Chapter 8) or set out by Hawksley (Section 10.1).

10.2.3 Common Methods Used in Managing Process Safety and Reliability

Many activities are commonly undertaken that are aimed at achieving a high level of process safety. They include:

- designing safety into a plant from the outset by such means as hazard studies, risk assessment, hazard and operability studies, and use of well-recognized codes of practice;
- detailed hazard analysis and risk assessment of existing operating plants;

- studying reports of incidents elsewhere and considering what can be done to reduce the chance of similar incidents occurring on a specific plant;
- inspecting the plant in the course of daily visits;
- implementing various procedures to control maintenance access and activities, to review proposed plant modifications, etc.;
- conducting audits; and
- training employees in the hazards of the plant and its processes.

All of these are valuable, and all have their place in a process safety program. But the questions remain:

"Is the standard of process safety high or low?"

"Is it rising or falling?"

"Where are the weaknesses?"

Without answers to these questions, management is flying blind and does not know whether the process safety program is inadequate, excessive, or about right.

What is missing is routine monitoring of the various components of process safety.

10.2.4 Differences between Monitoring and Auditing

Monitoring	Auditing
Must be undertaken by and for the line management responsible for the unit	Must be undertaken by people with no line responsibility for the unit being audited
Continuous	Periodic and brief
Comprehensive	Selective
Part of management control system, producing actionable results continually	Part of the protective system, producing actionable results only when a fault in the management system is detected

In effect, monitoring is part of the normal managerial control system. Without monitoring, the activity is not being controlled; that is, it is not being managed, as control is an essential part of management. On the other hand, auditing is a protective function. It is only really necessary when there is a weakness in control. The analogy with instrumented control and protection is very strong. It is important not to confuse the two.

10.3 SOME APPROACHES TO ASSURANCE OF EFFECTIVE PROCESS RISK MANAGEMENT

10.3.1 "Safety Management Systems"

"Safety Management Systems" (SMS) is a general term used to describe that wide range of procedures and activities that have an impact on safety. These procedures should also be designed to have a major impact on environmental performance and on reliability.

There are numerous ways in which such SMS can be structured, and most large organizations in the process industry have their particular structure.

However they are structured, it is important that they cover all the roles of management, as it relates to safety, environmental protection, and reliability. One way of approaching such a structure from first principles is to base it on the requirements derived from Hawksley (see Section 10.1), considering and defining whatever is needed to manage each requirement, that is, planning, motivating, organizing, and controlling (including defining the required performance, measuring, analyzing, and acting).

It is essential that such a structured approach to management of process risks of all kinds be adopted. This is the aim of the requirements in OSHA 29 CFR 1910.119 and EPA 40 CFR 68.10 (see Appendix E).

10.3.2 The "Safety Case" Approach to Safety Assurance

In essence, the "safety case" approach to safety assurance has arisen from the change, in recent decades, from safety legislation being prescriptive, that is, defining specific requirements to be met by operating organizations, to such legislation being performance-based.

Where legislation is performance-based, it requires organizations to operate safely and usually requires the organizations to be able to demonstrate, not just avoidance of accidents, but also a sound structure of management systems that can be expected to result in safe operation.

This requirement often calls for organizations to present, in a documented form,

- a list of the hazards;
- an assessment of the risks arising from the hazards;
- a list of hazard reduction initiatives;

- the preventive and protective measures to minimize the risks from the residual hazards; and
- the systems and procedures used to manage the residual risks.

The regulatory authority then uses this document, after approval, as the basis for routine inspections. Instead of using prescriptive regulations as the basis for assessing whether the organization is complying with the legislated requirement for safety, the organization's compliance with its own defined safety systems and procedures, etc., is used.

In principle, this approach recognizes the difficulty of keeping detailed legislation updated to match the safety requirements of changing technology. It enables each organization to define, for approval, what those requirements are, and for the regulatory authority to carry out its responsibilities by comparing the actual practice with what the organization has defined as being necessary.

Although the emphasis in the regulatory field has been on the "safety case," the same principles can be, and often are, applied also to environmental protection (e.g., by the US EPA) and to reliability.

In practice, however, problems have arisen. Quite often the safety case becomes just another document to be prepared by consultants or a specialist team in the organization, then filed, while operations carry on as before as if the safety case had never been prepared.

10.3.3 The "Highly Reliable Organization" Approach to Process Safety Assurance

Because of the identified weaknesses that have crept into application of the safety case approach in many instances, interest has grown in what it is that makes an organization "highly reliable."

In effect, it calls for a high degree of understanding and commitment throughout the organization to the requirements for process safety (Section 10.1) extended to cover reliability and environmental protection, quality, etc.

But, as with any other system, there is no assurance that a future change of key staff, for example, at board level or elsewhere, or economic pressures will not cause the focus to change to short-term survival, with a reduction of emphasis on activities that have a long-term impact, such as maintenance, training, and recruitment and retention of technical specialists.

10.4 DESIGN OF A PROGRAM FOR ROUTINE MONITORING OF PROCESS RISK AND RELIABILITY

10.4.1 Important Principles

In developing an approach to monitoring the risks to people and the reliability of a plant, there are a number of principles that should be applied:

• If the level of process safety of a plant is as important as the output, quality, and cost, then the management of it, including the monitoring and reporting, should be as systematically designed and as thoroughly implemented.

No one waits until the end of the year to find out how much of the product has been made. It is not good management to wait until the product is rejected by the customer before checking on the quality. No one waits until the company annual accounts are prepared and published before looking at manufacturing costs. Therefore, to be consistent, no one should wait until a plant has a major incident before instituting a systematic approach to management of process safety, reliability, and environmental protection.

• Process safety performance, as monitored, should be reported through the organization following the same management reporting structure as the production performance in the fields of output, quality, and cost and should be subject to the same management review process.

• The bulk of process safety monitoring should be done by those responsible for the safety of the process.

• The operators, tradespeople, supervisors, and professional production staff all have responsibilities for the safety of the process, and a personal stake in it. Further, the plant operators and first-line supervisors spend more time on the plant than anyone else and are in an excellent position to monitor process safety on a continuing basis. Therefore they should be equipped to perform that role.

• The nature of the program for monitoring the process safety of each plant needs to be designed to focus on the particular hazards of that plant, and on the relevant safeguards.

• The emphasis will need to change with time to give more attention to those areas that are recognized as in need of improvement, so as to enable management to assess the effectiveness of corrective action and programs.

• The manager of each plant, and the manager of each factory, should have a program for monitoring process safety such that he is not relying on audits by people external to his organization to provide him with the fundamental information he needs to manage process safety. (An audit should be a second check on an activity that is already being managed; it should not be the main means of managing the activity.)

- Both "hardware" and "software" need to be monitored.

10.4.2 Approach to Performance Measurement

Management theorists draw an important distinction between "key performance outputs," and "key performance drivers." The *key performance outputs* are the dimensions of success of an enterprise, and in the case of a private enterprise typically include its profit. But it is not possible for the directors of a company to take action directly on the profit; they need to take action on a range of other variables that influence the profit. These variables are termed the *key performance drivers*. Different terms are used here to reflect process industry terminology: "key safety outputs" and "controllable safety inputs," where the *key safety outputs* are analogous to the production quantity, quality, and cost of a process plant, and the *controllable safety inputs* are analogous to the process flows, temperatures, pressures, levels, compositions, etc.

In management of process safety, the desired safety output is a zero accident rate. However, as noted earlier, a period of even 30 years is not sufficient for a zero accident record to be a statistically significant indication of a low historical accident risk. Such a past history is even less an indication of a low current or future risk of a major process accident affecting safety, environmental performance, or production output.

Formal quantitative risk assessment is limited to assessing the quantifiable components of the risk; it is laborious; and it can only be a "snapshot" of the risk at the time of the study. Further, it is not yet possible to include the effect of important intangible components (such as the safety "culture" or "climate") into such assessments in a way that is theoretically sound or empirically validated.

It is commonplace in management studies for the outputs, such as profit, not to be directly controllable but to be measurable. But in process safety, there is an extra dimension of difficulty: the key safety output is not directly measurable in a manner suitable for routine monitoring. Thus, in a sense, those responsible for process safety management are "flying blind": controlling selected inputs without any direct indication of the effect their actions are having on the safety output. The closest approximation to a measurable process safety output is the pattern (not just the frequency) of "near misses" and unusual occurrences. This pattern is, in some respects, a surrogate measurement of the key safety output. See Figure 10-2.

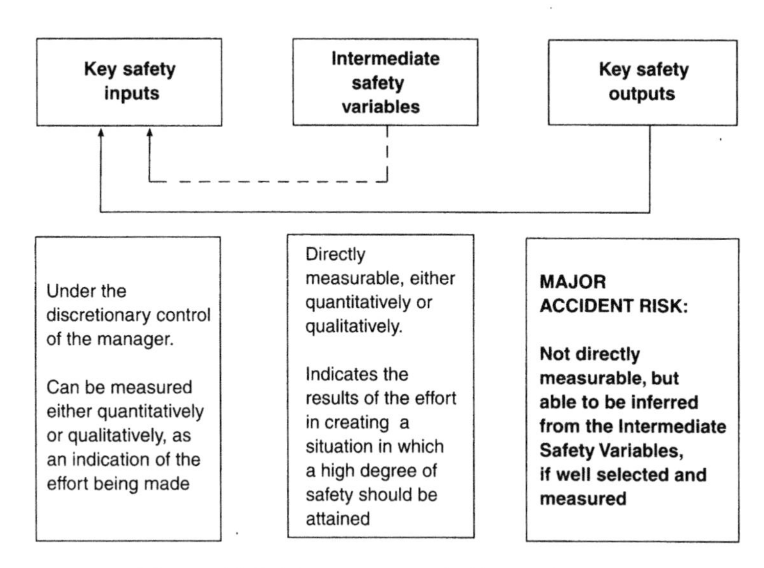

Figure 10-2. Measurement of effort and results.

As it is not possible to measure and monitor the ongoing level of the process safety output with confidence, it is necessary to adopt a different approach to performance measurement.

The linkage between any controllable safety input and one or more key safety outputs passes through several levels. For example, the quantity and quality of safety training on major hazards are directly controllable by management. (These are, of course, only two of the controllable safety inputs that aim to contribute to a good safety output, i.e., a low risk of a major incident.) As discussed above, the risk of major incident is not directly measurable and has to be inferred. But there are indicators at the intermediate levels between the maintenance and the training on one hand and the accident risk on the other.

Examples include:

- the physical condition of the plant and the equipment;
- the level of understanding of major hazards and the relevant safeguards, which may be measured in a number of ways, including formal testing and informal conversation; and
- the attitude toward safe operating practices, which could possibly be measured by surveys but may need to be assessed qualitatively by observation on the job.

These quantities (e.g., plant condition, understanding, attitude) may be termed *intermediate safety variables*. In some respects they are surrogate key safety outputs.

Thus, in view of the inability to measure the principal safety output (low risk of a major accident), it is necessary to place most reliance on measuring a combination of the controllable safety inputs and the intermediate safety variables.

10.4.3 The Concept of "Hazard Warning"

Where there is a very small probability of occurrence of an event that would have a very serious impact on the surroundings, for example, a bulk LPG road tanker crashing, leaking and catching fire, it can be very difficult to assess the risk. Even if this can be done mathematically, the assessed risk will depend on many assumptions about the condition of the tanker and the roadway, the driver's skill, the effectiveness of any emergency response, etc.

Whereas it is possible to monitor the frequency of small accidental spills at a tanker loading bay, for example, and to form an opinion about whether the procedure is improving or worsening, it is very unlikely that there are sufficient major tanker accidents and spills to determine whether performance is improving or worsening.

At the high-frequency–low-consequence end of the risk spectrum, the frequency of incidents is often sufficiently high for management to gauge the success of the risk management efforts by comparing the number of incidents in successive years.

For example, suppose a company has had, in successive years, the following numbers of small spillages at tanker loading bays:

9 13 11 16 18 15 23

If the number of tankers unloaded and loaded has remained constant, it is evident that performance is deteriorating.

At the low-frequency–high-consequence end of the risk spectrum, the frequency is (one would hope!) too low for any such comparison. For example, it is not usually possible to compare the number of explosions this year with those last year. This can lead to an "out of sight, out of mind" problem in the management of risks of major accidents.

Major accidents tend to have different causes from minor accidents. It is unsound to conclude that a falling frequency of minor accidents shows that the risk of major accidents is also falling.

Lees (1982a) postulated the principle of "Hazard Warning." He noted that, while the frequency of major incidents is very low, the precursor events that can, in some circumstances, lead to a major incident are much more measurable.

For example, a chemical plant may have a critically important pressure control system. Because of the importance of preventing overpressure, the plant may be fitted with a high-pressure alarm, an extra high-pressure automatic shutdown system, and relief valves. The combination of the instrumentation and careful attention by the operating, maintenance, and technical staff may result in a very low frequency of the plant suffering damage through uncontrolled overpressure. The frequency of vessel rupture from overpressure may be assessed as one in 10,000 per year (i.e., 10^{-4} p.a.). This frequency is too low to be monitored. There is only a 1% chance of it occurring in any 100-year period.

However, the assessment of the very low frequency of rupture due to overpressure may be based on one failure per year of the control system, with the failure being detected and acted on by one of the protective systems (probably the alarm system). By monitoring the frequency of activation of the alarm, or of any of the other protective systems, it is possible to get a feeling for whether the risk of rupture is increasing or not. If the high-pressure alarm were to sound three times in any one year, there would be cause for vigorous investigation, as it could be assumed (all other factors being unchanged) that the risk of rupture had increased by a factor of three.

Lees (1982b, 1985) and subsequently Pitblado and Lake (1987) have analyzed the mathematics of the Hazard Warning principle. An important practical lesson to be learned from the principle is:

Lesson: *Where the frequency of a major hazardous incident is too low to be monitored as an indication of the effectiveness of the risk management program, a program should be implemented for monitoring the precursors of the incident.*

10.4.4 Selection Principles

It is necessary to adopt a two-pronged approach when planning to measure performance in managing the risks inherent in process plant operations, namely:

1. measures to avoid identified scenarios and
2. measures to develop and maintain good practice.

(These are discussed more fully in Section 10.10, described as the "top-down" and the "bottom-up" approaches).

The selection of the variables to be measured should thus be partly plant-specific (based on the specific hazards and potential hazardous scenarios of the plant) and partly general (based on the requirements for good practice).

Because of the need to be plant-specific, it is not possible to set out a general-purpose checklist of what should be measured for all plants.

To maximize the reliability of the measurement of the performance of the risk management system, it is desirable to use redundant approaches of diverse types. This is one reason for using both the top-down and bottom-up approaches together, and for both proactively measuring and manipulating the controllable safety inputs, and reactively analyzing accidents and unusual occurrences.

10.4.5 Tangible and Intangible Variables

One of the limitations of quantitative risk assessments is that, naturally, they assess only the quantifiable components of the risk. There are many components that cannot yet be quantified in any objective manner. These include the effect of:

- inadequate design or physical condition of the plant;
- badly designed systems and procedures;
- inadequate organization, command structure, communications, and training;
- disinterested management, and poor safety climate and safety culture.

When a quantitative risk assessment is undertaken, the effect of strengths and weaknesses in the intangible variables on the quantified risk needs to be considered descriptively in parallel with the quantitative assessment, and not mathematically integrated with that quantitative assessment in some pseudoscientific manner. Serious intangible weaknesses invalidate a quantitative risk assessment, rather than simply requiring it to be adjusted.

Similarly, when routinely measuring the ongoing safety performance, it is necessary to report on some variables by describing them, or by ascribing subjective scores to them, or a combination of the two, rather than using objective numerical measures. In particular, it is important not to limit measurements to those variables able to be quantified objectively.

10.4.6 Key Safety Outputs

Although it is not normally possible to measure the frequency of major accidents, or their interaccident period, a valuable substitute is to monitor the "near misses" or "unusual occurrences," However, it is important not to aim to reduce the reported frequency of such incidents, as this will act as a disincentive to staff to report them. Instead, it is preferable to regard the frequency of such reports as a positive indicator of the interest in safety, and to use an analysis of them, their potential seriousness, and their root causes as a way of recognizing weaknesses in the intermediate safety variables, and of any need to attend to the controllable safety inputs.

Any judgment formed about the safety performance of a plant that is based on the frequency and potential seriousness of near misses and unusual occurrences covers the historical period studied up to the present. For such an analysis to have statistical validity, it will probably need to cover a substantial time period, and so may not represent current conditions.

On the other hand, a judgment formed from analysis of the root causes of any recent near misses and unusual occurrences can present a very up-to-date picture of the state of health of the intermediate safety variables, and thus provide an indication of the likely level of the risks at present and into the near future.

A major value of analysis of near misses and unusual occurrences lies in the insights it can provide into the quality of the routine monitoring of the intermediate safety variables.

10.4.7 Examples of Intermediate Safety Variables

The intermediate safety variables to be monitored should be selected in a structured way, matching the structure used in analysis of accidents, near misses, and unusual occurrences. (It is very desirable for the task of selecting the intermediate variables to be monitored routinely, and to be used when investigating those incidents, to be undertaken as a combined study so that the lessons learned in the investigations can be directly related to the monitoring program.)

As set out earlier, there are many different forms that the structure could take. Examples include those developed by Hawksley, CCPS, API, or the Safety Balance (Chapter 8).

Examples of intermediate safety variables that can be monitored in a quantitative manner include:

- number of leaks of process materials detected at month-end (indicating the level of containment in relation to minor leaks);
- number of defective work permits detected during the month—compared with the number inspected (indicating the level of understanding and commitment of operators, tradespeople, and their supervisors);
- number of control instruments in a failed state at month-end (indicating limited ability to control);
- number of actuations of critical alarms during the month (indicating difficulty in control);
- number of alarms and automatic shutdown systems that failed test during the month—compared with the number tested (indicating the reliability of the automatic protection system); and
- number of safety-related jobs in the job queue at month-end (indicating the ability of the organization to maintain the level of safety required).

There are numerous other intermediate safety variables that should be monitored in a nonnumerical manner.

10.4.8 Examples of Controllable Safety Inputs

Selection of variables to be measured should use a process structured around one of the established conceptual frameworks used to define the requirements for process safety. Suitable frameworks include those designed by Hawksley (referred to above and used below), CCPS, API, and others. Insights from the Safety Balance should also be incorporated.

Once a selection has been made of the intermediate safety variables that are to be monitored, the related controllable safety inputs should be listed. Some examples are listed below, grouped according to broad categories of intermediate safety variable based on the structure developed by Hawksley.

Understanding of Hazards:

- Formal training sessions for staff involved
- Informal discussions with staff involved
- Fuller involvement of staff in incident investigation

Facilities and Equipment:

- Systematic inspections of plant condition by managers, supervisors, operators, and tradespeople
- Preventive maintenance expenditure
- Reviews of the need to redesign the preventive maintenance program

Systems and Procedures:

- Revision of procedures (with or without a critical analysis of the opportunity for fault)
- Formal training of those who are to use or supervise the procedures
- Informal discussions with those involved
- Fuller involvement of staff in reviews of faults in execution of procedures

Organization, Staffing, Communications, Training:

- Review of the program of management meetings, and the selection of those who attend
- Review of the style of formal management meetings and informal management practices, and the way in which actions are decided and followed up

Emergency Capability:

- See the above headings, but relating specifically to emergency capability, such as understanding of emergencies, emergency equipment

Active Promotion of Safety Climate and Culture:

- Visits to the plant by senior management
- Participation by all involved in design of safety practices and in safety monitoring, investigations, improvements
- Full communication by management of reasons for decisions that may be interpreted as placing profit before safety.

These are only examples. A more comprehensive list should be developed by those responsible for the safety of each plant so as to:

- include plant-specific information,
- help develop understanding, and
- help build commitment.

10.4.9 Displaying the Assessed Safety Performance

The example displayed in Figure 10-3 (using hypothetical data for the assessed level of performance for the current period, e.g., year) is derived from the "Process Safety Footprint" used by British Petroleum.

It is tempting for managers to simplify the footprint to a single number; the average or the total of the individual scores. But this loses critical detail and suggests that a weakness in one area (e.g., a poor work permit system) can be compensated for by a strength in another (e.g., good design of instrumentation).

10.4.10 Putting the Principles into Practice

10.4.10.1 Introduction

A systematic approach to involving plant employees at all levels in identifying and monitoring hazards is described by Tweeddale (1985), and a strategy for process safety assurance by Hawksley (1987).

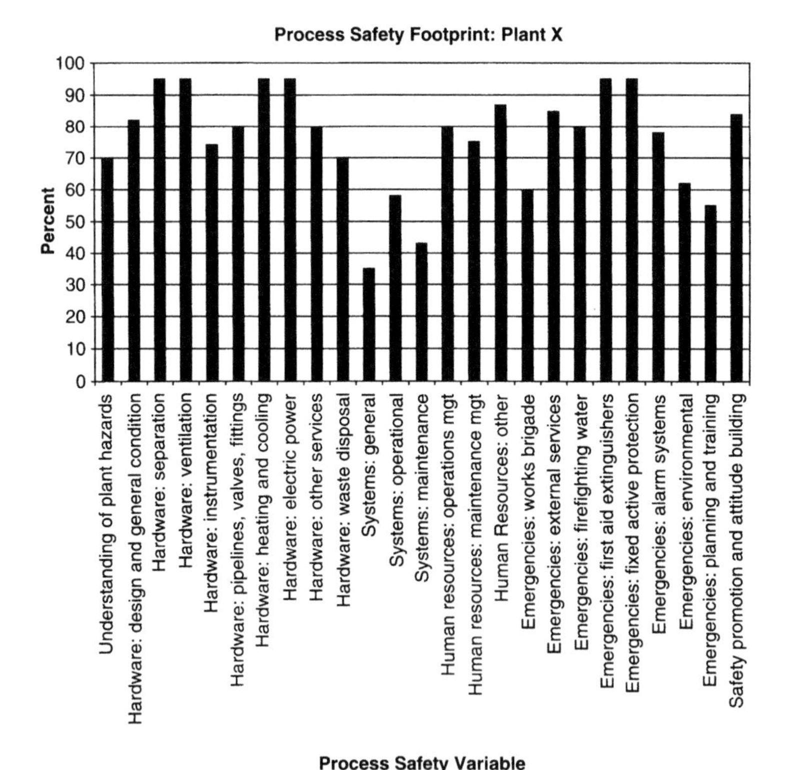

Figure 10-3. Example of Process Safety Footprint.

The two approaches outlined here are exploratory and are recognized by those involved as still a long way short of the degree of development reached by management systems for production volume, quality, and cost. They are described here primarily to illustrate the principles, and to provide a basis for others to build on.

10.4.10.2 An Approach in a Chemical Factory

Simpson and Smith (1988) set out a systematic approach to developing a process safety management system. They embarked on a three-step program for setting up process safety monitoring.

1. They set up easily measured indices of performance for each plant, for monitoring and reporting monthly by plant employees at all levels, covering such topics as:
   - number of permits-to-work checked and found to be defective (compared with the previous month),
   - number of leaks of process materials found on inspection at the end of the month (compared with the previous month), and
   - percentage of operating procedures prepared and up-to-date (compared with the previous month).

   The aim was to get started at once on a preliminary monitoring system, then to improve it by more thorough and systematic study, as in the next step.
2. By use of worksheets, they identified the critical features of each plant in relation to containment of materials and control of processes, and defined an appropriate monitoring program. The program could include such items as:
   - frequency of actuation of critical alarms compared with the previous month (this can be a simple number to collect on a computer-controlled plant),
   - number of alarms and trips due to be tested each month, and the number that failed the test,
   - number of control instruments recalibrated and the number that were out of calibration by a defined significant amount,
   - the number of points to be inspected for corrosion or thickness, and the number found to be in need of attention.
3. By undertaking Hazop studies of each plant in turn, they further refined the understanding of critical features and procedures, and

updated the monitoring program accordingly. Concurrently, a program of internal auditing was set up, involving only people from within the factory from other plants.

These steps were aimed at progressively improving the routine monitoring system for process safety. As they were implemented, there were to be changes in the nature of the audits involving nonfactory staff. The objectives of such audits would become:

- to review the quality of the monitoring and reporting systems and internal audits,
- to review whether revisions were occurring to account for accident experience on the plant or elsewhere,
- to probe one or two selected areas in detail to check the internal audit system, and
- to review progress on major hardware improvements.

10.4.11 Managerial "Breakthrough"

Juran (1964) discusses how an organization that is "breaking through" to a higher level of performance is not characterized by calm, order, and everything being under control. That is typical of an organization that is controlling performance at an existing level, or that is improving very gradually.

An organization that is making great strides in improving its performance is characterized by debate, arguments about personal roles, tension, and a degree of apparent disorder. (The converse does not apply: an organization with these characteristics may not be improving, it may simply be badly managed!) In an organization that is improving rapidly, unexpected things are happening, and people are responding to needs according to their capabilities, rather than just carrying out their responsibilities as set out at some previous time.

There is an important lesson here. If an organization is attempting to make a rapid improvement in its reliability and process safety performance, then it may be unwise to place too much emphasis on the documentation. Some documentation is essential, but it will be more important to get action to improve performance, than to document the situation in great detail. It is necessary to strike a balance. Because all the documentation will not be complete in an ideal degree of detail, there will be misunderstandings and debate, but this is part of what Juran calls "breakthrough."

10.5 AUDITING

10.5.1 Application of Accounting Audit Principles to Process Safety and Loss Prevention

Because the accounting profession has been conducting audits for much longer than engineers and production managers have, lessons can be learned from accountants' experience.

The most common understanding of an accounting audit is called the *attest audit*, which is undertaken with the aim of providing an opinion that enhances the credibility of a written assertion by those responsible for the subject of the assertion. (For example, an accounting auditor may be engaged to express an opinion about the validity of a set of annual accounts prepared by an organization—a common form of audit, required by the taxation authorities and the regulatory authorities concerned with the probity of corporate governance.) In undertaking an attest audit, the auditor does not recalculate all the incomes and expenditures, but checks a sample of the calculations undertaken by the organization, and checks the adequacy of the accounting systems.

The same approach can be followed in process safety and reliability audits. Instead of the auditor aiming to check every detail of the plant being audited (which is impracticable), he or she can require the plant manager to provide a simple written assessment of the level of process safety, using some systematic structure such as Hawksley's six principles. (See Figure 10-4, later in this chapter.)

A number of examples are listed below of lessons learned while auditing, and making use of the principles of auditing in the accounting profession. They are drawn from experience in a number of countries and cultures. Some of the more remarkable examples are from the more industrialized countries and high-profile companies. (See Tweeddale, 1995.)

- Accounting audits are based on sampling, rather than on studying all accounts, but the sampling is designed to test all significant contributors to the overall result. Although process safety audits cannot be expected to check all details of all equipment and all procedures, etc., they must check a sufficient sample of each for the conclusions about their overall standard to be confidently expressed and accepted. Both high-risk and high-consequence scenarios need to be covered.

Lesson: When undertaking an audit (as distinct from an audit-related service) of process safety, all the requirements for process safety should be examined, such as the six principles set out by Hawksley or the requirements specified by CCPS or API. This includes management awareness of hazards and safeguards, and the extent of promotion of risk management.

• Public corporations are required by law to engage external auditors, and the auditors are entitled to charge "reasonable" fees. During an audit, if they identify an area that needs more detailed examination than was initially expected, they can do that examination confident that their extra fees will be paid.

Lesson: External safety auditors need to be sure, if they need to examine some area more fully than was initially expected, that they will not be inhibited by a limitation on the fees able to be charged. This is necessary to protect both the client and the auditor. This may require an appropriate clause in an auditing contract.

• A process safety auditor visited a warehouse in which were stored a large variety of materials, including combustible liquids in drums, pesticide and herbicide active ingredients, combustible materials, and general chemicals. He inspected the first fire extinguisher he encountered. The inspection tag showed that it had been checked during the previous month. Yet the pressure gauge showed zero pressure. He then inspected the firefighting water reservoir, which comprised a brick-walled tank, around 15 meters in diameter, and around 1.5 meters high. It was a quarter full and was being filled when inspected. He was told that it had just been emptied so as to clean it of accumulated rubbish. Judging by the rubbish still visible, it had been empty for some time, and an attempt was being made to fill it before the audit, but the managers had underestimated the filling time.

• In a similar case, a process safety audit was conducted, on behalf of a regulatory authority, of a manufacturer of specialist lubricants. At the time of the audit, the auditor was impressed with the cleanliness of the surroundings, with little evidence of oil spills. The audit report was favorable. A few months later, the manufacturer sought a fire safety report as part of the approval process for a proposed extension to the factory. The same consultancy that did the audit was engaged to undertake the fire safety study. The consultant thus had the opportunity to visit during normal operations. The bunds around the storage tanks and the operating area around the blending and filling operations were all very oily, indicating to the consultant that the normal operations resulted in substantial spillages, and that such spillages were not cleaned up at once. It seemed very likely that the plant had been cleaned just before, and especially for, the statutory audit.

Lesson: If an audit is expected, it is quite possible that long-standing weaknesses will be corrected just beforehand, in an attempt to present a better picture than is normally the case. Accounting auditors frequently conduct spot-checks without giving notice. Similarly, some parts of a safety audit should be undertaken without giving advance notice, if the audit findings are to be representative of the normal situation.

• A safety auditor inspected the firewater pump at a factory. It was a gasoline engine mounted on a stand, coupled to a centrifugal pump. He checked the oil level and the battery acid level, both of which were satisfactory. He had to climb onto the stand to inspect the radiator, but because of the limited roof clearance he could not see into the radiator. He folded a strip of paper and inserted it, to find that the radiator appeared to have no water in it. The manager could not explain this but, after some inquiries, it was learned that there was a leak in the radiator, and that a large container of water nearby would be used to fill the radiator if the pump were needed. It was clear both that the radiator had been faulty for a long time, and that the management did not know of the problem.

• In another instance, a safety auditor was told that all emergency isolation valves were tested regularly, as were all alarms and shutdown systems. He was shown the test records, which supported this statement. While personally inspecting the factory he noticed some LPG vessels on the far side of a factory railway line, over a footbridge. He asked to visit them. The walkway from the bridge led to a platform at the top of the tanks. There was a ladder descending to the valves at the base of the tanks. The auditor climbed down the ladder and noted that the grease and dirt on the shafts of the emergency shutoff valves showed that they had not been operated or tested for a long time.

Lesson: The components that are most difficult to reach, and the operations that are most difficult to perform, are the ones that reveal how committed the responsible people are to high standards of safety. They are often also the most difficult components or operations for the auditor to check. Therefore an auditor should pay particular attention to those components and operations that are least easily checked.

• External accounting auditors give close attention to ensuring that they report at an appropriate level in the organization, because of the potential importance to the organization's viability of the findings and recommendations arising from the audit.

Lesson: Similarly, because process safety, or the lack of it, has the potential for serious impact on the commercial viability of the organization (e.g., effect on production, share price, credit rating, solvency, etc., after a major incident), process safety auditors need to pay similar attention to where they report in the organization, and what they include in that report.

• A consultant was engaged to undertake a risk analysis of an oil and gas facility. The scope was defined as covering the equipment (i.e., the "hardware") and not the "software." In the course of gathering background information about the maintenance of the equipment, and the effectiveness of the emergency response, it quickly became apparent that there were major deficiencies in many types of procedure. For example, there were no written operating procedures for a large part of the plant, and the emergency team, although well equipped, was not respected by the operating staff and was not involved in the emergency planning. The company, which had been expecting a very favorable report, requested the consultant to remove all reference to "software" from his report, on the grounds that it was outside the scope of the study. The consultant replied to a meeting of the senior managers that he had a legal responsibility to draw any observed deficiencies in safety to the attention of management even if they were outside the scope, since to fail to do so would be held as negligent, or worse, if an accident were to happen. Although this particular assignment could have been seen as a safety equivalent to an accounting "agreed-upon procedure" in which the auditor has a limited accountability for the accuracy of the findings as the scope is defined by the client, nevertheless the legal implications of not reporting or of suppressing a concern about the safety of the software are very serious.

Lesson: Even though an engagement may not specifically include auditing, many safety and loss prevention studies have an audit component as part of data collection. The responsibilities of the person doing the study are similar to those of a safety auditor.

• An accounting internal auditor was visiting an accounting office. He wished to check the administration of the system used for payment of invoices received from suppliers. As well as examining the system and how it was operated by the various staff, he decided to undertake a realistic test. Having notified the office manager of what he was planning to do, he inserted into the system a fake invoice prepared on a supplier's blank invoicing form, and kept a careful watch at the final stage: the authorization for payment by the responsible accountant. Normally an invoice should not be paid unless:

the corresponding order is on the file;

the ordered goods or services are certified, by the person who ordered them, as having been received;

the invoiced amount is the same as on the order; and

a nominated accountant checks that all is in order.

• The fake invoice passed through all the checks, and the auditor prevented payment by intervening after the accountant authorized payment but before computer processing.

Lesson: A system may appear satisfactory, but often the only means of really knowing how well it is understood and operated is to test it. In safety and loss prevention, this could be done by such means as inserting a poorly completed work permit into the checking procedure to see whether those undertaking the checks actually find the weakness; or inserting an ill-considered modification proposal into the "change management" system to see how far it progressed before being challenged.

• Internal accounting auditors are often used by management as troubleshooters and additional accounting staff, rather than purely as independent reviewers of the work of others. They cannot then comment in a detached manner on the work in which they have themselves been involved. Further, the position of internal accounting auditor is often used as a career move for a promising staff member to acquire experience. This can also limit the expertise available to the internal audit department.

Lesson: An internal safety auditor should watch very carefully for conflict of interest, and for being in a position where he or she is being asked to comment on a situation whose design he or she has been involved in. One cannot audit one's own work.

• A safety auditor was inspecting a marine terminal for a number of large international oil companies, noted for their high safety standards. The terminal was characterized by eroded earthen bunds, oil contamination of the ground inside and outside the bunds, rusted and decrepit road tanker loading facilities, and a generally derelict appearance.

• In a related example, an auditor was inspecting a warehouse on a factory site operated by a major high-profile international chemical company. It was noted that the contents of the warehouse included paper sacks of potassium nitrate, activated carbon, and powdered sulfur. They were not in adjacent stacks, but nobody had considered the possibility of spillages becoming mixed on the floor.

Lesson: The size of an organization, and its reputation, are not reliable indications of the standard of the safety of all its facilities, particularly the low-technology facilities and activities such as storage and warehousing, which often are neglected.

• In discussion with the safety manager of a factory before starting an audit, an auditor formed the view that the manager and his team were very thorough in their own inspections of plant safety. During an audit, a plant manager said that he personally checked all maintenance work permits before filing them, and also conducted frequent checks on the plant while maintenance work was being done. The safety manager added that he also carried out frequent checks, and the work permit system was working well. The auditor concluded that the system was being well operated and managed. Some months later there was a dangerous occurrence, in which a tradesperson was nearly killed. A detailed check found that the work permit system was, in fact, being operated and managed very poorly in the factory.

Lesson: It is important not to rely on reports from others, no matter how reliable they may appear. There is no alternative to personal inspection.

• Before being taken around a plant, a safety auditor was taken into the control room. The shift supervisor asked him to sign a new page in the visitors' book, and explained that the reason was to keep a record of who was on the plant in case of an emergency. The auditor noted this good practice, and went on with the inspection. Another member of the audit team inspected the book and found that the previous entry was over a year earlier.

Lesson: Even with personal inspection, an auditor can gain an entirely incorrect impression.

Specific lessons for auditing of risk or reliability of process plants can be learned from court cases involving accounting audits. For example, a benchmark ruling in an Australian court case (AWA Limited v Daniels t/as Deloitte Haskins & Sells & Orrs, 1992), discussed by Bottomley and Tomasic (1992), has highlighted a number of lessons. In this particular case, a manufacturing company that bought many of its components from overseas, and sold many of its products overseas, became active in foreign exchange trading, initially as a means of protecting itself against fluctuations in exchange rates. The young staff member leading that activity initially appeared quite successful and was given a free hand. The accounting auditors reported on the very high-risk exposure the company was facing because of the large sums of money involved, and the speculative nature of the foreign exchange field. However, no action was taken for a number of reasons. When a large loss was made on foreign exchange dealings, the company took legal action against the audit company, claiming that the auditors had not pressed their concerns adequately or in the correct quarters. The judge found that, although the manufacturing company shared the blame, the audit company should have taken more action to ensure that their findings were acted on.

Among the numerous points made in the finding are the following:

• The company claimed that the auditors were entirely responsible for detecting and reporting to the company any deficiencies or inadequacies in the company's system of internal control and record keeping. The auditors' reply to this claim was that "the primary responsibility for putting in place a satisfactory system of internal controls and accounts rested with management." While the judge found it remarkable that the company admitted having inadequate management systems, he held that both the auditor and the company shared the responsibility, with the auditor being very substantially responsible and therefore correspondingly liable.

Lesson: It should not be assumed by an auditor that line management is solely responsible for the consequences of any weakness in understanding, equipment, management systems, etc. The auditor may be held liable for having failed to detect such a weakness.

• The close working relationships between the auditors and senior managers of the manufacturing company, which included social contact, may have inhibited the auditors in pressing their views elsewhere in the company. The "relations between directors, managers and outside professionals (are all too often) structured according to the dictates of commercial convenience and personal loyalty, rather than those of legal duty and obligation" (Tomasic and Bottomley, 1992).

Lesson: The auditing role should be taken very seriously. A detached viewpoint should be adopted, regardless of the closeness of commercial or personal relationships.

• There was doubt in the auditors' minds about the authority for the particular staff member to engage in foreign exchange dealing with the nature and scale involved. The judge found that "in case of doubt, an auditor is required to inquire."

Lesson: If in any doubt, inquire.

• The judge found that an important role for management was to report to the board on any "gross breach" of guidelines. In safety and loss prevention, this may imply that where a board has a safety policy expressed in sufficient detail to be a useful guide to staff (and where it does not, the auditors should comment on that), the safety auditors should ensure that major weaknesses in safety management and performance in relation to that policy are reported to the board, and not, as so often happens, filtered out by middle management.

Lesson: The process safety and reliability auditor should ask: "What controls are there? How are they communicated to staff? What is reported back to the board? How accurate and representative is that reporting?"

• The judge noted that, at one board meeting, the external "auditor remained silent and as a result misled the board regarding (the situation)."

Lesson: Auditors must make their views plainly known at the appropriate level.

• The judge also made comments on the failure of the auditor to react to the knowledge that earlier warnings to senior managers had not been acted on. He found that an auditor's role goes beyond identifying and reporting on shortcomings; it also includes following up to ensure that management responds adequately, in terms of both the nature of the action and its promptness.

Lesson: A safety auditor should check carefully that the findings of earlier audits have been appropriately responded to. Where this has not happened, the auditor should not merely include a comment on this in the next report, but should actively follow up with more senior management.

• It was noted in the AWA case and elsewhere that managers sometimes insufficiently scrutinize the work of technical specialists. The reasons for this can include:

a reluctance to reveal their ignorance of that specialist field;

a faith in the expertise of their specialists; and

a reluctance to appear to lack faith in the specialists, some of whom may be sensitive to criticism and scrutiny.

Lesson: An auditor should check the nature of managerial review and penetration of the work done by safety specialists in the organization.

Perhaps the most important of the above lessons, depending on the legal situation in particular countries, is that a safety and loss prevention auditor carries a large responsibility, and either the legal or moral responsibility, or both, to ensure that the findings of the audit are communicated to an appropriate part of the management structure of the audited organization, and that the significance of the findings is fully understood. The best way of judging the extent of understanding is possibly to check the extent of implementation of previous audit findings, and the priority given to that implementation.

Summarizing the above points:

- To improve "ownership" of process safety by line management, require line managers to document brief written reviews of their level of process safety and reliability, with their strengths and weaknesses, and compare these with the audit findings as an indicator of the extent of the understanding of the requirements by line managers (i.e., use this as a check on Hawksley's first requirement).

- In an external audit, check *all significant* factors, and do not just rely on designing a good sampling pattern.
- Visit without appointment on some occasions.
- Pay particular attention to those safety activities that are difficult or laborious to perform or difficult to check or audit.
- Beware of limitations to the scope of an audit that may be imposed for financial or other artificial reasons.
- Test systems and procedures with practical examples.
- Do not be impressed by the size or reputation of the organization being audited.
- Do not overlook the simple or "low-technology" operations.
- Check personally, regardless of how reliable the source of information.
- Be skeptical.
- Where there is doubt, inquire further.
- An organization may rely heavily on the audit report as the principal source of information about weaknesses, and regard any failure to identify a significant weakness as a failure of duty by the auditor.
- Recognize the duty to follow-up to check on the implementation of their recommendations.
- Pay particular attention to selection of the organizational position to which the audit findings are reported. It may even be necessary to bring to the attention of the board of the organization significant matters that are contrary to the safety policy of the organization.

Perhaps the most important of the above lessons, depending on the legal situation in particular countries, is that a process safety and loss prevention auditor carries a large legal and moral responsibility to ensure that the findings of the audit are communicated to an appropriate part of the management structure of the audited organization, and that the significance of the findings is fully understood. The best way of judging the extent of understanding is possibly to check the extent of implementation of previous audit findings, and the priority given to that implementation.

10.5.2 Design of an Audit Program for Process Safety and Loss Prevention

Noting that an audit program is intended to be a backup for the ongoing routine monitoring program operated by line management, and not a replacement for it, it is important that the program encourage line management to undertake their own monitoring programs, rather than encourage them to rely on the audit program.

One approach to encouragement of ongoing monitoring by line management is to ask them to complete a pre-audit self-assessment sheet, either by mailing it to them before arriving to do the audit, or by leading them through it at the start of the audit. Such a pre-audit self-assessment questionnaire is shown as Figure 10-4.

To the Plant Manager:

In preparation for the forthcoming audit of the safety of your plant and process, we would be grateful if you could complete the following questionnaire, and give us a copy at the start of the audit. We will then use it as the starting point for our discussion with you and the plant inspection.

	Poor Fair Good
1. Understanding	
How do you rate the understanding, by yourself and those working on the plant, of the nature, scale and possible causes of process plant mishaps?	1.....2....3....4....5
2. Facilities and Equipment	
How adequate is the design and condition of the plant and equipment, from the process safety viewpoint, in relation to:	
a) Siting, layout etc.?	1.....2....3....4....5
b) Vessels, pipework, machinery etc?	1.....2....3....4....5
c) Control instrumentation?	1.....2....3....4....5
d) Alarms, trip systems, other protective equipment?	1.....2....3....4....5
3. Systems and Procedures	
How adequate are the systems and procedures for:	
a) operation (including modification control)?	1.....2....3....4....5
b) maintenance (including work permits)?	1.....2....3....4....5
c) monitoring performance of equipment and procedures?	1.....2....3....4....5
d) supervision and management?	1.....2....3....4....5
e) third-party auditing of performance?	1.....2....3....4....5

Figure 10-4. Example of pre-audit questionnaire.

(figure continued on next page)

4. Organization, Staffing, Communication and Training

How adequate are the following:

a)	the organization and the way people work together?	1.....2....3....4....5
b)	the level and competence of the staff for their roles?	1.....2....3....4....5
c)	formal and informal communication channels?	1.....2....3....4....5
d)	training and retraining programs and courses?	1.....2....3....4....5

5. Emergency Preparedness

How adequate are the arrangements and preparedness or emergencies:

a)	understanding of what is possible?	1.....2....3....4....5
b)	emergency facilities and equipment?	1.....2....3....4....5
c)	emergency systems and procedures?	1.....2....3....4....5
d)	organization, staffing, communication systems and training?	1.....2....3....4....5

6. Promotion of Process Safety and Reliability

How well is process safety and reliability promoted by senior management:

a)	by visibly spending time and attention on process safety?	1.....2....3....4....5
b)	by authorizing expenditure to improve it?	1.....2....3....4....5
c)	by giving it a high priority in routine management activities?	1.....2....3....4....5
d)	How do those working in the area rate the management commitment?	1.....2....3....4....5
e)	How do you rate the commitment of those working in the area?	1.....2....3....4....5

Thanks for your help.

Figure 10-4. Continued.

Initial Audit

An initial audit of a plant may be designed to be quite comprehensive, to enable the auditors to get a good overall feel of the state of safety management. But time constraints will probably prevent all topics from being explored in depth.

A typical checklist that can be used as the basis for a comprehensive audit protocol is shown as Appendix D. It is structured around the six requirements for process safety assurance derived from Hawksley. Other structures can be used, designed to probe the specific hazards and safeguards of particular plants or operations.

Subsequent Audits

In subsequent audits, it is common to undertake a shallow examination of all six requirements, and then to undertake a much deeper study of a selection of specific topics in one or more of the six requirements.

Because the aim of the audit is to assist the line managers in improving their ongoing monitoring, it is important that the audit findings be discussed with them, with emphasis placed on helping them identify what they should monitor themselves, and what they should aim for.

The scope of the two types of audit is shown conceptually in Figure 10-5.

10.6 CRITICALLY IMPORTANT PROCEDURES

There are numerous procedures that have a major effect on plant risks and reliability. Six particularly critical procedures, that are also difficult to keep in top condition, are:

- "Standard Operating Procedures" (SOP);
- training of new operators, and retraining and refresher training of experienced operators;

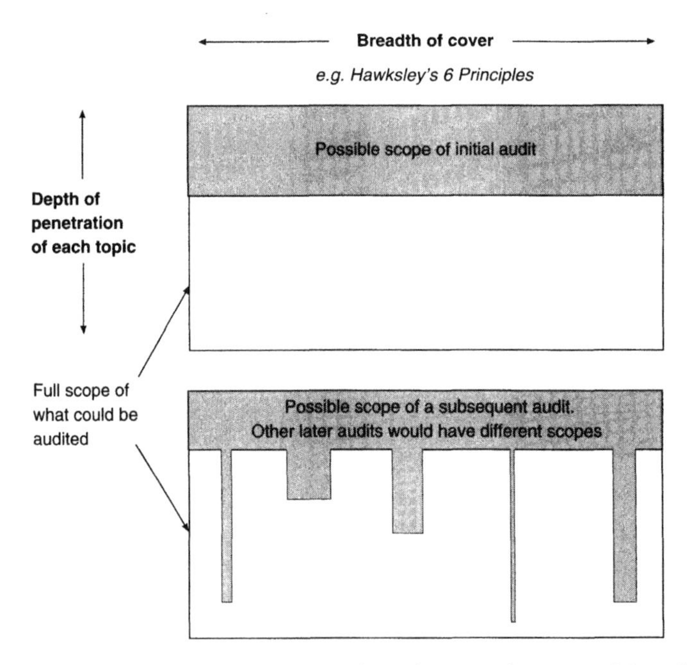

Figure 10-5. Conceptual representation of scope of process risk audits.

- "Permit to Work" procedures;
- control of modifications or "Change Control" procedures;
- condition monitoring; and
- testing of protective systems.

These are discussed in more detail below.

(The critically important task of investigating accidents, near misses, and other unusual occurrences is discussed in detail in Section 10.9.)

10.6.1 Standard Operating Procedures

Objective of Standard Operating Procedures

To define those methods of operation that will enable the production requirements to be met safely and without damage to the environment.

Important Principles and Features

- They must be written.
- They must be comprehensive.
- They must include means of recognizing hazardous situations, and the appropriate corrective action.
- There must be a routine procedure for updating both operating instructions and reference documents, drawings, etc.:

  as hardware modifications are made, and

  as decisions are made to change process materials, conditions, procedures, etc.
- They should be accessible to and usable by the plant operators and supervisors, that is:

  they should be located in the control room and other accessible points, and

  both full detail and quick summaries (e.g., batch sheets) should be available.
- There should be a mechanism for initiating and ending short-term modifications (at very least a note incorporated in the shift change logbook). This is especially important in research or developmental trials.
- The operating instructions should be accompanied by training and a supervisory environment such that operators do not follow the instructions blindly; they should recognize the need for judgment.

Comments

- One approach to updating is for specific sections to be updated as changes are made, with a periodic (e.g., annual) review to ensure that nothing has been overlooked.
- Pages in the operating instruction manual should be marked with the date.
- In some cases, there are brief operating instructions, backed up by detailed training modules.
- One method adopted in batch operation with frequent changes of process is for a computer listing of the batch sheet with instructions to include the hazard details of all materials, precautions to be taken with them, etc.

Common Faults

- Instructions for the more hazardous situations of startup, shutdown, "hold" positions, which are less frequently practiced, or cleaning or maintenance by operators are too often skimpy or nonexistent.
- Operating instructions are often too bulky, that is, detailed instructions not accompanied by working summaries or checklists.
- Operating instructions are often years out of date.

10.6.2 Training and Retraining

In view of what operators and supervisors are expected to know and remember, and the infrequency of some circumstances that call for special care (e.g., shutdowns and startups of complex plants that may be less frequent than annually), it is not surprising that many people regard training as one of the most difficult fields to be confident about.

This is a very big field, including training for operation, maintenance, supervision, management, etc., plus design. However, some pointers are summarized below.

Objective of Operator Training

To enable operators to perform the required operations to achieve production requirements safely and without damage to the environment. (This includes startup, shutdown, and abnormal situations, as well as normal operation.)

Important Principles and Features

- It should be based on written, updated, etc., operating instructions.
- It should be conducted by a person trained and competent in both:

plant operations, and

techniques of training.

- It should comprise a balanced combination of:

understanding of principles (i.e., why?) and

knowledge of procedures (i.e., what?).

- It should include the understanding of potential mishaps, how to recognize early signs, what situation is indicated, what the situation requires, and how to take the required preventive, corrective, and emergency actions.
- It should involve both "classroom" and "hands-on" supervised training.
- It should follow a defined (written) syllabus.
- It should include normal operating conditions, expected special circumstances (shutdowns, startups), and foreseeable abnormal and emergency situations.
- Competence should be formally tested both in the classroom (knowledge and understanding) and on the job (skills) using a defined list of questions or other performance criteria.
- It should include:

chemicals and their hazards,

processes and their hazards,

personal precautions in handling the chemicals,

principles of preparing the plant for maintenance (including Permit to Work system),

appreciation of electric supply, instruments, mechanical equipment, and hazardous failure scenarios and their identification and control.

- Refresher training is required, including training to cover changed circumstances, to maintain peak competence and the ability to cope with unusual circumstances.

Comment

Where the emphasis is on teamwork with operators helping each other, each operator should know his or her place in the team, and understand his or her specific responsibilities. Otherwise some critical checks and tasks can be overlooked in the belief that someone else is doing them. (Checklists can be helpful—for example, the preflight checklists used in aviation.)

Common Faults

- Relying almost solely on absorbing knowledge by working alongside an experienced operator. (This leads to inadequate and incorrect training.) However, working alongside an experienced operator is often the only way of getting hands-on experience.
- Management inaction where training and retraining shows an inability by an individual to acquire the necessary competence. (A probationary period of 3 months for new starters is one way of making it easier to avoid being stuck with an unsuitable person.)
- Perhaps the most common fault—the absence of a formal written syllabus, or any formal testing of the operator after training.

10.6.3 "Permit to Work" Procedures

There have been numerous major accidents and incidents of serious plant damage through slack procedures for permitting trades work in operating plants.

Objectives of "Permit to Work" Procedures

To safeguard people, the environment, and the plant by ensuring:

- that work is done on the plant only if it is properly authorized,
- that the plant is in a condition where it is safe to be worked on, given the information included on the permit,
- that the correct job is done on the correct equipment,
- that the tradespeople know and understand the hazards and the safe procedures,
- that the job is started, carried out, and completed safely,
- that the operators know that the plant is being worked on, and
- that the plant is in a safe and environmentally safe condition to be restarted.

Important Principles and Features

- Identification of work allowed without permit—for example, instrument checks, some routine running repairs.
- Positive identification to the tradespeople of the equipment to be worked on. (Neither tagging the plant nor physically showing the tradesperson is a foolproof method.)
- Clear definition of the work to be done.

- Comprehensive identification of the plant hazards, plus external hazards such as traffic and emissions.
- Authorization of procedures by appropriately senior people.
- Allotted time span of permit.
- Comprehensive making-safe, and means of preventing unauthorized alteration of the safe conditions before the work is finished.
- Identification of isolations to be made.
- Provision made for work by contractors and for their control. Contractors should be subject to the same disciplines as employees. They may require additional explanation of the need and the procedures.
- Reference to special procedures, for example, for excavation, hot work, use of explosive-powered tools, use of radioactive sources, entry to confined spaces, scaffolding, handling of asbestos, or access to roofs or other places where temporary means of access needs to be provided.
- Preparatory work should be recorded.
- After the job is completed, there should be formal acceptance of responsibility by production staff.

General Comments

- Routing of the permits to the file via the plant manager is a good way to enable monitoring of the procedure. Alternatively, a periodic audit can be made of filed permits.
- There is a need for checking when one shift prepares the isolations ready for work, and the next shift issues the clearances.
- There is a case for issuing separate clearances for slip-plating.

Common Faults

- Time pressure on the person issuing the permits.
- Issuing a permit without actually checking that the plant has been made safe, and that the tradesperson has a clear understanding of what is required.
- Incomplete or hasty change from one shift to another.
- Problems that arise when a permit on one job cross-references to an earlier job that established the isolations.

- Failure to remove identification tags and labels when the job is complete, or other sloppy signing-off practices. (Signing off is as important to process staff as is signing on to maintenance staff.)
- Difficulties with multitrades activities.
- Nonadherence by contractors.

For example, there was a potentially serious "near miss" incident on an offshore oil and gas platform operated by an international oil company. (The incident involved maintenance work on a pump and was very similar to that on *Piper Alpha*—see Chapter 13.) In the investigation, it was recognized that a major contributory factor was a poor work permit system.

A team was formed to make recommendations for improvement of the work permit system. The team composed of a platform supervisor, two junior engineers, and a contract maintenance supervisor. They recommended that the three forms of work permit (normal, hot work, confined space entry) be replaced by a new set of six forms, and that the forms be redesigned with a range of new boxes to be checked.

A consultant was engaged to review the recommendations. He noted the following features of how the existing system was operated.

- The contract maintenance supervisors regarded the work permit system as just paperwork to be completed before starting work.
- The night shift operators were given the task of putting equipment into safe condition for maintenance to be undertaken.
- The day shift platform supervisor allocated the work orders to the contractors, who started work directly without any further checks by the operators that the correct equipment had been made safe, and that it was, indeed, in a safe condition.
- The completed work permits were destroyed at the end of the day.

The consultant also noted that a feature of the "near miss" was absence of positive isolation of the pump before work was started, with reliance being placed on a closed remotely actuated isolation valve.

The consultant recommended that the emphasis be directed toward training on effective isolation, improved change of work to maintenance, and handback after maintenance, and toward training of everyone involved on the need for a tight work permit system, its objectives, its features, and typical incidents that had happened because of loose systems. He also recommended that the existing work permit documents be retained unchanged to avoid diverting the emphasis from how the system was operated toward the details of the design of the paperwork.

10.6.4 Control of Plant Modifications ("Change Control")

10.6.4.1 Introduction

It is fine to design a plant to be capable of environmentally friendly reliable, and safe operation, and to prepare operating instructions and procedures for maintenance, supervision, and management that will complement that design. But if the plant or the procedures are subsequently modified, the integrity may be lost.

There are numerous instances where a well-intentioned modification, with laudable objectives, was implemented without realizing that there would be very serious side effects.

A number of the accidents described in Chapter 13 illustrate the dangers of making modifications to plant or procedures without thorough investigation of the possible side effects and impacts. It is clear that modifications to the correct operating instructions, if inadequately considered, have the potential to result in hazard.

For example, a reactor was being started up. It was necessary for a minimum temperature to be reached (using steam heating) with only one of the reacting materials being recirculated in it before admission of the second reacting material. On one occasion, the second reactant was admitted before that temperature was reached. As a result, by the time the reactor reached the temperature required for the reaction to occur, the concentration of the second reactant was much higher than normal. The reaction started, then accelerated rapidly and overpressured the reactor, resulting in emission of irritant materials from the pressure relief devices. This caused public outcry and litigation.

See Chapter 13 for a number of other examples.

10.6.4.2 Design of a Modification Control Procedure

In the process industry, it is now common for a standard procedure to be followed before a proposed modification is approved.

This raises the question: "What is a modification?" A modification is any action that results in the plant equipment or procedures being changed from their original condition. Note that changes to procedures are modifications.

Kletz (1976) has postulated three requirements for plant modifications to be carried out safely. In essence, they are as follows:

- There must be a rigid procedure for making sure that all modifications are authorized only by competent people who, before doing so, try to identify all possible consequences of the modification and then specify the change in detail.
- There must be some sort of guide sheet or checklist to help people identify the consequences.
- Instruction and aids are not enough. People will carry out the instruction and use the aids only if they are convinced that they are necessary. A training program is necessary.

Mant (1981) describes such a procedure and illustrates one such form of a checklist. IChemE (1994) describe another such system in some detail.

There is a conflict between the wish to submit all modifications to the checking procedure (resulting in a great commitment of time, delays in getting modifications implemented, and a tendency to be superficial) and the wish to streamline the procedure by only putting selected modifications through the procedure (resulting in those modifications that bypass the procedure possibly introducing hazards).

A good way of getting a balance between thoroughness and efficiency is to adopt a two-stage approval procedure, as follows:

1. All modifications are described on a standard Modification Request form. A checklist of questions (set out on the back) is answered by the proponent. The questions cover a wide range of possible side effects and prompt the proponent to consider unwanted consequences. In considering possible side effects, the potential for the modification to affect adversely any of the following may be studied:
   - process conditions,
   - operating methods,
   - maintenance and engineering methods,
   - safety equipment,
   - environmental conditions, and
   - a wide range of design and equipment features.

   The form is then circulated in turn to several senior technical and line managers with responsibility for the plant or operation. They consider the modification and the checklist, and if they are satisfied, they sign the modification request form.

2. If any of the senior managers is unsure about the potential for unforeseen problems, then the modification is sent for a full Hazard and Operability (Hazop) study, by a suitably selected team. (The method for carrying out a Hazop study is described in Appendix A.)

It is common for such reviews to result in potential problems being identified that would otherwise probably not have been found until they occurred on the plant. In such cases, the proposed modification may be revised to avoid the problem without affecting the original aim, or the proposal may be dropped.

It is important that the procedure be audited periodically to ensure that it is not being bypassed. One aid to that is to require any trades job request that is to implement any change to the plant to have stapled to it a copy of the approved Modification Request form, and for the job requests to be checked each month by appropriate managerial staff.

The above may be summarized as follows.

Objective of a Change Control Procedure

To ensure that proposed modifications to both equipment and procedures will produce the benefits sought without unforeseen and undesirable side effects.

Important Principles and Features

- The procedure must be documented.
- It is essential that there be a definition of what a modification is: that is, what needs to be submitted for review. (Note especially the dangers of inadequately reviewed modifications to computer programs used for process control.)
- The authority needed for a modification to be implemented must be defined, and the authority must be at a sufficiently senior and experienced level.
- All modifications to hardware and defined relevant software must receive at least an initial screening.
- Both the screening, and also the more detailed reviews must be undertaken by more than one experienced, informed, and responsible person.
- Clear guidelines are needed for when a proposed modification is to be submitted to the second stage (a more detailed review).

- Screening and the more detailed reviews should use a standardized, comprehensive, and agreed checklist.
- All agreed modifications must be agreed, and drawings, P&I diagrams, operating instructions, etc., must be updated.

Comments

- To reduce the workload and elapsed time, it is common for such procedures to have two stages:
- a preliminary screening to determine whether a modification has sufficient potential for undesirable side effects to warrant a full study, and
- a full study, often using Hazop methods (see Appendix A), of those modifications that do not pass the first screening.
- Special care is needed with proposals to modify the logic of programs used by programmable electronic systems.
- One method of controlling unauthorized modifications to equipment is to require all job orders that will result in a change to equipment to be accompanied by a completed modification review form, and to audit job orders periodically to check that no such jobs are bypassing the system.

Common Faults

- The responsible manager may alone decide whether proposed modifications are worthy of study.
- The pressure to approve modifications may result in superficial study.
- The people who make modifications (supervisors, planners, etc.) may not realize that the change they authorize should be the subject of a study.
- Small modifications may be implemented (e.g., by a nightshift fitter) without being studied or considered for study.

10.6.5 Condition Monitoring

Condition monitoring is a form of preventive maintenance that aims to prevent loss of containment, loss of control, or loss of function of equipment.

Objective of Condition Monitoring

To identify early degradation of plant conditions to enable corrective action to be taken before a potentially hazardous event occurs such as breakdown, loss of containment (i.e., leak), or loss of control (e.g., reaction runaway).

(This is a field where there are often both statutory and corporate requirements, e.g., for pressure vessel inspection, but much more is required.)

Important Principles and Features

- The monitoring should be based on a regular, scheduled, systematic, and comprehensive review of where loss of containment, structural strength, process control, etc., could produce a potentially hazardous incident, and the likelihood of it.
- The monitoring should be designed around identification of specific hazards, rather than around the monitoring techniques available.
- The monitoring methods and results should be documented and filed.
- The monitoring schedules should be defined and followed on a routine, and the results reported routinely to management.
- The adequacy of the monitoring procedures and schedules should be periodically reviewed in the light of improvements in technology and accumulated plant history.
- The incidence of failures should be recorded.

Comments

Condition monitoring typically aims for early detection of future

- loss of containment,
- loss of control,
- machinery breakdown, and
- structural failure.

A wide variety of techniques exists.

Common Faults

- A technique-oriented approach is adopted, rather than a plant- or equipment-oriented approach. That is, the available monitoring equipment determines what is done, not the requirements of the plant.

- There is inadequate or inaccessible storage of historical information about previous monitoring carried out and the results of such monitoring.
- It is often based only on past incidents or near misses, rather than on an active study of possibilities.

10.6.6 Testing of Protective Systems

Objective of Testing Protective Systems

To ensure that the required level of reliability of protective systems is maintained.

Important Principles and Features

- It is important to define a list of all protective systems, active and passive (such as bunds), and not just instruments.
- Where there is no defined statutory frequency or formal company policy, the frequency of testing of each protective system should have been determined from a (possibly qualitative but preferably quantitative) review of the required reliability based on the following:

  possible magnitude of the hazardous event,

  likelihood of the hazardous event being initiated (i.e., a "demand") and thus protection being required, and

  likelihood of the protective system lapsing into a failed state between "demands" or tests.
- A formal test schedule should be established, for trips and other protective equipment (bunds, RVs, vents, flame traps, drains, etc.).
- Each system should be tested as realistically and as fully (i.e., from sensor to actuated protective equipment) as possible.
- Where it is necessary for the final actuated equipment to be disconnected or bypassed while the rest of the system is tested, the hardware should be designed such that it is clearly evident whether, on completion of the test, the system has been reactivated (e.g., trip-abort lock and warning lights on the front of the panel, rather than behind it, with access to the key suitably controlled).
- The authority to bypass the trip systems should be defined.
- Records of test results should be designed such that recognition of poor or decreasing reliability of individual systems is facilitated.

- Procedures should cover the period when protective systems are out of action for testing.

Comments

- Protective systems are those that are required to operate in the event of a potentially hazardous situation arising, to prevent the hazard from being realized, or to limit its magnitude or effect.
- Such protective systems include:

  instrumented alarms and trips,

  throttle bushes outboard of pump seals,

  relief valves,

  flame traps,

  vents,

  drains,

  nonreturn valves (pump delivery NRVs can be conveniently removed for testing as a routine whenever pumps are removed for repair or overhaul),

  bunds (including the possibility of leaks through bund bases),

  steam curtains and water curtains,

  fire systems, and

  emergency procedures.
- Untested protective systems cannot be relied on, as by the time a demand occurs, there is a high likelihood that the protective system has lapsed into an inoperable condition.
- Trip tests need to be carefully designed as, in some cases, routine testing has been a cause of plant interruption.
- It is helpful to classify the purpose and importance of alarm and trip systems.

Common Faults

- Testing confined to instrumented trip systems, not other protective systems.
- Definition of test frequency, and review of test results, being seen as the province of the instrument engineer alone.
- Keys being left in trip-abort switches, enabling unauthorized bypassing to prevent "troublesome" trips. It is prudent to have a system for authorizing and logging use of the key.

- Trip-abort facilities being designed such that it is not clearly evident whether a trip has been reactivated after testing.

10.6.7 Auditing Process Risk and Reliability Management Procedures[11]

The above six procedures are only a few of those that are needed. It is not usually practicable for an auditor to check all the required procedures on each occasion. What is needed is a short-listing process to minimize the time required in the audit.

One such approach is successively to ask four increasingly searching questions about each procedure. If one of the questions is answered unsatisfactorily, the auditor can conclude that the procedure is deficient either in design or in implementation. No further questions need to be asked about that procedure, and the auditor can go on to the next procedure. If all four questions are answered well, then the auditor can decide whether to undertake a detailed probe into the procedure and how it is used.

This approach is similar to a steeplechase with four hurdles, of increasing height, followed by a long, boggy stretch of ground. Falling at any hurdle disqualifies the athlete, but jumping them is not sufficient, as there is the boggy stretch to be covered.

The first question is: "Do you have a procedure for... ?" (for example: maintenance work permits).

If the answer is yes, then the first hurdle has been jumped. If the answer is no, the auditor's task is to explain the need for such a procedure and how it works, so as to prepare the ground for such a procedure being introduced.

The second question is: "What is the objective of that procedure?"

If this question is answered convincingly, then either the manager answering it has a reasonable understanding of it, or he or she is a quick thinker and persuasive speaker. But, the hurdle has been jumped, and the auditor then asks the third question. If the answer is not satisfactory, the task of the auditor is to explain the objectives, noting that the procedure is not likely to be well used in view of the lack of understanding of its objectives, and moving on to the next procedure to be audited.

The third question is: "How do you know that the procedure is being used correctly?"

1[1]Adapted from Tweeddale H. M. (1993): *The safety audit steeplechase*. The Chemical Engineer, No. 553, November with permission from IChemE.

If the answer shows that the manager personally has a good understanding of what is required, and personally checks it periodically, then the hurdle has been jumped and the auditor asks the fourth question. If, however, the question is poorly answered, either showing a lack of understanding of the requirements or a lack of personal checking, then the manager has fallen at that hurdle and the auditor's task is to explain the need for personal checking and the sorts of things to look for.

If the questions have been well answered so far, the fourth, final, and most difficult question is asked. It is: "Do you have to report to anyone higher in the organization about how well the procedure is operating?"

If the answer is no, then it is likely that the procedure could lapse and may have already done so, as its performance is not integrated into the management system. If the answer is yes and an explanation offered about how it is reported, then the manager has jumped the highest hurdle. The task for the auditor is now personally to check the procedure in detail. It may be necessary only to check one or two such details. For example, in the case of a work permit procedure, the auditor could ask, when inspecting the plant, to see the work permits being used by several of the maintenance tradespeople encountered on the way.

This approach is a valuable way for the auditor to reduce the time spent auditing procedures, and also for helping line managers understand their responsibilities for procedures.

10.7 LEARNING FROM ACCIDENTS AND "NEAR MISSES"

(See Tweeddale, 1999a.)

10.7.1 Approaching Accident and Incident Analysis

It is common for there to be an apparently obvious "cause" of an accident. This can lead to "tunnel vision," in which action is taken to treat that cause without treating the circumstances that led to the cause arising. To act in this way is to leave open the way for recurrence, either of the original incident or of other incidents with similar root causes.

A chlorine plant manager was away from his plant at a distance of around 1 kilometer. He smelled chlorine and hurried back to the plant where he found that no one was aware of the problem. He discovered that chlorine was escaping from the elevated vent from the purge gas scrubbing plant. The caustic soda solution used in the scrubber was fully saturated with chlorine and had not been renewed. The first reaction was to blame the operator for not checking and renewing the batch of caustic soda solution. In the investigation it was found that in recent weeks the batches had been becoming saturated much more rapidly than normal, but this had not been formally reported or investigated. It was then found by a member of the investigating team that the cause of the rapid saturation was poor performance of the chlorine liquefaction plant, due to a control problem. A recorder chart showed that the problem had started more than 3 weeks earlier. The chart abnormality had not been noticed by the shift operators, the shift supervisors, or the professional staff. When the chart abnormality was pointed out to the supervisor during the investigation, he asked what it indicated, what could cause it, how one would determine whether there was actually a problem or simply a defective recorder, and what one would do to rectify the liquefier problem if one existed.

Clearly, to blame the operator for not checking and renewing the batch of caustic soda solution would be to miss the opportunity of identifying the actual initiating incident (the liquefier control fault), as well as numerous serious weaknesses in understanding, training, management systems, etc.

For example, a pump handling anhydrous liquid ammonia was being overhauled. The tradespeople isolated the pump by closing a manual ball valve upstream of it and putting a "Do Not Operate" tag on it. They went away from the job for their coffee break. As it happened, an external auditing team was visiting the site and came across this inadequately isolated and deserted work site where a careless movement could have resulted in the ball valve being bumped into an open position, releasing anhydrous ammonia at a high rate.

It is worth considering the types of weakness of which that single incident could be an indication. The incident was not even a "near-miss," as there was no suggestion that an escape nearly happened. On its own, that example of poor isolation practice has little significance, apart from presenting the potential for a major escape. But the audit team also found the following weaknesses elsewhere on the plant:

- The anhydrous ammonia storage tank was of the atmospheric-pressure design, with an external open-air bund. Safety of such a tank relies on high integrity of control of a number of variables, including the tank level (to prevent overflow through the relief valves, etc.). Normally such a tank has three independent level indicators, using diverse measurement principles. During the audit, it was found that only one of the indicators was operational; one had been out of service for more than 12 months, and the other for around 3 months.

• The tanker bay at which anhydrous ammonia was filled into road tankers was fitted with booms, designed such that a tanker could be filled only if the boom was down (to prevent the tanker driving away while still connected), with the boom interlocked such that, once lowered, it could be raised only when the filling hose had been replaced on its stand. During the audit it was noticed that one tanker was being filled with the boom raised. It was discovered that there was a defect in the boom, so the operator had left it in the raised position and had bypassed the interlocks to permit loading to continue. It was also discovered that the plant manager did not know of the defect.

Each of the above weaknesses is indicative of a number of management defects, each able to be classified under the headings of the six general requirements for safety assurance.

Further, the above examples illustrate the importance of investigating, not just accidents and near misses, but also any circumstances that are unusual or unsatisfactory. They also illustrate the need to penetrate below the immediate weaknesses to identify the underlying causes.

It is important that the structure used for the analysis of root causes and contributory factors matches that used for safety performance measurement, so that the findings from such analyzes are in the most useful form for managing improvement. (See Tweeddale, 1999b.)

10.7.2 A Conceptual Framework for Accident and Near-Miss Investigation

In proactive safety management, it is well recognized as good practice to:

- actively identify hazards;
- identify the types of incident that could arise from each hazard;
- manage the risks by first aiming to eliminate or reduce the inherent hazards;
- then aim to prevent any incidents, or strengthen the response to any incidents, which could occur due to any residual risk.

This is often described as a "top-down" approach (see Figure 10-6). The first step is to identify each hazard on the plant. This is usually a relatively simple task. In this illustration, the identified hazard is a tank of motor spirit. The next step, moving lower down the diagram, is to identify the types of hazardous incident that could conceivably arise from that

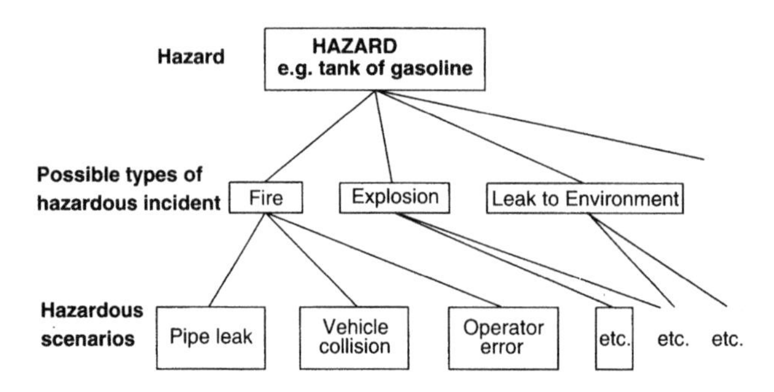

Figure 10-6. The "top-down" approach to risk identification.

hazard. This is also a comparatively simple task. In the example, the types of incident identified include fire, explosion, and environmentally damaging spillage. The next step is to define and understand the hazardous scenarios, that is, the ways in which those types of incident could occur. In the example, a fire is envisaged as possibly resulting from a leak from a pipeline connected to the tank, a vehicle colliding with the tank, an operator opening the wrong valve, etc. This is not a simple task, as there are usually numerous scenarios. For example, it is apparent that there are many other possible scenarios that could lead to a fire involving the inventory of motor spirit.

An important principle can be recognized if this diagram is continued further downward (see Figure 10-7) searching for the root causes of each of the identified scenarios.

One can classify the various types of root cause in a number of ways, for example, by using the structure of the six requirements for safety assurance derived from Hawksley. Each of the identified scenarios can be envisaged as resulting from a failure in one or more of those requirements. Then, proceeding one more level downward, the "universal root cause" is identified: "management failure."

Identifying the universal root cause as "management failure" is of little direct use: it has long been recognized that the responsibility for almost any problem can ultimately be laid at management's door. However, it is the starting point for the "bottom-up" approach.

Starting from "management failure" and then moving up the diagram, it is evident that management failure in relation to process safety can take many forms. Selecting just one, in this case "poor training of staff," and moving one further level up the diagram, it is evident that there are numerous other scenarios (shown diagrammatically here as X, Y, and Z) that had

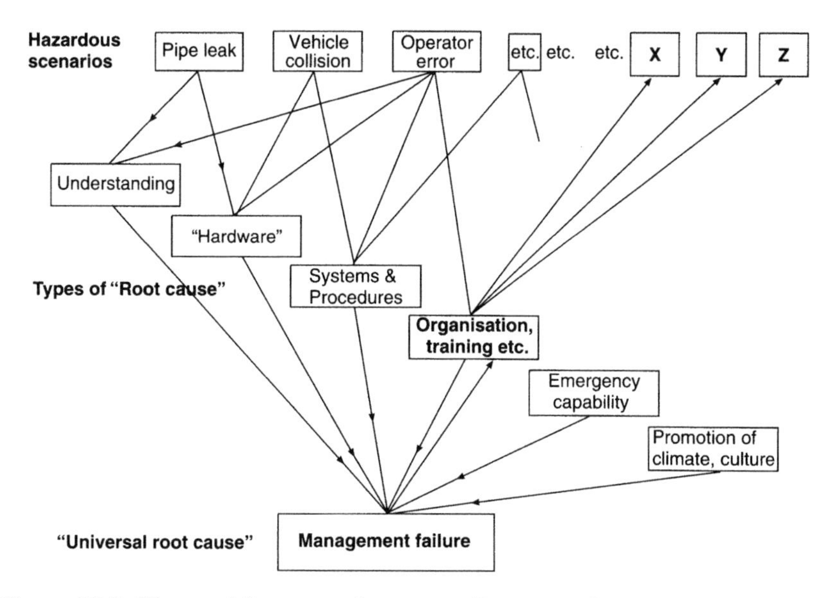

Figure 10-7. The need for a complementary "bottom-up" approach to risk management (continuation of Figure 10-6).

not been identified by the top-down approach. In fact, the range of mishap scenarios that could occur if staff are poorly trained defies imagination.

This clarifies an important principle for management of process safety and reliability, namely:

Lesson: *It is insufficient to base initiatives in a risk management program solely on identified hazards and hazardous scenarios.*

Because many scenarios typified by X, Y, and Z will not be identified in any top-down approach, and defy imagination in any case, they must be managed without being explicitly identified. This is done at the next level down, that is, at the level of the six requirements for effective risk management, by adopting "good practice" in each of them.

It is thus necessary, when investigating accidents, near misses, and unusual occurrences of any kind, to adopt a two-pronged approach to identification and design of process safety initiatives. The two prongs are:

- measures to avoid identified scenarios; and
- measures to develop and maintain "good practice."

In addition, accidents or incidents, which are reported to have occurred elsewhere, should be reviewed to determine whether they could reveal the potential for incidents that were not previously identified on one's own plant. A good approach is to challenge oneself or those managers responsible with:

- "Could that type of accident or incident occur here?"
- "If not, why not?"

Therefore analysis of accidents and incidents should aim to lead to two types of action:

- actions aiming to prevent that particular type of accident from recurring (note that exactly the same accident rarely recurs except in a negligent organization); and
- actions aiming at removing underlying weaknesses and thus preventing other (possibly undefined) accidents to which those weaknesses could contribute.

10.7.3 Methods of Analysis

There are many possible ways of investigating accidents and near misses.

Although many different approaches are possible for investigating accidents and incidents, there are several important principles. Some of them are very similar to those that apply to hazard identification (which is, in effect, investigation of accidents that are possible but that have yet to occur). They include the following:

- Every investigation should be undertaken with an open mind, constantly searching for possible contributory factors that have not previously been considered.
- Allocating blame and scoring points must be avoided, as they produce defensive reactions and closed minds.
- Because any one individual has "blind spots" (i.e., unrecognized gaps in his or her knowledge), the investigation should be undertaken by a team of people with a diverse variety of relevant expertise and experience.
- The opportunity should be taken to include in the team some of those who will benefit from an increased understanding of safety.
- The investigation should be systematic (so that important possibilities are not overlooked in the course of a vigorous discussion) and detailed (as it is the details that often prove or disprove a hypothesis about an initiating event or a contributory factor).
- Great care should be taken before a reported "fact" is accepted as truly factual.

- It is important to define the boundary of the problem; both where the problem is, and where it is not. Thus it is as important to define what were not initiating events and contributory causes as to define what were.

A form of fault tree analysis is often a helpful structure, starting from the top with the observed incident and analyzing down, progressively identifying the lower and lower layers of contributory factor or "cause." It is important, at each level, not only to show the obvious factor(s), but also to show other theoretically possible factors and to seek evidence to support or question their relevance in the particular incident under study.

A particular problem arises when the technical cause of an incident is not clear. This calls for especially careful analysis, imaginatively listing possible technical causes and contributory factors, and seeking hard evidence to prove or disprove them. However, this focus on the technical components of the incident must not be allowed to distract attention away from the nontechnical components, or to prevent further analysis to uncover the nontechnical root causes of the technical components. (In other words, referring to Figures 10-6 and 10-7, the importance of the technical components, which mainly appear in Figure 10-6, must not prevent the investigation from penetrating to the bottom of Figure 10-7.)

10.7.4 Incorporating Lessons from Accidents and Incidents into Routine Management Systems

Van der Schaaf (1991) describes a near-miss management system at a chemical process plant. The starting point is an "Incident and Near Miss Reporting Form." This leads to analysis of the contributory factors according to a classification scheme designed following extensive interviews with operators. There are two phases in the classification scheme: a coarse-scale grouping of factors into three main classes: Technical, Organizational, and Behavioral; and a second phase comprising a tree structure of possible causes and contributory factors in each of the above three main classes. In any particular investigation, the contribution of any of the possible causes is considered. The findings are stored in a database for statistical analysis, training, and providing a basis for improved safety monitoring.

Hale (1991) describes an interactive computer program for collecting and processing accident data. This program is designed to tackle the problem of having accidents investigated in sufficient depth. The person entering the accident details to the computer system is asked a series of questions, determined by the answers to previous questions, as a way of leading the investigator to search below the most obvious contributory factors to identify root causes.

Whatever approach is taken, it is important that the lessons learned from investigations of accidents and incidents not be lost. A very effective way of incorporating them into the ongoing management of operations is for the factors that are identified in investigations as contributing to the accidents and incidents to be added to the list of variables that operators, supervisors, and managers routinely monitor as part of operational control, and to the list of topics from which the program for periodic auditing is selected.

10.7.5 Case Studies

The following example illustrates the value of looking beyond the apparent boundaries of any particular incident.

There was a small fire at an isolated plant that processed raw gas from nearby fields for transmission to a major city and a number of important regional towns. The plant was located some hundreds of kilometers from the nearest town. It occurred at the glycol recovery heater, when methanol vented from the unit condensed, ran down the vent pipe, and was ignited by the burner. (Glycol is used to inhibit formation of hydrocarbon hydrate blockages, and methanol is injected to break up any blockages, which do occur. The diluted glycol is recovered in the heater by evaporating the absorbed water—and any injected methanol.) It appears that more methanol was being used to break up blockages than was realized. The fire was detected visually by a control room operator who happened to go outside the building briefly. It was extinguished within 5 minutes and caused no damage. The contributory factors leading to the incident were investigated and appropriate managerial and engineering action initiated. It was noted that, if this small item of plant had been put out of action, the entire plant would be shutdown.

This particular vulnerability led the directors of the company to consider the incident more broadly, wondering whether there were other possible incidents that could cause loss of ability to supply the market, and whether the appropriate safeguards were in place. This led to a wide-ranging review of the hardware of the plant. ("Software" was explicitly excluded from the review, as a major program for implementation of a package of safety-related procedures and systems was just starting.) The review identified such weaknesses as:

- very limited leak detection and fire detection equipment, particularly in view of the very low manning of the site;
- relief and blowdown facilities that, while meeting the requirement of the appropriate API Code of Practice, would not prevent a leak (and hence fire) continuing for a substantial time after recognition;
- very limited firefighting capability (again in view of the limited manning of the site) and actual faults in the installation of firefighting equipment;
- dubious practice when purging plant of air after maintenance;
- major layout weaknesses, with the main plant power and instrument cable runs very close to the pumps used for delivering hot oil to the plant, and hence vulnerable to pump fire;
- inadequate and poorly graded paving in areas handling flammable liquids, such that any ignited leak would result in major fire damage;
- inadequate fire protection of pressurized LPG vessels.

REFERENCES

API, "Management of Process Hazards, Production and Refining Departments." API Recommended Practice 750, 1st ed., January 1990.

AWA Limited v Daniels t/as Deloitte Haskins & Sells & Orrs. 10 ACLC 933, 1992.

Bottomley, S., and Tomasic, R., "Corporate Governance and Responsibility: the AWA Case." Australian Corporations and Securities Law Reporter Special Report. CCH Australia Ltd., 1992.

CCPS, "Guidelines for Implementing Process Safety Management Systems." Center for Chemical Process Safety, American Institute of Chemical Engineers, New York, 1994.

Hale, A. R., "IDA: An Interactive Program for the Collection and Processing of Accident Data." Chapter 7 in *Near Miss Reporting as a Safety Tool* (van der Schaaf *etal.*, Eds.) Butterworth–Heinemann, Oxford, UK 1991.

Hawksley, J. L., "Strategy for Safety Assurance for Existing Installations Handling Hazardous Chemicals." WHO Conference on Chemical Accidents, Rome, July 1987.

IChemE, *Safety Management Systems*. European Process Safety Center, published by The Institution of Chemical Engineers, Rugby, UK, 1994.

Juran, J. M. *Managerial Breakthrough: A New Concept of the Manager's Job*. McGraw-Hill, New York, 1964.

Kletz, T. A., "A Three Pronged Approach to Plant Modification." *Chemical Engineering Progress*, November 1976.

Lees, F. P., "The Hazard Warning Structure of Major Hazards." *Trans. IChemE* **60**, 211 (1982a).

Lees, F. P., "Hazard Warning: Some Implications and Applications." IChemE Symposium Series No. 71 (1982b).

Lees, F. P., "Hazard Warning Structure: Some Illustrative Examples." *Reliability Engineering* **10**, 65 (1985).

Mant, W. D., "A Stepwise Approach to Hazard Reduction in Process Plant from Process Specification through to Plant Operation." *Chemical Engineering in Australia* **ChE 6**(2), 31–35 (1981).

Pitblado, R. M., and Lake, I. A., "Guidelines for the Application of Hazard Warning." *Trans. IChemE*, July (1987).

Simpson, C. M., and Smith, D. A., "Development of a Process Safety Management System." CHEMECA Conference, Sydney, Australia, August 1988.

Tweeddale, H. M., "Improving the Process Safety and Loss Control Software in Existing Process Plants." Institution of Chemical Engineers, London. Symposium Series 94, Multistream 85, April 1985.

Tweeddale, H. M., "Upgrading Safety Management in Process Plants." Proc. International Conference on Safety and Loss Prevention in the Process Industries, Singapore, 23–27 October 1989.

Tweeddale, H. M., "The Safety Audit Steeplechase." *The Chemical Engineer*, No. 553, 15–16 November 1993.

Tweeddale, H. M., "Principles and Practices for Design of Process Safety Monitoring and Auditing Programs." European Federation of Chemical Engineering Process Safety and Loss Prevention Conference, Antwerp, Belgium, 6–9 June 1995.

Tweeddale, H. M., "Learning from Accidents and Near Misses." Proc. Conference Israel Institute of Petroleum and Energy, Herzlia, Israel, May 1999a.

Tweeddale, H. M., "Measuring Safety Performance in Process Plants." Proc. Conference Israel Institute of Petroleum and Energy, Herzlia, Israel, May 1999b.

van der Schaaf, T. W., Lucas, D. A., and Hale, A. R. (Eds.), *Near Miss Reporting as a Safety Tool.* Butterworth–Heinemann, Oxford, UK, 1991.

Chapter 11

Introduction to "Software" or the "Human Factor": Including Safety Culture, Safety Climate, and Human Error

When you have studied this chapter, you will understand:

- a way of classifying the elements of "safety software," or the "human factor";
- the importance of software;
- the effect of software on assessed risks;
- the importance of intangible components of risk;
- the impossibility (at present) of assessing the impact of intangible components on the total risk;
- a systematic approach to improving software (including safety "culture");
- ways of nourishing and poisoning the safety culture; and
- the nature and normality of human error.

This chapter addresses critically important elements of the risk management step "Reduce Risk" of Figure 1-4.

11.1 COMPONENTS OF "SOFTWARE"

11.1.1 Introduction

The term "safety software" includes essential but intangible factors such as the human element. Software can be classified in a number of ways, of which the following is a useful example:

- Organization
- Procedures and methods
- Knowledge, skills, and training
- Documented standards and methods
- Attitudes/"safety climate"/"safety culture"

These are discussed in turn below.

11.1.2 Organization

An organization chart aims to display the hierarchy and pattern of authority and responsibilities within an organization. Unfortunately, it is usually a very inaccurate indication of the unofficial but real hierarchy. The true organization is determined by:

- who decides what,
- who communicates with whom, and
- who defers to whom.

Because of the ways in which people interact in practice, this is too complex to be displayed on a chart without the result being confusing, yet the way the people in the organization interact is very important for the management of the risk and reliability.

For example:

- Are decisions about matters that affect risk and reliability being made by people with sufficient information for the decisions to be well made?

- Does the Risk and Reliability Manager, who may perhaps be shown as reporting to the General Manager, actually have an opportunity to communicate openly and with acceptance with the General Manager, or does he or she actually communicate only with people much lower on the official chart? Is this because of the diffidence of the Risk and Reliability Manager, or is the organization passively or actively resistant to the manager personally or to his or her role?

• Does the Engineering Manager, responsible (on paper) for ensuring that all plants are maintained in a safe and reliable condition, actually have any influence with the Production Manager, or is the Engineering Manager ignored with impunity because he or she does not take a stand and raise any concerns with the General Manager?

• Does the Production Manager have sufficient technical experience to understand the significance of requests from the production staff or from engineering staff for more capital to be spent on making the plants more reliable or safer, or is he or she an upwardly mobile "specialist manager" with a commercial background who is emphasizing cost reduction and short-term production maximization above all?

11.1.3 Procedures and Methods

The procedures and methods that affect reliability and process safety may be grouped under the following general headings:

- operating procedures and methods,
- maintenance procedures and methods, and
- supervisory and management procedures and methods.

The main objectives of formally documenting the procedures are:

- to facilitate discussion of them, as an aid to recognizing weaknesses needing correction,
- to facilitate communication and training of people in what is required,
- to encourage uniformity of practice, and
- to enable formalizing of decisions to modify and update practice in the light of experience.

It is one thing for a Safety Policy to be documented. It is another for it to be supported by documented "Arrangements for Implementing the Policy," which specify who is responsible for what.

There are numerous procedures and methods with an important bearing on reliability and safety in relation to major hazards or other major potential losses (with particular reference to a hazardous plant). They include:

- employee selection procedures,
- operating instructions,

- operator training,
- "Permit to Work" procedures (in the case of hazardous or critical plant and equipment: to precede and follow maintenance work, to safeguard people and the plant),
- maintenance procedures,
- training of maintenance employees (especially on specific plant hazards),
- modification control or "Change Control" (to prevent well-intentioned modifications—to equipment or procedures—from introducing hazardous or loss-producing side effects),
- condition monitoring (to check that the standard of containment in plant handling hazardous materials, control of critical operations, and reliability of the operating plant remains high),
- testing and inspection of protective systems, whether active or passive, whether based on equipment or procedures (to check that they are in an operable condition),
- control of electrical equipment in classified flammable areas (to ensure that flameproof equipment remains flameproof, etc.),
- safety inspections/process safety audits,
- "Unusual Occurrence" investigation and reporting (i.e., investigation of incidents that may, on investigation, be found to be precursors of serious accidents if circumstances change),
- accident investigation,
- emergency plans and procedures,
- safety and fire training,
- security procedures,
- arrangements for accessing and using technical resources,
- safety policy and arrangements for implementing it, and
- routine management auditing of procedures (to ensure that they remain appropriate and are complied with).

Numerous serious accidents have resulted, not primarily from equipment failures, but from inadequate procedures, or from failure to follow defined procedures. Several particularly important procedures are discussed more fully in Chapter 10.

11.1.4 Knowledge, Skills, and Training

It is not sufficient to have a good organization and well-designed procedures, if the people in the organization do not have sufficient knowledge and skills to carry them out. This requires training.

Common deficiencies of training include (see also Section 10.6.2 in Chapter 10):

- inadequate specification of the objectives and of what is to be learned, especially the "key points";
- inadequate identification of the base knowledge needed for the key points to be understood in depth, and not just rote learned;
- undue attention to narrowly defined "competencies," that is, focusing on the "key points" to the exclusion of essential understanding of the background and underlying principles;
- focusing primarily on normal operation, with too little attention to recognition, diagnosis of the causes of, and response to abnormal operating conditions;
- excessive formalization of training syllabuses and presentation, resulting in inflexibility and lack of attention to updating in the light of experience;
- inadequate attention being given to the means of training: often new employees are simply taught by an experienced operator, and so pick up old established bad habits and "folklore" rather than the correct methods and correct understanding; and
- inadequate testing that the trainee has in fact acquired the required knowledge and skills.

11.1.5 Documented Standards and Records

Documented standards are important in a number of fields. One field that is especially important in reliability and in avoidance of major hazardous incidents is engineering standards: the design approaches to be used in various applications, materials of construction to be selected to withstand various conditions of temperature, pressure, corrosion, reactivity, etc. In organizations where no such standards are adopted, or where they are not documented, there tends to be a variety of approaches adopted at any one time, dependent on the whims of individuals, with the result that many are inappropriate.

It is very important to have records of the design details of the plant when considering any change to the plant or its operating conditions. For example, it may be desired to operate a stock tank of a compressed liquefied gas at a lower temperature (by refrigeration) to reduce the internal pressure, and the inherent hazard in the event of a leak. But if there are inadequate records of the material of construction, it will be difficult to determine whether the material will be suitable for low-temperature operation.

In considering a modification to a section of a plant, it is important to have records of what hazards were envisaged in that section, and what the design philosophy was, to ensure that the modification does not unwittingly remove a safeguard.

Documented records of all sorts constitute the long-term memory of the organization, outlasting the employees who may change on average every few years for career advancement or other reasons. Organizations that do not keep records of accidents and near misses or other unusual occurrences, and of the lessons learned in the investigations conducted following such incidents, are doomed to repeat them. As Trevor Kletz says, "Organizations have no memory," and numerous examples can be quoted where the same type of breakdown or hazardous incident has been repeated in the same organization after several years because the people involved in the first incident were no longer around.

11.1.6 Attitudes, "Safety Culture," and "Safety Climate"

No matter how well an organization is structured and run, how good the defined procedures and methods, how knowledgeable and skillful the people, and how comprehensive the documented standards and records, if the attitudes in an organization are bad, the reliability and safety performance will be bad.

Other terms are sometimes used for *attitude*. These include *culture* and *climate*. Care should be taken in using any of these terms, as they can convey the impression that the fault is all with the other person or people, and not at all with the person speaking.

It is not unusual for people to think that safety is an optional extra, something that is given attention once production continuity, quality, and cost have all been taken care of. This attitude is reinforced by the common practice in which safety management is divorced from the organization structure and the management procedures used for managing the more obviously commercial objectives of the organization.

Positive attitudes toward safety can be developed if:

- people see that it is to their advantage,
- people see that improvement is possible, and
- people see that improvement will not cause excessive loss of some other benefit.

Rather than embark on a program aimed primarily at improving attitudes (which would probably be ineffective), it is probably much more effective to design a program to identify what is needed to improve reliability and safety in both a hardware and software sense, and to conduct it with clear organizational commitment and involving all those with a role and responsibility for reliability and safety (i.e., everyone). A well-managed program will lead to improved attitudes.

11.2 MEASURING THE STANDARD OF SOFTWARE

The standard of software is measured by inspection against a checklist of important features. The measurement can be purely descriptive, which enables management to determine the nature of the action needed to make improvements.

The measurement can be "semiquantitative," that is, numbers being assigned subjectively or even with guidance from quantitative criteria, to represent the standard of each feature. For example, some insurers specializing in the process industry have developed premium rating systems based on inspection of equipment (hardware) and a wide range of software factors. A great deal of effort must be put into developing consistency of scoring if those scores are to be used for such purposes as premium determination.

Such schemes are popular with management, which sometimes attempts to use them for intersite comparisons, or for determining whether there have been improvements. This can be very misleading.

11.3 EFFECT OF SOFTWARE STANDARDS ON QUANTITATIVE RISK ASSESSMENTS[11]

There is ample evidence that management standards that are better or worse than average will result in risks that are lower or higher

[11] Adapted from Tweeddale, H. M., "*Balancing Quantitative and Non-Quantitative Risk Assessment.*" Trans IChemE Vol **70 (Part B)** May 1992, 70–74 with permission from IChemE.

than average. However, currently there are insufficient data to enable quantification of the extent of the lowering or raising of the risk level. This is explained below.

The total frequency of hazardous incidents is the sum of those incidents that arise from:

- failure of equipment (mechanical, electrical, structural, etc.), and
- human error (in design, construction, operation, maintenance, supervision, management, etc.).

Both classes of cause are affected by the standard of management systems and how well they are administered.

There is no doubt that the standard of management can affect the frequency of equipment failure. Some consultants undertake a management audit and multiply the generic equipment failure frequency by a factor derived from the audit score. There are serious weaknesses in this approach, both in principle and in application. In outline, the weakness in principle is that hardware and software failures should often be added rather than multiplied (as failures of management systems can lead to hazardous incidents quite independently of the nature and condition of equipment), and the weakness in practice is that the magnitudes of important factors are not currently able to be determined.

The frequency of hazardous incidents is the following sum:

Hazardous incident frequency $= H \times F_{ENG} + E \times F_{OP}$

where:

H = Generic failure frequencies for equipment, based on some loosely definable level of management, etc.;

F_{ENG} = Engineering management factor for equipment failure, ranging below or above 1.0 depending on whether the management related to equipment failure is better or worse than the undefined level of management incorporated in data H;

E = Basic frequency of operational or maintenance errors for the particular type of plant (e.g., based on level of complexity or required closeness of control);

F_{OP} = Operational management factor for human error, ranging above or below 1.0 depending on whether the operational management related to human error is better or worse than the undefined level of management incorporated in data E.

Failure data H are available from data banks in generic form (expressed as a range in the better data banks).

Engineering management factor F_{ENG} is determined, in principle, by standards of design, maintenance, and a number of other aspects of supervision and management, that is, those that directly affect the condition of equipment and its probability of failure. Although it is possible to form a view as to whether management is better or worse than average, it is not at present possible to estimate the magnitude of F_{ENG}, beyond deciding whether it is less or greater than unity. In a poorly maintained plant, the magnitude could be of the order of 10, 100, 1000, or greater.

Human error frequency E for any particular plant requires assessment of the various ways in which human error in design, construction, operation, maintenance, or supervision could result in a major incident other than by physical failure of plant components. This is very difficult to do comprehensively because of the wide variety of forms that human error can take, compared with the finite variety of failure modes of physical equipment.

Management factor F_{OP} is a field in which a great deal of research is currently being done, but the results are mostly qualitative at present, taking the form of guidelines for reducing the likelihood of human error.

Note that the two management factors (F_{ENG} and F_{OP}) may relate to different components of the management role and management systems. For example, factor F_{ENG} may include the adequacy of inspection of equipment, maintenance training and supervision, etc., while factor F_{OP} may include the standard of the operating instructions, operator training, shift supervision, etc.

A great deal of research has been in progress internationally aimed at developing a method for incorporating management and attitudinal factors into quantitative assessments of plant risk. This research has involved staff from statutory authorities, psychologists and safety consultants. Hurst and others (1996) describe one approach in some detail. It relies on separately rating "software" including systems and attitudes in eight key audit areas such as design, maintenance activities, maintenance supervision, operation, and operational supervision. The generic failure rates for equipment are multiplied by a modifier factor calculated from the scores for each of the audit areas and a weighting factor derived for each audit area from an analysis of the causes of a large number of past accidents. The modifier factor has a range from 0.1 to 10. The developers of the method believe that their method has a sound theoretical and statistical basis, but note that it is demanding on those who audit the management systems and attitudes. They also report that the results correlate well with the lost-time injury rates for a number of plants.

As a pragmatic aid to risk management, the approach appears to have much to recommend it, except for the apparently large effort needed to implement it. It provides a way of studying and reporting on practices and attitudes in an explicit way, allowing a more focused improvement effort. But whether it can validly be used to calculate the absolute level of risk arising from both equipment and human failures will probably only become clear after more thorough peer review. Based on experiences such as those described by Tweeddale (1992), issues that need resolution include:

- whether it is theoretically valid to base assessed risks due to operator error and inadequate operational supervision on a foundation of equipment failure frequency (i.e., if a hazardous material is released because an operator opens the wrong valve, equipment failure rates are of little relevance);
- whether it is sound practice to base the risk assessment for a specific plant on factors averaged over many plants;
- whether it is theoretically valid to use a modification factor in which good scores for some audit areas can balance bad scores for other areas (e.g., whether close supervision of operations compensates for severely corroded pipework—the situation appears analogous to a chain in which some strong links are expected to compensate for one weak link);
- the confidence that can be placed on the views formed during an audit (for discussion of some problems when auditing, see Tweeddale, 1995);
- whether there can be a sound basis for defining the magnitude of the modification factors; and
- whether, in view of the demonstrably poor correlation between lost-time injury rate and process safety, it is sound practice to validate a method of assessing process risks by correlating risk scores and lost-time injuries.

The appropriate approach to inclusion of software in probabilistic risk assessments of the process plant appears at present to be as follows:

- Where specific human errors can be defined, then those errors should be incorporated in the fault trees and event trees, and included in the frequency estimation process.

- In general, generic failure data should be used for equipment (unless there is clear reason to use higher or lower frequencies and a clear basis for selecting the extent of the deviation from the generic data), but clearly with a statement in the written report broadly as below.

  "A risk assessment attempts to quantify the risks from a defined set of failures, using generic data (possibly adjusted to take account of the perceived condition of the specific equipment).

  The frequency of those specified failures could be appreciably higher or lower as a result of various management and human factors, but by an amount that cannot be assessed at present.

  There would be additional risks that cannot be assessed arising from defective human performance from the design stage onward. Some of these risks may be included implicitly in the generic data used for equipment failures, but this cannot be stated with confidence.

  Where defects in the management systems (etc.) are identified in the process of undertaking a risk assessment (and such defects should be searched for), then they should be rectified, rather than included in the risk assessment."

One helpful way of visualizing risk is to see risk as having two dimensions: quantifiable and nonquantifiable. (This concept was suggested by the Argand Diagram, where one axis has "real" dimensions and the other has "imaginary" dimensions.) Often the unquantifiable risks are intangible: subjective, emotive, and relying on value judgments. These can be represented as in Figure 11-1.

The total risk displayed in Figure 11-1 could be expressed as:

Total Risk = 1.6 Tangible + High intangible

A subjective numerical scale could be derived between Low and High on the Intangible axis, but because the units on the two axes are different, it is not possible to avoid having two terms in the total risk expression. Therefore, a quantitative value for the vector sum cannot be determined.

As the aim of risk assessment is not principally to know what the risk is, but to aid risk reduction and risk management, this is not a serious problem. When the tangible and the intangible risks have been analyzed in arriving at the assessment, they can be examined and action taken to reduce them.

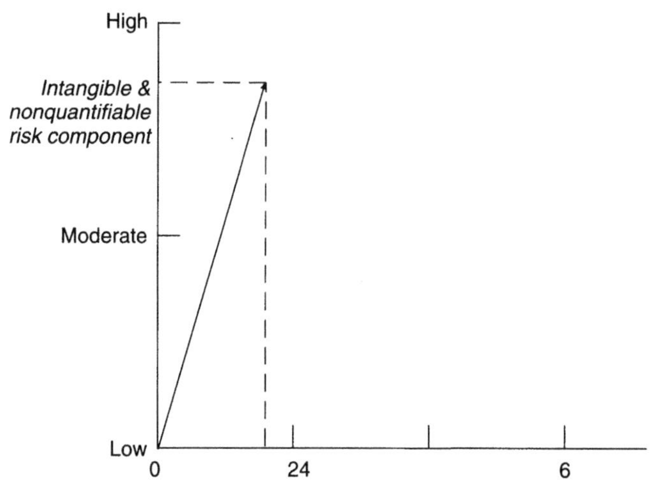

Figure 11-1. Illustration of combination of quantifiable and nonquantifiable assessments of risk.

11.4 "SAFETY CULTURE" AND "SAFETY CLIMATE"[21]

Adapting the description of a strong "safety culture" proposed by the IAEA (1991), *safety culture* may be defined as "that assembly of characteristics and attitudes in an organization and the individuals in it that determines the extent to which safety issues receive the attention warranted by their significance."

A safety officer for a major trade union once remarked that the term "safety culture" had unfortunate connotations. He heard it being used in a manner that implied that the safety problems stemmed from the union members not being "cultured." He preferred to use the term "safety climate" of an organization; the climate being determined by senior management, with his members responding to that climate. There is a case for both terms being used, but with no suggestion that "safety culture" is linked with the meaning usually associated with the word "culture" in the general community.

Thus the concept of safety culture can be helpfully subdivided into two complementary components:

- safety culture (relating to the characteristics and attitudes to safety at the operating site); and

1[2]Adapted from Tweeddale, H. M. (2001): *Nourishing and Poisoning a "Safety Culture."* Trans IChemE Vol **79** (**Part B**) number B3, May with permission from IChemE.

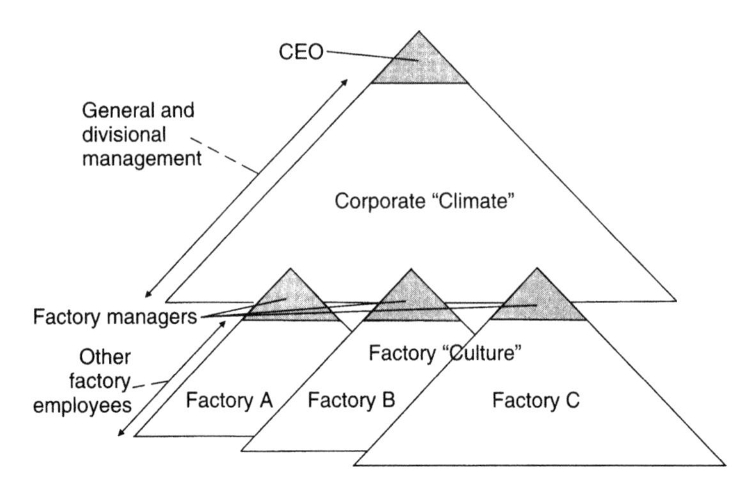

Figure 11-2. Safety climate and safety culture.

- safety climate (relating to the characteristics and attitudes to safety from board level down to the most senior manager of the operating site).

(The border between those responsible for safety climate and those responsible for safety culture is debatable, but one convenient location is at the level of the operating site manager. This manager has responsibilities in both fields: upward into the realm of the safety climate to ensure that sufficient information is forwarded to senior managers for them to assess the safety needs and any initiatives they should take to strengthen the safety climate, and downward to create as strong a safety culture within the site as the safety climate permits.)

Diagrams do not always illustrate concepts particularly well, but Figure 11-2 may help illustrate this separation between climate and culture.

11.5 SENIOR MANAGEMENT ROLE IN SAFETY CLIMATE AND CULTURE

11.5.1 Introduction

The climate of an organization is determined by senior management within constraints imposed by outside influences.

Kletz (1977) explores the reasons for innovation or major steps forward in process safety. He lists them as:

- a major accident: leading to investigation and understanding of contributory factors and causes, and the corresponding approaches to prevention; also causing management of the organization concerned to attempt to "shut the door after the horse has bolted";

- legislation: motivating management to comply with the letter of the legislation, if not always with the spirit (legislation being able to compel action, but not attitude—one cannot "legislate for virtue");
- insurers: through research by a few leading industrial insurance companies wishing to be able to advise their clients on risk reduction;
- a "champion" within the organization concerned, with sufficient seniority and influence to push successfully for adoption of good standards, and the leadership skills and standing to encourage people to change their attitudes and follow his or her lead.

Given that the upper limit to the safety culture in the workplace is determined by the safety climate, and that the safety climate is determined by the commitment and effort of the CEO, the task facing those concerned with raising the level of process safety throughout industry is how to motivate and educate the CEOs of such organizations.

Those with that concern include:

- people working in high-hazard plants, who are personally at risk, and who are aware of those risks;
- production and engineering line managers of such operations, who need the support of more senior management to make the required improvements;
- safety managers with a functional responsibility in such operations; and
- legislators who, whenever a serious accident occurs, are held accountable by the public for failing to ensure safety.

As was outlined in Chapter 1, the principal classes of motivation for safety are:

- legal,
- commercial, and
- ethical.

These are considered in turn.

11.5.2 Legal Motivation

Attempts to impose a higher level of safety through legislation have had a marked effect over the years, from the first factory safety legislation in the 1800s to now. Prescriptive legislation (which specified required safety features—such as machine guards—and procedural requirements, even organizational requirements such as safety officers) has been falling out of favor because it has been seen as incapable of keeping up with changes in technology and management theory. In its place, "performance" or "goal-oriented" legislation is currently widely preferred by regulators and management (but not always by those at personal risk in the workplace), requiring management to provide a "safe place of work." Yet this is accompanied by a prescriptive element: instructing management on how to provide a safe place of work and requiring a pro-forma demonstration that they have done so (e.g., by means of a quantitative risk assessment; provision of "competence-based training"; preparation of a comprehensive "safety case"; safety audits; etc.). It is likely that, in a few years, this approach will be found to have been inadequate, as was the old prescriptive legislation.

There is a second dimension to the legal requirement for safety. Not only does management have statutory obligations as required by legislation, but they have a more general requirement to take care.

Over the years, it appears that senior management have not been greatly motivated toward plant safety by the potential of their organization being fined for breach of legislation, or being liable for damages because of a failing under common law, although they are motivated to some degree by the embarrassment of the publicity given to legal proceedings. They are motivated to some degree by the potential for severe economic loss due to loss of production from unreliability or the cost of damage to production facilities by a damaging incident. Certainly they are highly motivated by the potential for imprisonment, where that is provided for by legislation.

However, this risk of personal imprisonment can lead to senior management acting with the main objective of protecting themselves personally from prosecution by going through the motions of managing safety, while at the same time not really providing the climate in which those junior to them can build the required level of safety. This is clearly evident to the junior staff whenever it happens, and it destroys the safety culture.

There is a case, whenever a member of an organization is being prosecuted for an alleged failure to meet his or her safety obligations, for automatically including the CEO of the organization in that prosecution, for failing to provide the required climate and failing to ensure that all employees comply with their obligations. Although it is unlikely that a sufficiently strong case will be made against the CEO for a custodial sentence, the stress of having to mount a legal defense, and the embarrassment of the publicity, would be an incentive to CEOs in general to minimize the risk of such a situation.

11.5.3 Commercial Motivation

The commercial motivation toward safety has limitations. If plant safety is measured only in monetary terms, then its priority is allocated among all the other activities that affect profit. This overlooks the human value of life, and the outrage of the public with its consequent political (and hence commercial) impact on not just the company involved, but the whole of the industry. But in times of economic rationalism, this is common, and is unfortunately accepted implicitly by some naïve risk consultants who, for simplicity of prioritization of risks or because of pressure from their clients, reduce the consequences of all risks to monetary terms. This is a case of believing that one can provide a simple one-dimensional solution to a complex multidimensional problem.

11.5.4 Moral or Ethical Motivation

Of these three forms of motivation, the most durable (where it exists) is the ethical concern. Where a CEO feels a moral or ethical commitment for safety, this can be spread down throughout the organization provided that the CEO understands the requirements for safety (e.g., the six requirements postulated by Hawksley—Section 10.1 of Chapter 10) and has the leadership ability to drive that commitment down through the management hierarchy (see Case Study 1 in Section 11.6.1 of this chapter).

A child psychologist once spoke of the importance of parents and counselors avoiding what he called the "slight smile" response when hearing of some adventurous but unacceptable behavior. If the child notices a slight smile, or other comparable nonverbal body language that suggests admiration for the bravado if not the behavior, then no verbal reprimand will be regarded by the child as being sincere.

Similarly, no amount of promotion and apparent commitment will withstand lack of real interest or decisions that site employees perceive as placing commercial or other interests ahead of safety. It is necessary for senior management, when considering issues such as organizational changes or the size of maintenance budgets, to inquire actively into the safety implications of those issues, to make whatever decision is appropriate with a full awareness of those implications, and, where the decision could be interpreted as contrary to the best safety practice, to ensure that the basis of the decision is clearly conveyed to the relevant site employees, demonstrating that the safety concerns have actually been taken care of and that management is keenly aware of the requirements.

How, then, do those concerned with the level of process safety in an organization, where the CEO is not particularly committed to safety, raise that level of commitment?

This has challenged legislators, safety professionals, and others for many years, so unfortunately there is no generally applicable answer. Perhaps the main options are (a) to demonstrate the link between safety and reliability, and between both of them and productivity and profit; and (b) to show the economic consequences to the organization and the personal risks to senior management from major plant upsets that are associated with loss of safety and reliability.

So, just as it is necessary for senior management to demonstrate personal commitment to safety in tangible ways, it is equally necessary for the operating site manager to present information that enables senior management up to and including the CEO to understand the risks and needs of the operating site.

11.6 MEASURING THE CLIMATE AND THE CULTURE

The key elements of climate in relation to risk and reliability management are:

- management commitment, and
- management style.

The key elements of culture in relation to risk and reliability management are:

- interest in risk and in its minimization, and
- willingness to act accordingly.

Climate is being constantly measured in an ad-hoc, qualitative, and subjective manner by the employees. Suitable measures or indicators include:

- the extent to which management devotes time to risk and reliability management;
- the extent to which management calls staff to account for risk and reliability management performance and efforts;
- the extent to which risk and reliability management is included in the normal planning, budgeting, and reporting cycles;

- the extent to which management can be seen by the workforce to be giving attention to risk and reliability management; and
- the willingness to spend money, to forego production, or to exert effort, to improve risk and reliability.

Culture can be measured, using ad-hoc qualitative and largely subjective indicators in a similar manner.

As in the case of F_{ENG} and F_{OP} (Section 11.3), numerical scores can be given to such indicators as an aid to communication, but any attempt to add the scores to find a realistic total, or to use the numbers in a quantitative risk assessment, would be unsound and misleading.

11.6.1 Case Studies of Nourishing the Safety Climate and Safety Culture

Case 1

A chemical company had for many years regarded itself as a leader in safety. Then a new CEO was appointed from a company with a record of excellence in safety. Soon after arriving, he announced that the company was to reduce the lost-time injury rate (LTIR) by 20% each year. Senior managers regarded this with some incredulity.

Then he asked when the next safety audit was being conducted within the company. He was told that a new safety audit protocol had just been developed and was to be trialed for the first time in the following week in a particular factory. He said that he would take part in the audit. During the audit, when a safety weakness was identified, a plant manager remarked that he had sought approval for the required expenditure to remedy it, but that his application had been unsuccessful. The CEO remarked to the works manager that if it were resubmitted, no doubt the money would be found. Naturally it was resubmitted and naturally the money was found.

At that time, a feature of the way in which the company was managed by the board of directors was for a series of 6-monthly review meetings, at which each divisional general manager (in effect the CEO of each major division) would present to a meeting of the executive directors an analysis of performance over the previous 6 months, together with the business plans for the future. At the first of these during the incumbency of the new CEO, his first question was the level of the LTIR in that division. The general manager had never been asked that before and had no information about it. He was told in no uncertain terms to find out. The next morning, when the next divisional general manager attended his panel meeting, he had been forewarned and knew the LTIR of his division. When he was asked, and gave the answer, he was told by the CEO that it was far too high, and would he please present his work plans for reducing it. He had no such plans. Before long, every manager in the company was preparing safety improvement plans.

Then the CEO began a series of staff communication sessions, starting in the head office. Everyone was invited to attend, leaving only one person per floor to answer the telephones. He gave a talk about the company's position and his plans for the future, and then had an open question time, in which anyone could ask any question and he would answer. This session was then repeated at each factory, typically in a warehouse with chairs brought in, with staff including plant operators, maintenance workers, and management attending. He used these sessions to discuss all major topics, including his requirement for safe plant and safe work practices.

He totally changed the safety climate in the company in only a few months. In spite of the initial skepticism of local managers, the LTIR did, in fact, fall by over 20% for each of the years during which he was CEO. He insisted on safety being seen as a line management function that required planning, organizing, controlling, and leading, just like production volume, quality, and cost, and one that was managed through the normal line structure, rather than being largely delegated (or abdicated) to the functional stream of safety managers and officers who, as a result of this changed emphasis, became more in demand in a specialist role to assist the line management, rather than as a "ginger group" trying to get the line managers to act.

Case 2

A company chief medical officer learned that a plant was to be built on a site that had been severely contaminated by leaking drums of a moderately volatile toxic liquid. He explained to the responsible managers that the people undertaking the excavation for foundations and drains would be exposed to dangerous concentrations of the liquid and vapor, and the site would need to be cleaned before construction began. He could not get his concern acted on. So he went to the director with responsibility for safety and the environment, explained the problem, and said that, having exhausted the normal management hierarchy, professionally his remaining option was to alert the director to the problem. The next day the director called a meeting of all the stakeholders—managers, engineers, health and environment staff, and union representatives—and said that the problem was to be addressed and solved properly, with the required full communication and approval by the statutory authorities and community groups. This proved expensive, but it was undertaken successfully and led to a strong feeling of pride that, when serious issues of safety and health arose, the company would face them properly and professionally, and not "fudge" a solution.

Case 3

A chemical company planned an important expansion of a plant that manufactured a highly reactive liquid. Because the site was already extensively developed, there were limited options for the siting of the expansion in a location sufficiently close to the existing plant to be operable by the same team. The only really practicable location for a particular distillation column was next to a large stock tank of a compressed liquefied toxic gas. As such columns had a worldwide history of explosion, the safety manager said that the location preferred by the production staff was unacceptable. Eventually the General Manager called a meeting of those involved. The factory manager said that, although he fully supported the company's safety policy and standards, where so much was at stake commercially, surely some flexibility was appropriate. In response, the General Manager said that if the factory manager would be embarrassed to explain the basis of his recommended siting to a coroner's inquest after a fatal accident, then he should not make the recommendation. This led to a renewed investigation of design options and then to successful identification of a satisfactory solution involving a different design that avoided the inherent hazard. In turn, this led to reinforcement of the understanding that the company's safety policy was to be implemented, not just paid lip service.

Case 4

A divisional general manager of a large company read an account of the Flixborough explosion and noted that a mineral-processing plant currently being designed for his division was superficially similar in appearance. He sought an independent review of the safety of the design. Even though the process was entirely different from that of Flixborough, the review found that there were major inherent hazards that had not been adequately addressed. By taking this action, and following up on it, the General Manager effectively gave notice that safety was critically important in the design of the plant.

Case 5

There was a fatal accident in the mineral processing section of a metalliferous mine operated by a diverse international resource company. The general manager of that division visited to learn how the managers planned to avoid any further fatalities. He said that he was prepared to close down the operation altogether, but, in view of their comprehensive presentation to him, he would give them one more chance. This led to a major effort being directed toward improvement of safety.

The principle illustrated by the preceding five examples is that visible demonstration of commitment by senior management has a powerful effect on the seriousness with which safety issues are addressed elsewhere in the organization.

Case 6

The environment manager of an extensive chemical manufacturing site had a practice of visiting each production manager four times per year, to discuss the environmental performance of their plants, and to agree the appropriate action to remedy problems and to improve performance further. He was experiencing difficulty in getting the production managers to take the agreed actions. At one meeting they would make a commitment to implement agreed actions, but at subsequent meetings would offer excuses why no progress had been made. The manager to whom the environmental manager reported arranged for the quarterly reviews of environmental performance to be conducted at a higher management level, himself accompanying the environmental manager to those meetings. The more senior managers then held their production managers accountable for the agreed actions and ensured through their routine management meetings that progress was made as planned.

Although this is an environmental example, the principle applies equally in safety. This principle is: for a strong safety culture, safety must be managed alongside the other priorities of the business, by the same management structure. Where there is insufficient management commitment at a lower level, more senior management must be brought in.

Case 7

A company with potential for major process incidents that could cause multiple fatalities, serious plant damage, and environmental releases found it difficult to interest the operators and the unions in a safety improvement program. One union representative asserted that unless the company erected many safety signs, he would not believe that they were serious. So the company started a program of hazard identification and risk ranking, using a large team of operators, maintenance tradespeople, supervisors, professional staff, and senior managers, working together. The systematic discussion and the indicative quantification of the consequences and likelihood undertaken by discussion in the group between technical, operating, trades, and supervisory staff led to a changed and greatly improved understanding of the risks, their possible causes, the potential severity of the consequences, and the required engineering and operational safeguards. This led to consensus about the most important risk-reduction actions to be taken, built a strong feeling of teamwork between the various groups in the organization, and showed clearly the extent of management commitment.

The principle illustrated by the preceding example is: to build commitment to safety among any particular group in an organization, for example, engineers, production staff, supervisors, operators, or tradespeople, it is important to involve them in the safety management program as fully as practicable.

Case 8

A very large and geographically dispersed works processing a wide range of hazardous chemicals had a number of maintenance workshops, each with its own team of tradespeople, and each dedicated to a small number of plants. It was believed that a greater efficiency in the use of these maintenance resources would be achieved by establishment of a single central workshop and single maintenance team. Concern was expressed by the production staff that this could lead to a loss of "local knowledge" by maintenance tradespeople, supervisors, and engineers, and a decline in the condition of the plants, leading to more downtime and lower levels of safety.

A series of discussions were held between production staff and engineers to list the physical features of well-maintained plants, and the other variables and conditions (many not able to be quantified but needing to be described in words) that would be monitored. Then inspection worksheets were prepared, the plants were inspected, and the worksheets were completed, showing the defects identified. When the maintenance reorganization was undertaken, quarterly inspections were instituted using the same worksheets to observe any change in plant condition. This approach was regarded by all those involved as effectively managing the change in maintenance organization, and as evidence of management commitment to reliability and safety.

The principle illustrated by the above example is: where it is commercially imperative to downsize production or maintenance staffing at any level, or to outsource production or maintenance services, it is essential to determine beforehand whether there is any potential for a detrimental effect on safety, to identify those critical variables or conditions that indicate the level of process safety, then to implement a monitoring program examining those variables both before the downsizing and afterward. Doing so allows the effect on safety to be managed, and employees can be assured both that the level of safety is being maintained, and that management is really committed to safety, rather than simply paying lip service to it.

The consequences of downsizing and cutting maintenance budgets in the interests of profit without due regard to the needs of management for careful technical appraisal and review were well known over a century ago at least, and are described gleefully in a maritime parable by Kipling (1898), which also illustrates the need for technical people to stand up and be counted in the face of technically ill-considered commercial "imperatives."

Other approaches to nurturing and growing a healthy safety culture are suggested by Tweeddale (2001), Kean (1995), and many others.

11.6.2 Case Studies of Poisoning the Safety Culture

Case 9

The corporate managers of safety and environment performance in an oil refinery sought assistance from a consultant in polishing the documentation of their safety and environmental vision statement and the targets for the following year, in response to a requirement from their overseas parent company. They wondered how to get commitment to these targets at the refinery level. Noting that refinery management had not been involved in reviewing the vision and setting the targets, the consultant suggested that the director responsible for safety and environmental performance might be asked to discuss them at the refineries, and might ask to be shown the plants where the major safety and environmental upgrading would be needed. The consultant was told that the director would be too busy to be involved in that way.

The principle illustrated by the above example is: to poison the safety culture, senior managers should avoid personal involvement. (Staff assess the extent of commitment of senior management to safety, and the priority they really allocate to it, by observing the amount of personal time and effort they devote to it, and the extent of financial support they give it—not by policy statements, posters, memos, and public statements.)

The above example illustrates another principle: To poison a safety culture, fill the safety staff positions with people who have failed in other roles. Judging by the nonverbal response of the corporate safety and environmental managers when the suggestion was made that the director should explore the major issues at the refinery, it appeared that the two managers did not have the confidence of the director and were reluctant to approach him. This perception was strengthened when one of the managers left the company soon afterward.

Case 10

The CEO of a chemical company visited a factory that manufactured and formulated a range of herbicides and pesticides. The factory had a history of environmentally damaging releases to the stormwater drainage system, as the stormwater and trade waste systems had been initially installed years previously and had not been kept properly separate. The factory manager told the CEO that there was a major problem in that any major spillage of any one of a number of the chemicals handled would result in serious contamination of the open drain that passed through the factory. The CEO politely acknowledged the problem. The factory manager then explained that, because of the troublesome history of releases and the increasingly critical community and regulatory response, any further release on the scale suggested would result in the factory being closed down. At this the CEO became keenly interested. It was apparent that the CEO was not particularly interested in environmental performance, unless there would be an adverse impact on profit. This message quickly spread around the factory. This is another example of environmental culture rather than safety culture, but the same principle applies.

The principle illustrated by the above example is: to poison a safety culture, demonstrate that safety is only really important if it affects profit, that is, through unreliability and loss of production.

Case 11

(This is not a process industry example, but the principles apply universally.) Information was prematurely released by a financial institution to some of the interested organizations, enabling them to profit before the information was published generally. The premature release occurred because a computer operator, who was instructed to type an e-mail message with a 6-minute delay on its transmission, entered the required specification for the delay, but did not call up a separate menu that actually activated the specified delay. As a result, even though a delay had been specified, the message was transmitted at once.

An internal investigation identified numerous weaknesses in the systems used by the financial institution. As well as recommending many changes to the systems, it was recommended that the computer operator be punished by transfer to another position and a reduction in salary, and that some of the supervisors in the area should also be penalized.

The justice of this response is highly questionable. Kletz (2000) remarks to the effect that "to blame human error for accidents is about as helpful as to blame gravity for falls," on the grounds that just as engineers recognize gravity, and design to cope with it, we need to recognize that it is human to err, and we need to design systems to cope accordingly. One principle arising from that is that any system where a single human error will lead to serious consequences is a high-risk system, and that sooner or later the consequences will inevitably occur, without the person or people involved being other than human. Computers are particularly prone to this: most people who use computers have at some time made a serious error, such as corrupting or erasing a critical file.

When the response to the investigation was queried, the management of the financial institution asserted that, although there were certainly weaknesses in the system, the operator concerned had not taken the level of care appropriate to the consequences of error. This is a common and futile response to an accident: "tell the operator to take more care." There was no suggestion of any action being taken against senior management who had required such systems to be established, and had not arranged for them to be adequately reviewed for risk of failure, etc.

This is not strictly an example of a safety culture, but the principles apply equally to safety. Such selective punishment of junior people leads to a feeling that "the organization" does not understand the requirements for low-risk operation and is more concerned with the political priorities of being seen to have isolated the problem and taken decisive action.

Case 12

In a military exercise involving movement of troops by helicopter at night, a collision occurred between two of the helicopters. Several of the troops were killed, including the pilot of one of the helicopters, and there were many injuries. The official inquiry, conducted by officers of the same service, provided another example of blaming and punishing junior staff who are closest to the incident, or even its victims, as the inquiry led to the charging of the middle-ranking and junior officers involved in the exercise. Even though the report mentioned the budgetary constraints that minimized the amount of preparatory training possible, there was no criticism of, or action taken against, the more senior officers (some more senior than the investigating officers), civil servants, and politicians who defined the level of capability required from the armed services but withheld the necessary funds.

The principle illustrated by the above examples is: to poison a safety culture, when an accident occurs, conduct a witch hunt for the guilty and punish the innocent, all as a means of deflecting any aspersions cast in the direction of senior management.

Case 13

A factory handling large quantities of molten metal had a record of serious incidents, including fatalities, due to process causes. Their management engaged safety consultants specializing in occupational safety. The consultants did not tackle the inherent hazards of the plant or the adequacy of the hazard-specific safeguards (saying that those were the specialty of the client) but focused on reduction of lost-time injuries due to lifting, tripping and falling, etc. This approach was seen by the operators and tradespeople as not addressing the real issues, and they lost confidence and interest in the safety program being promoted by management.

This also illustrates the damaging effect on safety culture of management not recognizing the difference between process safety (addressing the hazards inherent in the materials handled and the processes used) and occupational safety (addressing the hazards inherent in work methods of those involved).

The principle illustrated by the above example is: if senior management demonstrate limited understanding of safety issues (when compared with their usually excellent grasp of economic issues), they demonstrate their lack of real interest in safety, and they cause loss of confidence in the safety programs that are in place and a worsening of the safety culture.

Case 14

A risk assessment study was undertaken of a plant manufacturing a highly toxic gas. The report, which detailed a number of concerns and a series of recommendations, was presented to the corporate director with responsibility for that factory. He pushed it away and said that he did not wish to see it. Subsequently, presumably after discussing his managerial obligations with his legal adviser, he asked to be fully briefed on the findings of the report, required presentation of the work plan for implementing the findings, and called for a monthly update on progress. He never sought any information on any other safety issues related to the plant that may have been identified as work progressed. Nor did he seek to initiate similar studies of other plants. Nor did he inquire whether any additional support was needed; rather, his manner implied that any such request would indicate incompetence by the local manager. It was clear that he was less interested in safety issues receiving the attention they deserved, than in having the identified items on the work plan completed. His manner throughout was less that of determination that the plant be made safe, than of protecting himself from liability.

The principle illustrated by the above examples is: if senior managers are principally concerned with minimizing their own legal liability for any future accidents, their insincerity will become evident to the rest of the organization, and their ability to inspire confidence in their leadership of the safety program will be undermined.

Case 15

A coal mine operated by a subsidiary of a major worldwide mining corporation had a history of serious accidents, partly due to the unusually hazardous nature of the coal seam being mined. Their management developed a comprehensive safety management program, which provided for substantial ongoing involvement of the mining unions and every level of management. The program was well designed, but because of the history of industrial relations problems at the mine, management imposed it on the unions as a finalized package. The paradox of imposing a program that was designed to be operated in a collaborative manner had not been recognized. After the unions refused to have anything to do with it, management remarked that if they had involved the unions in the design of the program, implementation would have been delayed by the discussion and they would still be designing it. That the program was now unable to be implemented was not seen as demonstrating the fundamentally incorrect approach they had adopted to their management of the design process. However, this approach was consistent with their view (apparently part of the corporate ethos) that management must manage. Although few would argue with that as a philosophy, there are various approaches to *how* management should manage.

The principle in the above case is: to poison a safety culture, or to stunt its growth, regard design of safety management systems and safety management programs as the specialist field of professional managers and technical specialists, and relegate the workforce to the passive role of those who do as they are told.

These cases lead to the following conclusions.

1. The two complementary fields of safety climate and safety culture each call for effort by those responsible for developing, strengthening, and maintaining them.
2. The required effort entails positive attitudes, resources, and initiatives; coupled with avoidance of negative indications able to be interpreted as lack of real commitment.
3. The strength of the safety climate determines the upper limit to the strength of the safety culture.
4. The safety climate can be nourished by such means as:
   - visible expenditure of resources on safety, including the time and attention of senior managers from the CEO downward, and availability of staff resources and finance to rectify safety problems;
   - senior managers clearly having made the effort both to understand the principles of good process safety management and also to learn of the safety problems in the organization;
   - positive recognition of improvements in safety performance; and
   - requiring safety to be managed simultaneously with and by the same management structure as are production, quality, and cost.
5. The safety climate can be poisoned by senior managers failing to understand safety imperatives as well as they understand economic imperatives, and by failing to manage safety as actively and with as much understanding and personal attention as they use to manage the other business objectives.
6. The safety culture on operating sites can be nourished most effectively by creation of a strong safety climate, including visible commitment by those at higher levels of the corporate management structure, coupled with involvement of all those on the operating sites in safety planning, safety studies, and implementation of the findings of those studies, and in monitoring safety performance.

7. The safety culture can be poisoned by a sick safety climate that demonstrates lack of interest higher in the corporate management structure and thus unduly limits the resources available to nurture safety. It can also be poisoned by perceived insincerity by management at any level, and by ignoring the potential for everyone to contribute to the safety program.

11.6.3 An Approach to Improving the Safety Culture by Increasing Involvement

It may be recognized that a plant needs improved safety software. This recognition may arise in one or more of a variety of ways:

- a poor record of losses, a major incident, or an unacceptable frequency of unusual occurrences or near misses,
- a hazard analysis and risk assessment revealing deficiencies in software, and a dependence on software for safety, or
- increased technical complexity of equipment, either for processing or for control.

Tweeddale (1985) sets out a stepwise approach to defining the hardware and software requirements of a plant, the approach being designed to involve all levels of the plant organization in such a way as to identify the requirements effectively and to build teamwork and positive attitudes—these being important components of software.

In outline, the steps are:

1. Brief all concerned on the importance of reliability and process safety, and the nature of the inherent hazards threatening them.
2. With a group, representative of all organizational levels, systematically identify all the significant hazards in the plant. The discussion could follow the general approach set out in Chapter 2, using the Chemical Hazard Checklist and the Plant Section Hazard Checklist.
3. With the group, postulate a range of hazardous incidents that could arise for each of the identified hazards, and identify how they can be avoided altogether or identified early and stopped.
4. With the group, define procedures or equipment to facilitate avoidance or early recognition and control of the potential incidents.
5. With the group, define the appropriate responsibilities for carrying out the procedures, and for periodic checks.

6. Define any further training requirements for those responsible to be able to carry out their responsibilities properly—in carrying out procedures, or in operating and maintaining any new equipment.
7. Implement.
8. Monitor and maintain.

Kletz (1992) says: "Don't try to change people's attitudes; just help them with their problems." Another way of expressing it is: "Don't try to change people's attitudes; just change their working 'environment.'"

After all, most people respond to their situation. If it is perceived that other people have a bad attitude, it is very likely that they have been put in a situation where that attitude is the likely response. Therefore, according to Kletz, the task for management is to change the situation so that the attitudes will respond. That will almost certainly first require a change in the attitudes of management, leading to a change in management behavior, style, climate, etc. In other words: a poor workforce safety culture is very possibly the result of a poor managerial safety climate.

By adopting the approach set out above, management illustrates a willingness to learn and to work in a team with the workforce. It also illustrates a positive commitment of management to safety. This type of approach has led to tangible improvements being recognized in the plant and procedures, leading to intangible but substantial improvements in the understanding of and commitment to safety by the workforce. Similar benefits have been obtained by involving people at all levels of the workforce in the rapid ranking process for identifying hazards and ranking risks. (See Chapter 3.)

Employees watch management closely to identify any signs of insincerity, or failure to put their money where their mouth is. Exhortation for improved performance is cheap. If a manager is seen to devote personal effort, and to require others to do so, that is a great encouragement. If the opposite is observed, such as reluctance to spend money, interrupt production, etc., the real priorities of management are immediately observed and noted.

11.7 HUMAN ERROR

11.7.1 Types of Human Error

Kletz has commented informally that to blame human error for accidents is as unhelpful as to blame gravity for falls. He notes that engineers are aware that the forces of gravity can cause structures to collapse, so they design them to be resistant to those forces. Similarly, it is well known that people sometimes do not follow rules precisely, that they have lapses of attention, that they do not make important decisions well when stressed and under pressure to do so quickly, that they are emotional, etc., so we should design our plants and procedures to minimize the likelihood or impact of such normal human factors.

Kletz (2001) comments that people should not be blamed for behaving as most people would behave, for example, occasionally making a mistake. He cites numerous examples where equipment has been designed such that human error is inevitable.

One common situation is a lecture room with both a whiteboard and flip-chart paper. Erasable pens are provided for the whiteboard, and permanent markers for the paper. Sooner or later a lecturer using a permanent marker on the flip charts will go over to the whiteboard, and cover it with permanent ink before discovering his error. Whose fault is it? Not the lecturer, but the person who put permanent markers in the same room as the whiteboard.

Similarly, if a machine is to be supplied with a particular specialized lubricant, and if drums of various lubricants are stored in the area, sooner or later the wrong lubricant will be used.

Case Study

An operator of a high-temperature fused-salt bath used for cleaning equipment used in spinning polyester fiber was instructed to go to the store to collect an additional supply of one of the salts used in the bath. He did so, but was given the wrong salt, which was clearly and boldly labeled with its identity. The supervisor saw the bags but did not pay attention to the labeling, and instructed the operator to empty the bags into the bath. He did so, and narrowly escaped serious injury when the bath erupted due to a violent chemical reaction.

Simply relying on checking by other people may not be sufficient.

Case Study

In a prestige book of around 700 songs, each page of proofs was checked by five people. When the book was published it was found that the wrong tunes had been printed for two of the songs, and that there were numerous other errors and omissions.

What may have been a better approach to proofreading is:

- preparation of a checklist of the more probable types of error (including typographical errors, layout errors, omissions, incorrect matching of tune and words, etc.);
- holding a briefing session for all those involved in checking;
- requiring that the boxes in the checklist be completed for each song checked;
- having only two people check each page, to focus their attention on the importance of their role.

On the other hand, consider the following case study:

Case Study

A commuter lived a couple of kilometers beyond the end of the railway line. His wife drove him to the station each morning and collected him each evening when he phoned. To save money, he adopted the practice of calling (a seven-digit number) from the public phone booth, waiting until he heard the phone at the other end being picked up, then putting his phone down. (He could hear that, but he could not speak to his wife or hear her.) His wife knew that, if the phone rang at around that time of the evening, but the caller put the phone down as soon as she picked her phone up, it would be her husband. However, if he dialed the wrong number, the following would happen:

- he would be left waiting for a long time in the incorrect belief that his wife was on her way to collect him;
- his wife would be wondering when he was going to call;
- some stranger would be wondering who had called and hung up.

In a total of around 500 occasions on which this procedure was used, that is, around 3500 digits being dialed, the commuter never dialed the wrong number: it was important to him to dial correctly, and he took great care. Because of the importance of dialing correctly, whenever he wondered if he had made an error, he quickly put the phone down and dialed again.

This illustrates that rates of human error can be extremely low in some circumstances (such as if the person performing the task understands the possibility of error and is personally committed to avoiding the error).

Kletz (2001, pp. 3–5) classifies human error into the following categories:[31]

1. those caused by a slip or a momentary lapse of attention;
2. those caused by poor training or instructions (not knowing what to do, or "knowing" the wrong thing to do);
3. those caused by lack of physical or mental ability, that is, by a mismatch between the abilities of a person and the situation;
4. those caused by a lack of motivation or a deliberate decision not to follow instructions, etc.; and
5. those caused by managers, perhaps resulting from a lack of understanding of their responsibilities or the role they should be playing.

Kletz then classifies accidents a second way: according to the means available to prevent them. They are:

(A) better design;

(B) better construction;

(C) better maintenance; and

(D) better methods of operation;

plus a special category:

(E) those related to computer control.

He discusses each of these in some detail.

11.7.2 Estimating the Probability of Human Error

Gonner (1913), writing about something else, said something applicable to the estimates of the probability of human error. "The best of them is but an approximation, while the worst bears no relation at all to the truth."

One approach to inclusion of human error in risk assessments is to undertake a *task analysis* in which the required activities of individuals are studied in detail, to identify the opportunities for error, and then the likelihood of error is estimated in each case.

Where there is a simple choice, such as "act" or "not act," for example, in response to an alarm, it may be possible to derive a reasonable estimate,

---

1[3] Adapted from Kletz T. A. (2002): *An engineer's view of human error* (Second Edition). IChemE, Rugby, England with permission from IChemE and the author.

taking account of the circumstances in which the person is placed at the time (stress, time pressures, etc.).

However, a major problem arises when considering the different forms of incorrect actions that are possible. One way of illustrating this is to consider a partly trained operator facing a situation in which a number of alarms sound at once. Perhaps the most likely action is to do nothing: to freeze for fear of doing the wrong thing. But if the operator does not take that course, there is an immense variety of incorrect responses that he or she could take—numerous buttons that could be pressed or controls that could be operated, in any of a vast range of different sequences and timings.

Again, it is easier, when constructing a fault tree in the usual manner, from a single top event downward, to identify many of the ways in which a human error on the part of an operator can lead to an event higher on the tree. But if one considers such factors as poor training, poor modification control procedures, and poor work permit systems (to name just a few), it is not practicable to identify all the ways in which these may lead to many different types of top event.

Various tables of human error probabilities have been developed over the years for various tasks and work situations. A number of them are listed by Kletz (1992). However, attempts to validate them have proved disappointing, with wide variations between the results and the real-life performance.

This gives further support to the view held by some practitioners that quantitative assessments of risk, although very valuable aids in developing understanding of risks and the options for improvement, have too insubstantial a basis for the absolute values to be used for much of the decision making for which they are intended.

REFERENCES

API, "Management of Process Hazards, Production and Refining Departments." API Recommended Practice 750, 1st ed., January 1990.

CCPS, "Guidelines for Implementing Process Safety Management Systems." Center for Chemical Process Safety, American Institute of Chemical Engineers, New York, 1994.

Gonner, E. C. K., in *Royal Statistical Society Journal*. Quoted by J. Boreham and C. Holmes in *Vital Statistics*. BBC Publications, 1974.

Hurst, N. W., Young, S., Donald, I., Gibson, H., Muyselaar, A., "Measures of Safety Management Performance and Attitudes to Safety at Major Hazard Sites." HSL, Liverpool University, Four Elements NL VROM. *J. Loss Prev. Process Ind.*, **9**(2), 161–172 (1996).

IAEA, "Safety Series No. 75-INSAG-4." International Atomic Energy Agency—International Nuclear Safety Advisory Group, Vienna, 1991.

Kean, N., "Developing a Risk Awareness Culture through the Implementation of Total Risk Management." Proc. AIC Conference "Total Risk Management," Sydney, July 1995.

Kipling, R., "Bread upon the Waters." Short story in *The Day's Work*. Macmillan, London 1898. (Reprinted by Penguin 20th Century Classics, 1990.)

Kletz, T. A., "What Are the Causes of Change and Innovation in Safety?" Loss Prevention Symposium, Heidelberg, Germany, September 1977.

Kletz, T. A., *An Engineer's View of Human Error*, 3rd ed. Institution of Chemical Engineers, Rugby, UK, 1992.

Kletz, T. A., *By Accident*, Chapter 10. PFV Publications, London, 2000.

Lees, F. P., *J. Loss Prev. Process Ind.* Butterworth–Heinemann, UK, 1980, 1996.

Tweeddale, H. M., "Improving Process Safety and Loss Control Software in Existing Process Plants." IChemE Symposium Series No. 94, 269–278, May 1985.

Tweeddale, H. M., "Balancing Quantitative and Non-Quantitative Risk Assessment." *Trans. IChemE* **70 (Part B)**, 70–74 (1990).

Tweeddale, H. M., "Nourishing and Poisoning a 'Safety Culture.'" *Trans. IChemE* **79 (Part B)**, 167–173 (2001).

Chapter 12

Role of the Risk and Reliability Manager

When you have studied this chapter, you will have a preliminary understanding of the role of a Risk and Reliability Manager, and of the way in which such a manager can start to approach his or her responsibilities. This has a major impact on how the elements of Figure 1-4 are approached, and the success achieved with them.

12.1 ELEMENTS OF MANAGEMENT

Although there is wide agreement that management can be described as "getting things done through people," as outlined in Section 10.2 of Chapter 10 there are many ways of defining the elements of management.

However, the following are commonly regarded as essential elements:

- planning,
- motivating,
- organizing, and
- controlling.

These are all interrelated and are briefly outlined below in relation to management of process plant risk and reliability.

12.1.1 Planning

The first requirement is a policy about process safety and reliability. Once this is available, the next step is to define targets for each. In the case of process safety, they may be in terms of risks to employees and the public, risks to property, or risks to the environment. In the case of reliability, they may be in terms of plant availability, or total cost of maintenance plus loss of profit due to unavailability, etc.

(There is much loose talk about safety, e.g., "Safety is our number one objective!" No organization is established with the principal purpose of being safe. Few people would invest money in such an organization. They are looking for something else from the organization. It may be that they wish to make a profit, or to provide a service. But in achieving the agreed aims of the organization, they do not wish to be unsafe. Thus they wish to operate in the region of safety, rather than "unsafety." That is, safety or "low risk" defines boundary conditions for their operations, not the objective of their operations.)

Then it is necessary to understand the present situation. For example, where do we have policies, and where are they in need of definition? Where have we defined targets, and where do we need to define performance targets? Where is performance consistent with the policy and the performance targets, and where is it failing?

Then the tasks needed to bring the present situation to a standard that complies with the policies and the performance targets are defined.

12.1.2 Motivating

This depends greatly on whether the risk and reliability manager is a "manager" or a "consultant." This is discussed in Section 12.2 below. See also Chapter 11.

12.1.3 Organizing

This entails assembling information about the resources needed to complete the required tasks, the human and material resources available, and the priorities of the various tasks. This can be aided by use of a form of rapid ranking (see Chapter 3).

This may be assembled into a ranked list of tasks or projects, each with defined requirements for resources. Then a decision must be made about the rate of progress needed. It is rarely possible to undertake all projects at once, or in any one financial year, so it is likely that the projects tackled in any one year will be limited by the human or monetary resources available. On the other hand, it may be that the urgency of tasks on the list calls for review of the resources available, and provision of additional people or monetary resources to the work program.

Although resources can be loosely classified as human, financial, etc., the human resources (at least) need to be structured into project teams, with leaders, specialists, support people, etc. Each project needs a defined objective, responsible project manager, staffing, cost limits, reporting requirements, etc.

A documented work plan, perhaps in the form of a Gantt chart, displaying all projects with start and finish dates, and the responsible people, summarizes the way the risk and reliability program is organized and provides a solid foundation for management of progress against cost and timetable requirement.

12.1.4 Controlling

As discussed in Chapter 10, the elements of control are:

- definition of a desired benchmark or standard level of performance;
- monitoring or measuring the actual performance;
- comparing the actual performance with the standard; and
- taking action to bring the actual performance into line with the standard.

12.2 AUTHORITY AND RESPONSIBILITY FOR PERFORMANCE

Authority is the power to decide and to act. It is the power to get things done. *Responsibility* is the liability to be called to account.

The authority and the responsibility of any position must match. Granting authority without responsibility is foolishness, as the person with the authority can do as he or she wishes. Imposing responsibility without authority is cruelty, by holding someone to account for something that they did not have the power to influence. (An ancient form of this was the practice of executing the bearer of bad news.)

It is important to understand "line" and "functional" authority and responsibility as they relate to performance in process safety and reliability.

A *line manager* is one who has the authority to authorize expenditure and to give instructions to people working in an area. Plant managers and production managers are examples of line managers. They have the authority to set budgets and define production plans. They are therefore liable to be required to account for their expenditure and for achievement of the production plans.

The risk of major incident (either of process safety or reliability) is only managed effectively if the primary responsibility is that of the line managers, as they and their staff at all levels are the people who have the authority to determine what is done on their plants and in their operations.

A *risk and reliability manager* may have one of two possible roles depending on the management structure and the defined authorities and responsibilities of the position. These are:

- adviser to line managers, responsible only for providing good advice when asked, but without being called to account if a plant does not perform well, or
- manager, sharing with the line managers some of the responsibility for the performance.

A simple test to determine the nature of the authority of the risk and reliability manager is to ask: "If there is an incident, although the line manager is the primary person to be asked to explain how it happened and to be held responsible, will the risk and reliability manager be regarded as partly responsible for allowing the conditions that led to the incident?"

Typical questions could be:

- "Did you know of the possibility of this incident occurring?"
- "If not, why not?"
- "If you did, why weren't you having something done about it?"

If those questions could be asked, then the risk and reliability manager has a responsibility for the "function" of risk and reliability performance. Therefore he or she has the matching shared authority to arrange for preventive action to be taken, that is, the authority not just to give advice, but also to:

(a) monitor performance to determine the need to give advice; and

(b) press for it to be adopted.

In other words, he or she is not just a consultant, but a "manager," that is, someone who "gets things done through people." A risk and reliability manager is usually a *functional manager*, that is, someone who "gets things done through line managers."

If a risk and reliability manager feels that those questions would not be asked, then he or she is not responsible, and are performing a passive advisory or consulting role, and not an active managing role. In such circumstances, the choice is either:

- to move into the role of active manager, with authority, and accept the matching responsibility; or
- to accept the passive role of simply offering advice that may be ignored.

It is possible for a risk and reliability manager to act as a consultant in a passive role, waiting to be consulted, and going away when the line manager has satisfied himself.

It is also important for line managers to understand that role the risk and reliability manager has.

12.3 SOME MANAGEMENT SITUATIONS AND TACTICS

12.3.1 The "Law of the Situation"

One of the early management theorists described a case where a laborer was taken to an open patch of ground by a supervisor and told to dig a trench in a particular direction. After digging for an hour, the laborer was told to go to a different spot, and dig in a different direction. After another hour, he was told to go to a third spot, and start again. When this was repeated a fourth time, he objected and refused. The supervisor then told him to be reasonable, saying that there was a drain pipe somewhere in the area that was not shown on any plans, and that digging was the only way to locate it. The laborer said "I wish you had explained that at the start," and recommenced digging.

From this the management theorist propounded the deceptively simple "Law of the Situation." In brief, people are usually prepared to do what is required by the situation, no matter how inconvenient, although they may be most reluctant to do the same thing if simply told to.

This principle can be very useful in developing or enlarging the role of a risk and reliability manager. It is common, when a new position is being defined, for the responsibilities to be set out in more detail than the authorities that go with them. This need not deter the incumbent. Given a definition of the responsibilities, it is relatively simple to define what needs to be done (in broad terms) to be able to carry them out. Then, the incumbent is empowered by the situation to get on with it.

For example, one approach is for the risk and reliability manager to take the initiative in going to line managers, seeking approval to:

- become familiar with the plant and operations,
- review the plant and operations to identify the nature and magnitude of any current or potential problems that are in his or her field, and
- to discuss these perceived problems with the line managers to agree on the correct courses of action, and on a program for implementation.

It is important to ask for approval before starting to look into an area controlled by a line manager, but the situation does not allow acceptance of "no" for an answer!

If there is opposition, or if the line manager asks the risk and reliability manager to stop the investigation before it has been properly completed, then the response determines whether the risk and reliability manager is going to be a manager with authority and responsibility, or a passive consultant. Someone who is going to be an active manager will find a way to get permission to continue, perhaps by explaining the situation—the responsibility that he or she carries for the "function" of safety or occupational health with its matching authority—and if that shows signs of failing, invoking the authority of more senior management (ideally in a way that allows the production manager a graceful way of changing his or her mind).

12.3.2 Establishing Influence in "No Man's Land"

In starting up a new position, or developing an existing one, there are always gray areas where the authority is disputed or unclear.

This is analogous to the land between opposing military forces. It is usual for that land to be patrolled by both forces, with skirmishes from time to time. If one side stands firm in the face of the opposing forces, it effectively establishes its control over that land, and part of the "no man's land" has been taken over.

For example, a newly appointed risk and reliability manager was in a large meeting with a project engineer and other staff working on a project. The safety manager queried a design feature that had a clear weakness. After discussion, the project engineer, who had agreed the feature with other engineers earlier and was presumably reluctant to have to lose face and ask for a change, asserted: "As the project engineer, I am responsible for safety performance and I say it is OK!" The risk and reliability manager replied: "Because of the creation of the position of Risk and Reliability Manager, the rules have changed, and the situation requires that I, too, be satisfied which, for reasons I have explained earlier, I am not!"

The others at the meeting watched this exchange with interest. The debate was not continued, and the meeting moved on to less contentious topics. The Risk and Reliability Manager thought it would be wise, in view of the rising tempo of the debate, to allow time for the project engineer to consider the situation. On the next day, at another review meeting, the project engineer said that he had held further discussions with the design engineers, and they had found a better design. The perceived authority of the Risk and Reliability Manager was increased as a result.

12.3.3 Resolution of Disagreement

The management theorist mentioned earlier suggested three means of resolving disagreement:

- *domination*, in which the winning party is gratified, and the loser dissatisfied, sowing the seed for further disagreement;
- *compromise*, in which neither party is fully satisfied, sowing the seed for further disagreement and renewed negotiation; and
- *integration*, or creatively finding a solution that fully satisfies all parties.

For example, a major extension was planned for a plant handling a potentially unstable and explosive material. The proposed type of extension had a bad history of explosions in plants around the world. The extension had to be located near the existing plant. The only available location was also very close to vulnerable stock tanks holding a liquefied toxic gas. There was a strong commercial motive for building the extension, and the line management was strongly in favor of going ahead. The safety manager said that location was unsuitable because of the severity of the consequences of rupture of the stock tanks and the credibility of the postulated initiating explosion (without any formal quantitative risk assessment being necessary). For that reason, he would have to object to the extension being built there. Gradually the discussions headed toward confrontation. Then, in the course of a planning meeting chaired by a senior manager, the need to insist on the highest safety standards when there was so pressing a commercial need was queried by a line manager. The senior manager emphasized the company safety policy and arranged for technical staff to make a further overseas trip, at some expense, to explore the safety issue further. The result was realization that all the plants that had exploded had a particular design feature, whereas those of an alternative approach had never exploded, for clear and definable reasons. Thus it was possible both to build the extension, and to maintain the safety of the stock tanks of toxic gas. This is an example of integration. In this instance it is difficult to envisage a form of compromise that would be technically sensible, while domination would have been possible at a cost—either commercial cost or increased risk.

An important feature of the approach was that the safety manager gave early notice of his concerns, the reasons for them, and the position in which he was placed, so that those involved were not confronted by a sudden nasty surprise. Thus there was no personal antagonism, and the process of deciding the correct course of action was taken rationally and calmly.

12.3.4 Confrontation

The only graceful way out of confrontation, for both parties, is not to get into it!

In establishing a new position, or developing an existing one, in the process risk, safety, or occupational health fields, there will be times when confrontation looms. The choices are as follows:

- Find a solution that satisfies all parties, as described above.
- Avoid it by backing off, thus surrendering a piece of no man's land to the other party.
- Walk around it in the pursuit of a higher priority objective, leaving it to be tackled at a later date when it is more convenient or when success is more likely.
- Make the safety position clear, stand firm, and give the other party the opportunity to decide whether to back off, or to head into the confrontation.

Tactically, it is important to go into a confrontation only if the situation really requires success (even at the cost of antagonizing the other party and possibly reducing subsequent cooperation), and preferably if there is a reasonable prospect of being successful.

In the case quoted above (the hazardous plant extension) the safety manager had clearly explained his concern very early in the project, and that he regarded the hazard as so serious that, in the event of it being decided that the project would proceed anyway, the situation in which he was placed by his responsibilities would require him to take his objections progressively to the highest level in the organization. (Such a step should be kept as an ultimate "reserve power" if at all possible, as invoking it is equivalent to admitting failure to resolve problems oneself.) By openly adopting that position, the involvement of the general manager and the positive solution adopted were facilitated.

12.3.5 Sources of Information

A common problem faced by any person in a functional role, such as Risk and Reliability Manager, is not being told about situations in which a contribution could be made. This is mostly entirely unintentional; the line manager responsible does not recognize the opportunity for the Risk and Reliability Manager to contribute.

It is very desirable for functional managers to cultivate informal lines of communication. Such an informal line of communication could be simply joining the production or engineering staff in the cafeteria at lunchtime. Quite apart from the general social conversation, one can gain an appreciation of the problems being tackled at the time, giving the opportunity to make an innocent approach to someone at another time, asking about a problem, and offering to help. After doing this a few times, line managers start to think of the safety or occupational health manager as soon as appropriate situations arise and approach them on their own initiative.

However, note that the perceived level in the organization that people ascribe to the safety or occupational health manager is determined as much by who he or she has lunch with, as by the position on the organization chart. Therefore, one should choose one's lunch table with care!

12.4 LINES OF INQUIRY FOR A RISK AND RELIABILITY MANAGER

A Risk and Reliability Manager is constantly taking stock of the performance of the operation compared with:

- the performance standards and targets defined by statutory authorities, the community, or the organization itself; and
- some absolute standards, as an indication of the potential for improvement of the normal performance standards.

12.4.1 Risk and Reliability Manager of a New Site

A person appointed to be Risk and Reliability Manager of a new site or facility has some special opportunities but also some special problems.

The special opportunities may include:

- being able to sample the state of the local community and environment before operations start (this is valuable as a benchmark for comparison with the situation later);
- being able to influence the locations of plants within the site, and the design of the plants, from the outset, to meet community concerns;
- being able to establish rapport with the local community to hear their concerns before the design is fixed.

Special problems may include:

- staff being preoccupied with project work of design and construction, and thus being reluctant to get involved in what they regard as a task for operations staff once the plants are in operation;
- being appointed too late to have any substantial influence on layout and plant design.

Note that determining the state of the surrounding community and physical environment before starting operations may necessitate careful statistical and analytical design, so that subsequent comparisons are valid. In undertaking this initial study, it is important first to identify the nature of the external impacts that are theoretically possible, as a result of the nature of the materials to be used and the processes involved. Then the study can be designed with assistance from the appropriate specialists, such that both the initial results have meaning, and subsequent changes can be properly sampled, analyzed, and compared with the initial data. This may necessitate assistance from a statistician, design engineers, a specialist analytical chemist, acoustic specialist, botanists, ecologists, etc.

12.4.2 Risk and Reliability Manager of an Existing Site

A person appointed to the position of Risk and Reliability Manager of an existing site is in a different situation from a manager on a new site. Much may have been done already. Therefore an initial review could comprise a quick overview of each of the six requirements defined by Hawksley (see Chapter 10) to determine the "state of play."

A person appointed to be Risk and Reliability Manager of an existing site may have a number of advantages and disadvantages. The possible advantages include:

- there may already be a recognition of the need to do better, and a commitment to doing so;
- there may already be resources available; and
- there may be systems already in place, for example, for auditing plant condition, sampling and analyzing waste streams, etc.

The possible disadvantages include:

- there may be a feeling that existing practice is satisfactory and that the appointment is a political gesture by "the head office";
- the existing systems may be seriously defective, and changing an existing system may be more difficult than implementing a new one; and
- the climate and culture within the organization may have become fixed, with a degree of complacency and a resistance to change.

Having conducted that initial review of the plants and the surrounding community attitudes and concerns, the task will then be to list the deficiencies, or the opportunities for improvement, to identify the means available for making improvements, to prioritize them, and to win support within the organization to undertake them. After that, it is a matter of project management and following the steps set out in Chapter 1.

12.5 DEALING WITH THE PUBLIC

12.5.1 Introduction

There is a still a tendency for the technical aspects of risk to be emphasized and the importance of community perception to be overlooked.

The following section introduces principles, rather than attempting to set out how to handle the full range of situations that can arise.

Any individual's view of the correct way of interacting with the community will be markedly influenced by a number of factors:

- the individual's view of how democracy should work;
- the individual's view of the community's "right to know";
- the individual's faith in the power of logical argument and technical facts; and
- the individual's experiences in dealing with the community.

These are discussed briefly below.

12.5.2 Different Understandings of How Society Works

One view of democracy is that members of the public periodically elect their representatives to form a government and permit themselves to be governed by them until the next election. If they are dissatisfied with the performance of their elected representatives, they accept their right to make decisions until the next election, at which time they replace them with others. A contrasting view is that every citizen has the right to make those decisions that affect himself or herself.

An intermediate view is that, between elections, members of the public should make their views known to their elected representatives, but abide by their decisions. A variation on that is that, if the elected representatives disagree with the expressed views of a community group, that group should escalate its opposition until the elected representatives give way.

There are also differing views of how the regulatory authorities operate or should operate. Some believe that they exist to implement the decisions of the elected representatives. Some believe that the regulatory bodies are primarily motivated by a desire to protect the public. Others believe that they are the real (unelected) government with their own "agenda" and plans, that the elected representatives are manipulated by them and act to legitimize the views of the regulatory bodies, and that the safety and interests of the public are relatively unimportant to the regulatory bodies.

In practice, all those views exist in the community, and there is evidence to support each of them.

The implications for organizations seeking to operate within the community are discussed later.

12.5.3 The Community's "Right to Know"

There is increasing legislation enshrining certain rights of the public to demand access to information. There is increasing expectation by the public, partly stimulated by the legislation, of being informed about anything that will affect them. There is pressure from community groups for the community's right to know to be enshrined in legislation.

Regardless of one's views of the public's right to know, the public has increasing expectations. Any organization operating in the public domain that ignores that expectation must expect difficulties arising from distrust.

12.5.4 The Inadequacy of Logical Argument and Technical Facts

From the Renaissance onward, there has been optimism in the wider community about the benefits of greater knowledge. There was enthusiasm for the discoveries of science and the achievements of engineering. The prevailing view was optimism—that humankind was advancing toward some sort of "golden age."

This is no longer the case. Possibly starting with Hiroshima, there is an awareness that there is a downside to scientific discoveries and engineering skills. The prevailing view is of pessimism (that we are wrecking the world with pollution and nuclear energy, that we are running out of fuel, etc.). And for many people, it is science and technology that are to be blamed.

There is disillusionment with the integrity of scientists (e.g., scientific fraud), and distrust of engineers (e.g., great engineering work at the expense of the environment). There is little faith in "experts," who have often been found fallible and who are constantly disagreeing in the press or as witnesses in court cases.

Logical developments have led to problems (possibly because of faults in the logic, or unsound "facts," or insufficient information. Technical "facts" are now seen as little more than technical opinions. (And often that is all they are and can be!) The public believes that the age of the irresponsible scientist and the rapacious philistine engineer is with us. So while we must be logical ourselves, and base our work as much as possible on established facts, we cannot rely on "logic" or technical "facts" to convince the public.

Two important relationships are:

1. Risk (technological) = potential consequence (of the occurrence of the hazard) × likelihood
2. Risk (sociological) = potential consequence + outrage (after Sandman, 1991)

This means that the community response to a risk is determined primarily by:

- the magnitude of the hazard (i.e., the possible consequences) rather than the magnitude of the risk (i.e., the consequences weighted by the frequency);
- the emotional stress or outrage associated with the hazardous facility.

The public is not convinced by assessments of likelihood. They are well aware of the fallibility of "experts." If it *can* happen, they feel it *will* happen.

The emotional stress component is often more important than the magnitude of the hazard. Factors influencing the magnitude of the emotional stress factor include:

- unfamiliarity of the nature of the hazard;
- mysteriousness of the hazard, such as difficulty of detection (e.g., nuclear radiation);
- involuntary exposure;
- untrustworthy sources of information, or managers of the risk;
- secrecy, reluctance to give information; etc.

Further, managers of companies are often perceived by the public as:

- patronizing,
- manipulating,
- "control freaks,"
- deceitful, and
- arrogant.

All of these generate, or add to, outrage. Therefore it is critical that corporate representatives and spokespeople, in their dealings with the public, be aware of these perceptions and conduct themselves in a way that does not reinforce those perceptions, and in fact counters them.

It is important to note the plus sign (+) in the second formula. It shows that the sociologically viewed risk can be severe even if the facility is physically incapable of generating incidents with a significant impact.

It is difficult to lower the emotional stress once it has been raised. It is of little use attempting to lower it by rationally explaining the nature of the risk. It is much preferable to prevent the level being raised, by addressing the emotional stress factor from the very outset of the project.

12.6 THE PRECAUTIONARY PRINCIPLE

12.6.1 The Principle

In 1992, The Rio Declaration on Environment and Development, Principle 15, set out the Precautionary Principle in a widely accepted form as it applies to environmental issues. It states:

> In order to protect the environment, the precautionary approach shall be widely applied by States according to their capabilities. Where there are threats of serious or irreversible damage, lack of full scientific certainty shall not be used as a reason for postponing cost-effective measures to prevent environmental degradation.

Applied to process plant risks, this may be understood as:

"In order to protect potentially exposed people, property and environment, the precautionary approach shall be widely applied by operating organizations according to their capabilities. Where there are threats of serious impact, measures for prevention, protection or mitigation of such impact should not be postponed because the reality of the threat is not technically proven."

However, the understanding of the precautionary principle in the minds of the public is not fixed, and it is the public that ultimately determines what is to be accepted. A common public view of the precautionary principle is of the form below:

"Unless it can be proved with certainty that a proposed new facility or activity is absolutely safe, it should not be undertaken. Similarly, any existing facility or activity that cannot be proved with certainty to be absolutely safe should be shut down."

In other words, in the minds of the some members of the public, the Precautionary Principle, which states, "If there is a risk, *do* take precautions," tends to become, "If there is any risk, *don't* undertake the activity."

This is, of course, inconsistent with allowing aircraft to fly over cities, or allowing towns to be built downstream of dams. It is inconsistent with the reality of risk of death being a fact of life. It is totally inconsistent with progress of any kind.

12.6.2 The Precautionary Effort

The extent of the effort to prevent or mitigate potential impacts, as seen by the public, can be derived from relationship 2 in Section 12.5.4:

Precautionary effort required = (envisaged impact in event of threat being real × perceived credibility of threat) + outrage

This formula has clear implications for management:

- It is important that the public understand the true nature of any threats, rather than relying on sensationalized versions propagated by the uninformed.
- It is important that the public understand how such threats could be realized, and the precautions taken to prevent occurrence. Their perception of the credibility of the threat will be greatly influenced by their view of the credibility of the organization responsible for the threatening facility or activity.
- But the most important factor is *outrage*.

All three factors—envisaged impact, perceived credibility and outrage—are interrelated and are determined by the manner in which the organization interacts with the public.

12.6.3 Dealing with the Public

In dealing with the public, it is important to have realistic expectations.

- Members of the public actively disbelieve political spokesmen of the party they did not vote for, and distrust political spokesmen of their preferred party.

- Members of the public distrust statutory bodies and believe them to be in league with "developers," "big corporations," and other perceived self-interested groups.
- The public does not trust industry, or any government utility, such as power generation, railways, port authority, etc., to do what is in the public interest.
- The community is sometimes told by specialist action groups: "You can't trust business and industry as they are ruled by short-term gain; you can't trust government because it is driven by its own desire for power and is in league with business and industry; but you can trust us!"
- The public appears to trust the integrity (e.g., impartiality) of the legal system, but to have reservations about its capability. Although people are prepared to trust their own legal advisers, there is widespread distrust in lawyers as a group, and in technical circles there is growing dissatisfaction with the adversarial system as a means of resolving technical disputes.
- En masse, the public is not always logical and rational in its decision making, partly because they lack a decision-making structure, and partly because people, whether members of the public or managers of an organization, are human and subject to both rational and irrational thinking.
- The public perceives (often correctly) that managers are not always logical and rational in their decisions or official viewpoints as they affect the public. They perceive that managers sometimes rationalize that their proposals are in the public interest, or at least are not to the public detriment, when their objectivity is clouded by commercial interests.
- Community groups often comprise people with a wide range of objectives and motivations, including:

  seeking betterment of the community;

  advancing the objectives of some other community group (e.g., a specialist action group may seek to advance its objectives by joining a community advancement group);

  advancing individual political objectives;

  advancing a perceived party-political objective; or

  getting personal publicity and attention.

No one likes being presented with a "fait accompli." Yet industry and government often delay opportunity for public comment until plans have reached the "glossy expensive public report stage," when it is obvious that the proponents are heavily committed to the proposal and would be very reluctant to make any significant change. True communication is not possible in such circumstances; all that is happening is information exchange without any readiness to think about what is received. The situation degenerates to confrontation.

It is difficult, for many reasons, to win over a hostile community group. It is much preferable to forestall formation of such groups by preventing situations arising of the kind that generate hostile community groups. The key to this is open communication, and communication of a kind that earns and creates trust. This requires:

- communication that can be understood (i.e., no jargon);
- communication that is found to be accurate;
- communication that does not conceal; and
- a "vulnerability," or openness and readiness to admit uncertainty, ignorance, and error.

The public cannot be expected to start this process.

The various styles of interaction with the public are illustrated by Arnstein's "Ladder," of which a modified form is shown in Figure 12-1.

Neither end of the ladder is ideal. At the bottom, the public is ignored and becomes hostile. ("We are going to build an oil refinery here, regardless of what you think.") At the top, the organization allows the public to determine the organization's future, which may be totally impractical and

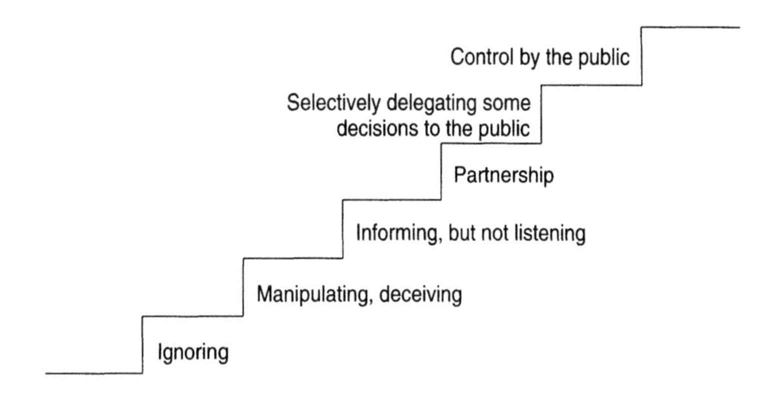

Figure 12-1. Heirarchy of control.

uneconomic. ("We have this large amount of capital to invest. What do you want us to invest it in? An oil refinery? A recreation facility and child care center? A hospital and a retirement home?")

It has been suggested that the approach with the greatest chance of success is to use an approach that is marginally higher up the ladder than the public has been accustomed to. If it is too low on the ladder, the public is offended by being given less involvement than they expect. If it is too high, they are (rightly) suspicious.

12.6.4 Handling Queries and Complaints from the Public

A query or even a complaint from the public can be a very positive opportunity to "build bridges" with the community. There are many simple principles:

- Be courteous (including cultivating a friendly manner of answering the telephone—the first contact often determines the nature of the continuing discussion, and you do not know when you pick up the handset whether the caller is a colleague, a member of the public, a regulator, etc.).
- Listen carefully, without interrupting.
- Be as open as possible in replying, and if literature is requested, etc., ensure that it is sent promptly.
- If the caller has the wrong extension, don't give the person another number to call; say you will get the correct person to call back (and make sure that that happens).
- If practicable, go around personally to see a person who complained, both to investigate the nature of the complaint more thoroughly than is possible by phone, and also to show that you, as the representative of the organization, care about the complainant and take the complaint seriously. Often a well-treated complainant becomes a strong supporter in the community.
- If the organization has made a mistake, or if the complaint is the result of any known fault of the organization, admit it. Discuss how it happened, what is being done to restore the situation, and to prevent recurrence.
- If the complaint needs investigating, or if the full cause, etc., of the incident is not yet clear, explain what is currently known and offer to get in touch when the inquiries have been completed (then do so!).
- In discussing incidents, outline what happened, but do not discuss responsibilities of individuals, etc.

Note that although there may be some legal restriction on discussing incidents, failure to discuss them can lead to the very situation that the legal advisers are trying to avoid.

12.6.5 Relevance for a Specific Organization

Where some activity of a specific organization may be expected to impinge on the community, it is important that the organization take the initiative to build communication with the community.

Where risks are involved, the organization should discuss the risks in a simple way, comparing them with risks that the community accepts without concern (not road accidents, which are correctly seen as a high risk), and without reliance on very small decimal numbers.

The organization would need to accept that low assessed risks do not ensure safety, because of the residual risk, but particularly because of the possibility of future changes that invalidate the risk assessment.

The organization may benefit from consideration of the principles behind the Responsible Care program of the chemical industry:

- defining standards for operation in areas of greatest relevance to the public (by working with public groups to define the areas of interest, and to define those standards);
- preparing and implementing plans to reach those standards (again in discussion with public groups);
- publicizing the standards, plans, and progress in achieving the objectives.

REFERENCES

Rio Declaration on Environment and Development, Principle 15. 31 ILM 874, June 14, 1992.

Sandman, P. N., "Risk Communication." Australian Institute of Petroleum Seminar, Sydney, Australia, 5 December 1991.

Chapter 13

Lessons from Incidents

When you have studied this chapter, you will understand:

- some of the typical causes of, and contributory factors to, serious incidents on process plants (including failures of both hardware and software); and
- some of the early warning signs to be watched out for on operating plants.

This chapter provides practical detail to illustrate the principles set out in Chapters 9, 10, and 11.

(The incidents described in the first four case studies were very influential in leading to development of regulations aimed at improving the safety of installations capable of generating serious off-site impacts.)

13.1 THE VAPOR CLOUD EXPLOSION AT FLIXBOROUGH, UK, 1 JUNE 1974

The Nypro site at Flixborough, UK, had a number of plants, producing materials used in the manufacture of nylon. One of the plants, the cyclohexane oxidation unit, had six reactors, arranged as in Figure 13-1. The reactors operated at elevated temperature and pressure. Cracks were

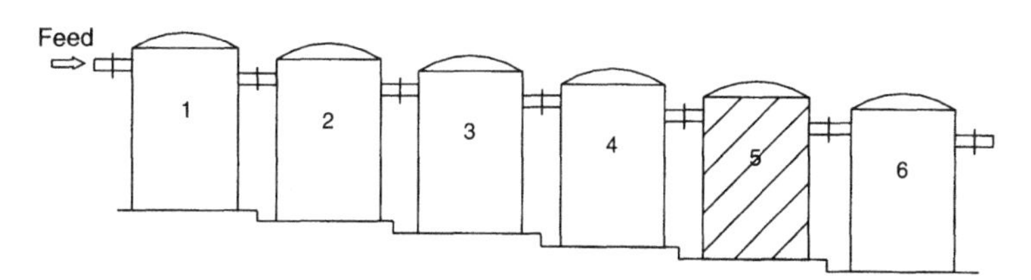

Figure 13-1. General arrangement of reactors.

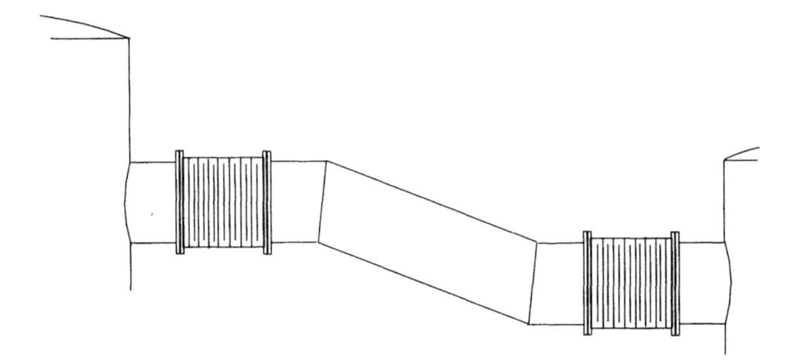

Figure 13-2. Configuration of the "dog-leg" pipe.

found in the fifth reactor, and it was decided to take it out of service for repair. To enable production to be continued during the repair, it was decided to bridge the gap with a pipe. The loss of performance due to using only five reactors was small compared with the loss of all production if the gap were not bridged. (The physical properties and hazards of cyclohexane are similar to those of refined gasoline.)

Reportedly the workshop foreman sketched out a design for a pipe, and it was installed as shown in Figure 13-2. It had a "dog-leg" bend at each end to enable it to fit, and a bellows at each end to cope with thermal expansion. It was supported on scaffolding.

The plant was started up and ran for some days, even though it was noticed that the bellows at each end were distorted for some reason not apparent to the people who saw it. (For the explanation, see Figure 13-3.) On a Saturday afternoon (1 June 1974), when the plant was largely unattended except for the laboratory staff, a few shift operators, and an unusually large number of people (26) in the control room, a bellows ruptured, and the pipe fell to the ground, allowing an estimated 50 tonnes of cyclohexane to escape. Because the cyclohexane within the plant was

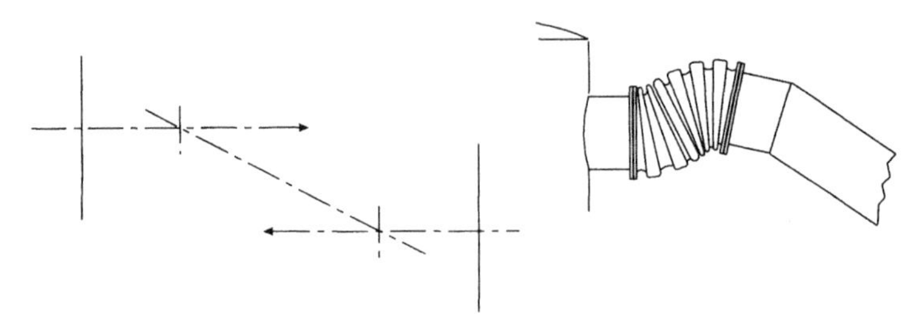

Figure 13-3. Rotational forces on the temporary pipe.

at a temperature above its boiling point at atmospheric pressure, but was held in liquid form by the high pressure, a high proportion flashed instantaneously into vapor as it escaped. After around a minute, the vapor ignited and exploded. The explosion destroyed much of the plant, including the control room, and started numerous fires as other flammable liquids leaked. All 26 people in the room were killed, as were the driver of a tanker at the plant and an operator with him. Extensive damage was done to houses at a distance of around 400 meters, and shop windows were broken some kilometers away. The local community was severely shocked and community life dislocated, and outrage was expressed throughout the country.

In the extensive inquiry that followed, the responsibility was not directed at the workshop foreman who designed the dog-leg pipe, and who did not understand the problems caused by the forces imposed on the pipe and the bellows by the shape of the pipe, but on management, who had not seen fit to replace the works engineer who had resigned many months before.

The workshop foreman did not know that he lacked the knowledge to design the pipe, and it was not to be expected that a person in that position would have the necessary knowledge. It was held that management should have recognized the importance of high engineering standards and appointed an experienced engineer whose training would equip him to recognize the need for proper design practice.

This example illustrates many points, including the following:

- the need to look for other process routes to a product;
- the need to minimize the inherent hazards in processes, by using different process materials, lower temperatures and pressures, lower inventories, etc.;
- the need to maintain a high standard of engineering at such plants, and the importance of professionally qualified staff in so doing;
- the importance of not making any modifications to the plant, hardware or software, without systematically checking for side effects;
- the importance of plant design and layout, especially control room location and design;
- the need for organizations that enter partnerships to run plants to decide which of them is responsible for the safety policy and standards; and
- the vigor of public response to a disaster, and the impact on the organization responsible and other organizations.

It was also noted at the inquiry that no one at the plant had thought to check whether any of the other reactors had cracks, when No. 5 reactor was taken off line. Inspection after the explosion revealed that the other reactors were, in fact, affected in the same way. If the pipe had not failed, it is quite possible that one of the reactors would have done so at a later time.

This accident led the UK government to plan controls on the location and safety of "hazardous" industry, a move that was overtaken by European Community initiatives following the Seveso release (see later).

(In the official inquiry into the accident, a so-called "two-event" theory was postulated, in which a small fire from a leaking flange led to weakening and rupture of a larger pipe, which led to accumulation of flammable vapor, which exploded, rupturing the dog-leg pipe. At the time, it was completely discounted by the inquiry, which held that for it to be correct, too many extremely low probability events would have had to occur. However, the initiating "trigger" for the Flixborough explosion is still debated, and it is possible that it was a smaller fire such as was postulated by the two-event theory.)

It was notable that, at the time of the explosion, the factory had a very good occupational safety record; another illustration of the unsuitability of lost-time injury rate as an indicator of the level of process safety. (See also Section 13.11.)

13.2 THE REACTOR RELEASE AT SEVESO, ITALY, 10 JULY 1976

A batch reactor was being used to manufacture the herbicide 245T (2,4,5-trichlorophenol) from 1,2,4,5-tetrachlorobenzene and caustic soda in the presence of ethylene glycol. In the reaction, TCDD dioxin is normally formed in minute quantities, but the concentration increases as the temperature (and hence speed) of the reaction increases. The reactor was heated with exhaust steam from a turbine, the temperature of the steam normally being too low to overheat the reaction.

In this instance, on 10 July 1976, the reaction accelerated, overheated, and ruptured the bursting disc. There was no catchpot to collect the discharge, and it sprayed over the countryside, contacting many people and contaminating a large area. There were no fatalities, but about 250 people contracted chloracne, a severe skin disease, and a large area of land was rendered unusable.

The circumstances leading to the reactor overheating were as follows:

- The reactor was heated by external steam coils, supplied by exhaust steam from a turbine.
- The reactor was shut down at the end of the reaction, but before the ethylene glycol had been removed. (This was because an Italian law required the plant to be shut down for the weekend.)
- The plant was not normally shut down at this stage of the cycle.
- The steam was reportedly isolated, but residual heat remained.
- The stirrer was turned off.
- The temperature of the wall of the reactor above the liquid level increased, leading to a slow exothermic reaction starting in the upper region of the liquid.
- After around 7 hours, a runaway reaction occurred.
- Because the runaway reaction had raised the batch to a high temperature, the dioxin level was very high compared with normal levels.
- There was no catchpot to collect the discharge from the bursting disc.

Lessons to be learned include the following:

- Modifications to normal operating procedures can introduce unexpected hazards.
- Beware of residual heat after completion of normal operations, as it can lead to overheating of materials and then to runaway reactions, auto-ignition, etc. (See also Section 13.7.)
- Where a batch of material can be released from a plant, a catchpot or some other means of preventing it from reaching the outside or the environment should be considered.
- There is increased danger of hazardous shortcuts being taken just before the end of the workday.

There was a major public outcry, leading to the European community issuing an edict that is now popularly titled the "Seveso Directive," requiring member countries to pass legislation requiring organizations operating inherently hazardous facilities, etc., to demonstrate their safety. This directive, and the legislation developed in member countries of the European community in compliance with the directive, made extensive use of the principles being developed by the UK legislation following the Flixborough explosion.

13.3 TOXIC GAS ESCAPE AT BHOPAL, INDIA, 3 DECEMBER 1984

On 3 December 1984, around 25 tonnes of methyl isocyanate (MIC) was released from a storage tank at a pesticide manufacturing plant at Bhopal, India, operated by Union Carbide India Limited. The vapor spread beyond the factory boundary to an adjacent shanty town, killing several thousand people and injuring many thousands more, a substantial proportion permanently. This was the worst chemical plant accident in history.

MIC is a very volatile liquid, boiling at around 30°C. The vapor is extremely toxic, having at the time a TLV[11] of 0.02 ppm.

It is relevant that the plant was not profitable, as the market was less than had been expected. There had been a high turnover of staff. There was a history of smaller leaks from the plant in recent years. An audit of the plant in 1982 by Union Carbide Corporation staff had expressed a number of important safety concerns.

The accident began when a stock tank of MIC, produced at the plant as an intermediate material in the manufacture of carbaryl insecticide, became contaminated with water. (Whether this was an operational error or a deliberate act of a disgruntled employee is still debated. The investigation sponsored by the Indian government held that it was design and operational error in which water being used to flush out a pipeline was able to reach the stock tank, whereas the investigation sponsored by Union Carbide Corporation concluded that that scenario was physically impossible and contradicted by observation, and that the most likely cause was an action by a disgruntled employee[2]—see Kalelkar, 1988). But it is agreed that entry of water led to the start of a runaway exothermic reaction. The refrigeration system intended to cool the tank was shut down at the time. The scrubbing plant that should have absorbed the vapor released from the relief valves was not in service at the time, having been shut down some months before as an economy measure, as the plant

---

1[1]TLV: The airborne concentration of a substance that most workers could be exposed to day after day without adverse effects on their health.

[2]In investigating the cause of a serious incident, sabotage is commonly suggested by the responsible organization (and seized upon by the media) and evidence found to support the theory. This may arise for a number of reasons, for example: (a) the operating organization cannot believe that their operation could produce such an incident without it being deliberately initiated; and (b) sabotage can be held to be beyond the powers and hence responsibility of management to prevent. Therefore an investigator or an observer should be cautious before accepting the sabotage theory. However, sabotage does sometimes occur, and in this case hard evidence was reportedly found to disprove the alternative theory and to support the sabotage theory.

was not profitable, the market being less than predicted. The flare system, which was the second line of defense after the scrubber, was out of service for maintenance. (But, in any case, it is reported that the scrubbing plant and the flare system were not sized to cope with such a rapid generation of vapor.)

The Union Carbide team members were prevented by officials of the Indian government from having access to many of the people who had been closely involved, and to critical documents, until long after the event. This hampered their investigation, and necessarily introduces an extra element of doubt about the findings of each of the investigations.

The lessons from this disaster include:

- Try to use processes that do not require hazardous raw or intermediate materials. (There was at the time another process for making carbaryl from the same raw materials that did not use MIC as an intermediate step.)
- Separate hazardous plant and sensitive land uses, such as residential areas or high-density population centers.
- Take care in the design of the plant, and in the Hazop studies, to identify the possible consequences and causes of contamination, and design and operate to minimize the opportunity for that to happen.
- Design the plant with the capability of coping with the identified events as well as designing and operating to maintain the reliability of the prevention and protection systems.
- Keep all equipment in good working order. Provide redundancy in protective systems in critical duty, and inspect, maintain, and promptly repair any failures, so as to avoid circumstances where all the protective barriers are out of service at the same time.

Kletz (2001) lists other lessons. See also Lees (1996).

The worldwide public outrage over this disaster affected the entire chemical industry, not just the company involved. A number of major international companies started or increased process safety surveillance of operations by their subsidiary companies in other countries, regardless of their previous attitudes toward local autonomy.

13.4 EXPLOSION AND FIRE AT PHILLIPS 66 PLANT, PASADENA, TEXAS, 23 OCTOBER 1989

At around 1:00 P.M. on 23 October 1989 there was a release at a polyethylene plant operated by Phillips Petroleum at Pasadena, Texas. This formed a flammable vapor cloud that exploded within a couple of minutes

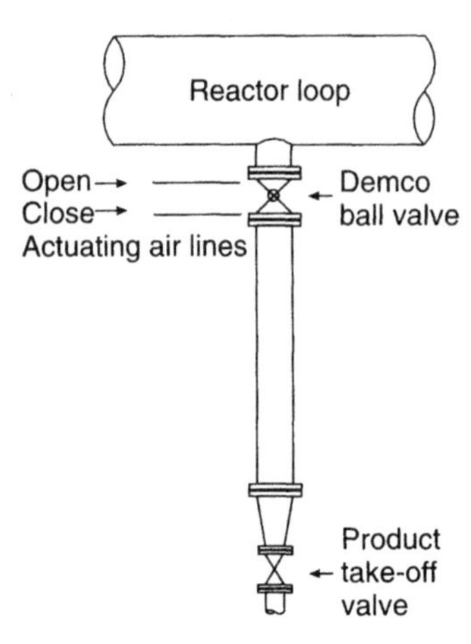

Figure 13-4. Simplified diagram of the setting leg of the reactor.

of the escape starting. There were several subsequent explosions and fires. The explosions and fires killed 23 people, injured more than 100 more, and caused extensive plant damage.

The process entailed injecting ethylene into a recirculating stream of isobutane, the catalyst carrier, operating at a pressure of around 5000 kPag and high temperature. As the ethylene polymerized, the particles settled out into settling legs and were removed from the process and separated from the catalyst carrier. See Figure 13-4.

On the day of the explosion, workmen were clearing plugs of polyethylene that had accumulated in one of the settling legs. The normal procedure was to isolate the air-actuated DEMCO valve and then to disconnect the air lines.

The maintenance team had partly disassembled the leg, and had cleared some of the plug, when the escape occurred. The reactor vented through the leg and, because of its elevated temperature, flashed instantly into vapor, forming a very large cloud that ignited between 1 and 2 minutes from the start of the escape and exploded with an effect estimated to have been equivalent to between 2.5 and 10 tons of TNT. The explosion, and the fires that resulted, then led in turn to two other explosions; two isobutane tanks of 20,000 US gallons and a nearby polyethylene reactor that failed catastrophically. It damaged the incident command center, disrupting telephone communications, sheared fire hydrants from their fittings, and disrupted the power supply to the electrically powered fire pumps.

It was found in the investigation that the DEMCO valve had been open at the time. The air lines to it had been crossed over, such that when it was actuated to close, it opened. Although it was a corporate rule (and recognized industry practice) that isolation for maintenance required either two isolation valves to be closed as a "double block," or positive isolation by fitting of a slip plate, this was not the custom at this plant. The plant did not have a dedicated firewater system, relying on use of process water for that purpose. One of the three diesel fire pumps was in maintenance at the time, and one of the remaining two quickly ran out of fuel.

The OSHA report (OSHA, 1990) lists a number of deficiencies in safety management at the plant. They include the following:

- failure to undertake a hazard analysis;
- inadequate consideration of separation distances between plants, etc.;
- poor location of the control room;
- failure to minimize populations of people in the more hazardous areas;
- lack of planning for escape routes;
- lack of any fixed flammable gas detectors;
- inadequate control of ignition sources;
- inadequately controlled maintenance work permit system;
- inadequate procedures for isolation for maintenance;
- inadequate integrity of the firewater system;
- unreliable fire pump condition;
- failure to follow-up on safety audits; and
- inadequate planning for emergencies.

See also OSHA (1990) and Lees (1996).

13.5 CONTAMINATION OF THE RHINE BY A WAREHOUSE FIRE, SWITZERLAND

A warehouse fire at Basel, Switzerland, on 1 November 1986, led to pesticides and herbicides being washed in spent firefighting water into the Rhine. As a result, the water life (fish, animals, etc.) and stock along the river were poisoned, in some cases as far away as the North Sea.

Because some villages along the bank of the river, throughout Germany, rely on wells extensively for water, the wells being supplied partly by groundwater that had been contaminated by the river, it was necessary to bring water by truck to those villages for months.

The lessons from this accident include:

- avoid warehouse fires (see Chapter 8), and
- provide means of collecting spent firewater at chemical plants and warehouses.

The response to this fire was very great throughout Europe, especially Switzerland, Germany, and the Netherlands, which were the countries directly affected. The company constructed massive catchment basins for spent firefighting water. New warehouses elsewhere are often required to provide similar catchment basins, and existing warehouses are having them retrofitted. This can be done by bunding the warehouses, or by damming up the lower parts of a site, and providing means of isolating or diverting the stormwater drains.

13.6 EXPLOSION OF A PRINTING INK TANK

A tank exploded at a printing ink factory early in the morning. The lid of the tank was thrown through the air, landing around 30 meters away, and narrowly missed employees arriving for work. The contents of the tank caught fire, but the fire was extinguished promptly with foam. See Figure 13-5.

The tank was around 4.5 meters diameter and 12 meters high. It was nearly empty, with the depth of the ink being around 1 meter. The roof was fitted with a 50-mm vent.

Printers ink is thixotropic, that is, it sets to a gel unless it is continually worked. The contents of the tank were continuously recycled by the

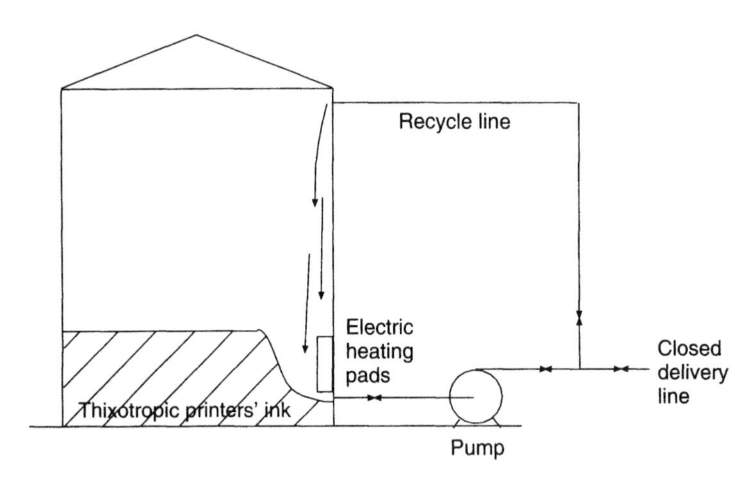

Figure 13-5. Configuration of tank and pumping circuit.

15-kW delivery pump. To keep the ink mobile in the vicinity of the pump suction, there were three electric heaters (two of 3000 W, one of 750 W) attached to the outside of the wall of the tank near the base. The configuration of the tank is illustrated in Figure 3-5. The recycled ink ran down the wall directly over the surface heated by the 750-W heater.

There appeared to have been an overheating of the 750-W heater and its supply cables where they were attached to the outer surface of the tank directly above the heater, and damage to its thermostat bellows, which was permanently distorted into the open position; but the other heaters appeared in good condition. They were set to control between 70 and 90°C. The heating coils were not attached directly to the tank, but to a stainless steel backing plate that was in turn attached to the tank.

At first it was suspected that overheating of the surface of the tank by loss of temperature control of the 750-W heater had led to vaporization and auto-ignition.

After investigation, it was recognized that the recirculation pipeline entered the tank directly over the suction line, and that the bulk of the ink was not mixed by the recirculation, but that a very small volume was being rapidly recycled. It is now suspected that this ink was absorbing the friction heat of the pump and being heated sufficiently to generate vapor and auto-ignite. A simple temperature-rise calculation confirmed the adequacy of the friction energy to heat the small recirculating volume of the ink to auto-ignition temperature.

There were ash deposits around the recycle return to the tank, consistent with previous local overheating and combustion of this stream before the explosion.

It was learned during the investigation that there had been several "pops" at this tank and others in the past, but that they had not ruptured the tanks, and so were not rigorously investigated.

A number of lessons can be drawn from this accident. Options for preventive measures include the following:

- Investigate unusual occurrences, so that any lessons can be learned and action taken before a serious accident occurs.
- Take care with recycling a pump delivery line to its suction. Ensure that there is sufficient cooling.

Options for protective measures include the following:

- Use a nitrogen blanket.
- Put a high-temperature sensor in the pump suction line.

13.7 EXPLOSION OF A "SLOPS OIL" TANK

An old riveted oil slops tank at a marine oil terminal exploded. The roof ruptured along a riveted seam, and two contractors working on the roof, installing new handrails, suffered extensive burns.

At the start of work that day (Friday), the contractors were issued with a hot work clearance after a flammable gas test was done around the tank vents. No further gas tests were done. The contractors' supervisor gave final instructions for completion of the job and left. The work would finish that afternoon, after which the contractors could go home. The weather was sunny, with a gentle breeze in the vicinity of the tank. Additional slops oil was pumped to the tank during the morning.

The tank exploded around midday. The tank roof peeled open along a riveted seam. There was a burst of flame, then a brief fire that extinguished itself after a few seconds. The two contract employees were burned, but managed to escape from the tank via the ladder, which was damaged but not destroyed.

The contract employees were confused about what they were doing and where they were immediately before the explosion. There are several possible scenarios:

- The tank, warming up with the heat from the sun, began to release vapor from the vents, and this vapor was ignited by welding or a spark from the welding equipment.
- Heat from welding on the roof of the tank locally raised the internal temperature to above the auto-ignition temperature for the vapor or the oily residue on the inside of the tank to ignite.
- Sparks from the welding entered the tank through a gap in the riveted plates, or an uncovered vent, igniting a flammable vapor mixture inside. See Figure 13-6.

Lessons to be learned include the following:

- Where hot work is to be done in the vicinity of flammables, there need to be frequent tests of the atmosphere.
- Contractors need careful training about hazards and safeguards.
- Contract work needs careful supervision.
- Many accidents occur near the end of work, when there is a temptation to take shortcuts.

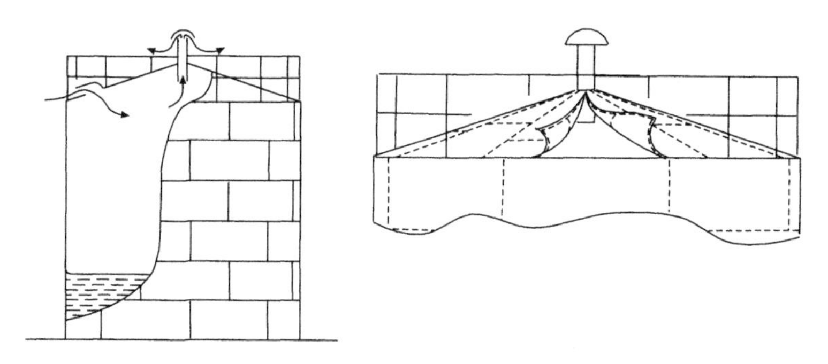

Figure 13-6. Slops oil tank.

13.8 REPEATED EXPLOSION OF A VARNISH KETTLE

A company manufactured varnishes and other surface coatings in a "kettle" (see Figure 13-7). The kettle was heated by direct gas firing in a firebox. There was an agitator, driven by an electric motor. The process was purely one of formulation, mixing resins and solvents; there were no chemical reactions involved. The solvents had a flash point of around 180°C, and an auto-ignition temperature of around 280°C.

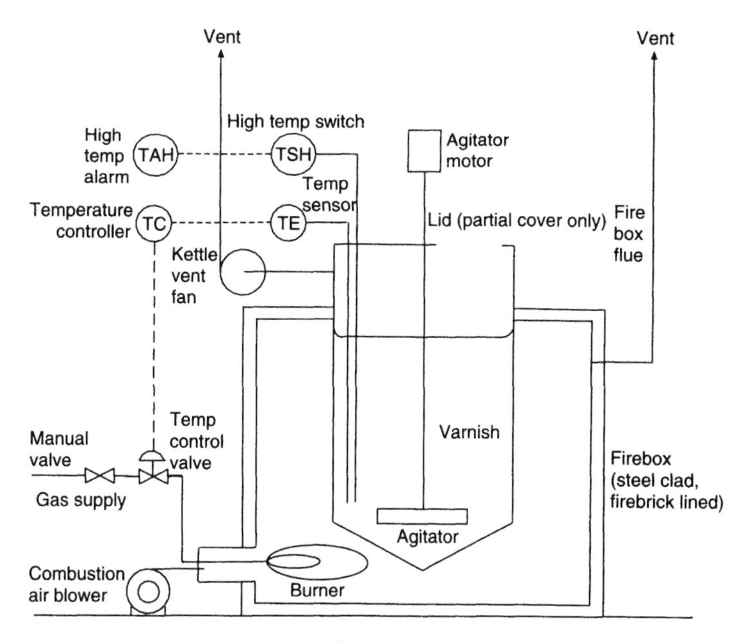

Figure 13-7. Resin plant.

The process entailed heating a charge of solvent to around 150°C, then adding resins and allowing them to dissolve over a period of around 3 hours. Further solvents and other ingredients were then added, and the gas heating turned off. After a further hour, the contents were then pumped to a holding tank.

One Friday afternoon, at 4.30 P.M., an operator turned off the gas and immediately started to pump the completed batch to the holding tank. That reportedly took about 5 minutes. The operator then left the building. About 10 minutes after the pumping was finished, there was an explosion in the kettle, which was almost empty at the time, having only around 50 mm of varnish in the bottom.

When other people in the vicinity came to the building they found the kettle, with its lid open, burning vigorously. They activated the piped CO_2 extinguishing system piped directly into the kettle and extinguished the fire. They then went up the stairs to the kettle platform and considered what to do. After around 10 more minutes, the kettle exploded again, resulting in one person suffering superficial burns to exposed skin.

The explanation was that, because the contents had been pumped away before the firebox was allowed to cool down, the residual heat of the refractory lining was sufficient to heat the varnish on the inside surfaces of the kettle to above the flash point and the auto-ignition point, resulting in explosion of the vapor in the kettle. After extinguishing the following fire, the CO_2 was removed by the vent fan, allowing reignition of vapor in dead areas (probably the bottom) of the kettle, resulting in the second explosion.

Lessons from this incident include the following:

- Direct firing, with its inherently high temperatures, introduces the potential for over-temperature, which other forms of heating can avoid.
- Beware of residual heat.
- The temptation to take procedural shortcuts near the end of work can introduce serious dangers.
- Modifying (or departing from) the approved operating procedures should be done only after careful study as part of a modification control procedure.
- Operators should be trained to understand the reasons behind the operating instructions, rather than just to follow them blindly. This will make them less likely unwittingly to take unsafe shortcuts.

13.9 EXPLOSION OF A "FUEL OIL" TANK

A storage tank for by-product oil, used as fuel oil for a boiler, was fitted with a breather pipe at the top, to allow air to leave or enter as the level rose or fell. It is possible that the breather was initially fitted with wire gauze to act as a flame trap, and to prevent birds from entering.

The tank was mounted on a level concrete base, on sloping ground. A small wall was constructed to catch any small overflow, but it was not a full bund.

The general arrangement was probably as shown in Figure 13-8, but this was never positively determined, as it had been modified some years before the explosion, and no records were available of the modification.

The precise sequence of subsequent events is not clear. However, at some past time, an extension had been fitted to the breather, presumably to prevent overflows from dirtying the side of the tank. It is likely that the gauze was removed at this time, to prevent the vent from blocking, which would lead to tank rupture, or being sucked in.

The tank was operated satisfactorily for years in this configuration.

One day the tank exploded, separating from the base around the bottom weld and being projected more than 100 meters. The oil in the tank ignited and flowed downhill. Several people were killed.

The reconstruction of the event is as follows. It was a crisp, still morning. The sun was warming the tank. The air and vapor in the tank (noting that the oil was a by-product from distillation and appears to have contained some light fractions) was expanding and was expelled down the breather into the catchment area between the tank and the wall, where it accumulated. There was a shutdown at the time. Many work permits were being issued. Supervisors regarded the work permit as a troublesome paperwork exercise. Flame cutting was being undertaken by contractors in the area close to the tank (which should not have been permitted by the work permit or the

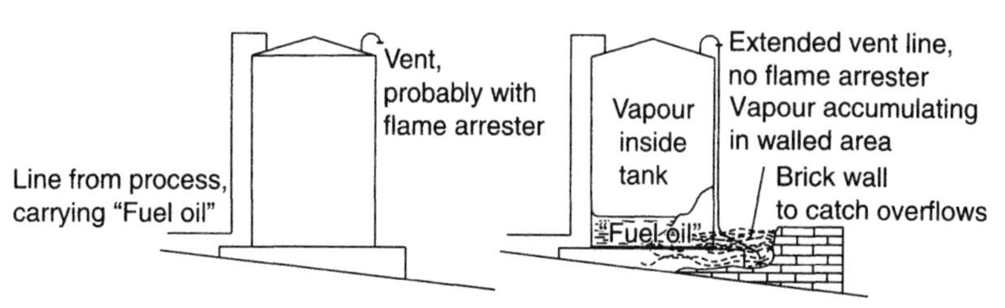

Figure 13-8. Fuel oil tank.

maintenance supervisors). Sparks from the flame cutting ignited this accumulated vapor/air mixture, which flashed back up the vent into the tank.

Lessons from this incident include:

- Modifications to plant should not be made without careful study as a part of a modification control procedure.
- A maintenance work permit procedure is critically important to safety and needs constant retraining, supervision, and monitoring.
- Contractors need safety training and careful supervision.

13.10 FIRE ON THE *PIPER ALPHA* OFFSHORE OIL AND GAS PLATFORM

On 6 July 1988, the *Piper Alpha* offshore oil and gas platform in the North Sea, standing in water around 75 meters deep about 100 km off the coast of Scotland, caught fire. The fire destroyed the platform, killing 167 people. The public inquiry held under the leadership of Lord Cullen, a Scottish judge, with three technical assessors to assist him, led to a greatly increased understanding of the hazards of such structures, and of what is required to manage them safely. This led to major changes in design philosophy, and in legislation.

The platform existed to separate oil, condensate (mostly propane), and gas from wells in the sea floor, and to pump the three phases to an onshore processing plant. It also acted as the collecting point for product from other platforms, so that the consolidated products could be pumped ashore together.

The general layout of the platform was as shown in Figure 13-9.

The sequence of events was determined to have been as follows:

1. In one of the two processing modules, there were two condensate export pumps, one operating, one on standby. The delivery pressure was around 7000 kPag. The standby pump had been electrically isolated that morning in preparation for a major overhaul, but not yet physically isolated with slip plates, etc. No work was done on that day. At around 9.45 P.M., the operating pump tripped and could not be restarted. So the operators decided to recommission the standby pump, by reconnecting the electrical supply, then remotely opening the valves. They did so.

At the time the operators did not know that a separate work permit had been issued for replacement of a relief valve on the condensate line. The valve had been removed, and because of the lateness of the time, the

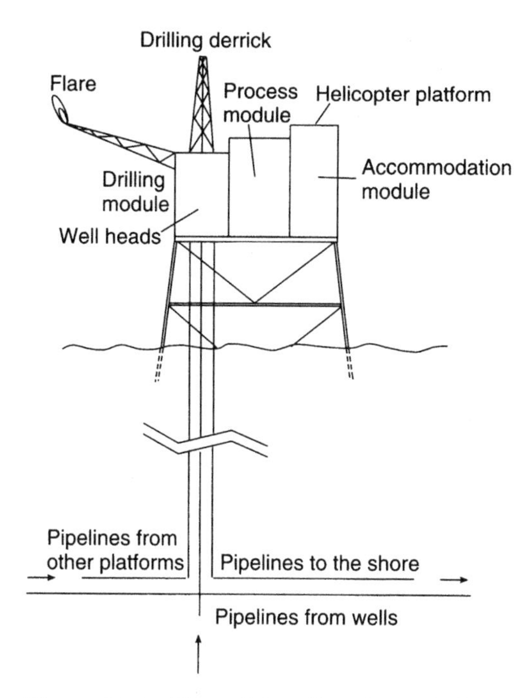

Figure 13-9. *Piper Alpha* oil and gas platform.

contractors planned to fit the replacement on the next day. The work permit was returned to the supervisor's desk, but was not discussed with him as he was very busy. So, at the time the condensate pump was restarted, there was an open nozzle on the condensate line, which allowed propane to escape. (There was no cross referencing of work permits. The contract supervisor had never been trained in the work permit system, nor had most of the process staff. This weakness in the operation of the work permit system was found not to be an isolated incident, but the normal standard.)

The floor-level gas alarms from that section came on several times.

At just after 10.00 P.M., there was a gas explosion at that section.

The explosion killed perhaps two people. It probably damaged some of the firefighting water mains, and the electrical supply system to parts of the processing section.

2. The gas explosion caused damage to the firewalls around the condensate module and started leaks in the heavy oil module alongside. This led to a large oil fire.

Although the potential for fire had been considered when designing the platform, no consideration had been given to the possibility of explosion.

The fire continued for a very long time, long after the oil inventory on the platform would have been consumed.

Although the other platforms supplying oil to *Piper Alpha* knew that there was a serious fire, they continued to pump oil, as they assumed that *Piper Alpha* would be able to control the fire. All communication equipment on *Piper Alpha* was damaged in the initial explosion.

The emergency diesel firefighting pumps (intended for automatic startup if firewater were needed and the electrically driven pumps did not start) failed to start, so no cooling was available to the exposed plant.

Because of the risk of divers working underwater around the platform being sucked into the pump suction lines, the platform manager had instructed that all pumps were to be switched to manual start whenever a diver was in the water, not just when a diver was near the pump suctions. The weakness of this ruling had been listed in a safety audit report long before, but not acted on. As divers were in the water for 12 hours per day in summer, that meant that auto start was bypassed for 50% of the time.

Because of the use of salt water for firefighting, the water pipes had corroded and tended to block the spray nozzles. Flushing the pipes had been tried, but this increased corrosion. Larger nozzles had been tried, without success. It had been decided to replace the pipework with corrosion-resistant pipework, and this had been done on one module. But this was all that had been done in the 4 years since the problem had been identified.

3. The oil fire weakened the main gas pipelines (typically 400–500 mm NB, and around 14,000 kPag) which came up from the seafloor through the center of the processing module (in the middle of the oil fire) from wells and other platforms. A gas pipeline ruptured. The resulting fireball enveloped the platform. The continuing fire could not be isolated.

The possibility of such an escalation had been identified by a young engineer 12 months earlier when doing a study of the case for continuing to keep the standby vessel alongside the platform. He had noted in the report that if such a fire occurred, the standby fire ship would be useless, and there would be immense loss of life. During a management discussion of his report, that section was not raised. The pipework could have been fire-insulated (to buy time), or provided with a deluge.

4. People arrived and left the platform normally via the helicopter platform on the top of the accommodation module. In the fire, people fought their way to the module, where the management was. The module had been designed with some fire resistance, but no toxic gas resistance.

Most of the people who died in the accommodation module did so from carbon monoxide poisoning, as the doors were being constantly opened and windows broken.

Although it would have been evident that no helicopter could land, no command was given to evacuate the module and jump into the sea. Presumably the managers were shocked and unable to think logically in the circumstances. (Such circumstances are those for which emergency plans are designed and training should be undertaken, so that there is a conditioned response, rather than reliance on quick and calm rational thought.) The eventuality had not been considered, and no relevant emergency plans prepared or training had been given.

Some people did leave the module and jumped into the sea. It is not known how many tried, but 28 did so and survived. (The accommodation module collapsed and fell into the sea, being recovered in due course, enabling the situation during the fire to be determined.)

Numerous weaknesses and faults can be identified when studying this disaster. They include:

- missing management systems, or poorly operated management systems;
- lack of commitment to safety (e.g., very slow attention to firewater pipe corrosion);
- poor quality audits and little follow-up (although numerous audits were conducted);
- inadequate training; and
- a reluctance by management to face up to the possibility of the "worst credible accident" and take preventive and protective precautions.
- a belief by management that, because they never heard of any safety problems, everything was satisfactory. (There are always problems with safety. No news is bad news!)

(A comprehensive list of the lessons is set out by IChemE, 1990.)

13.11 BLEVE AT AN LPG RAIL TANKER UNLOADING BAY

On 17 August 1987, a 22-tonne railcar of LPG was involved in a BLEVE at a terminal in Cairns, Australia. One person was killed; he had refused to leave his nearby home, where he was hosing the wall of his house to prevent it from catching fire.

At around 3 P.M., a tanker arrived at the terminal. The operator connected the liquid discharge hose and the vapor return hose to the tanker. At around 3:20 P.M., very soon after the valves were opened, there was a major failure of either a hose or a coupling, etc., resulting in a large LPG leak. The tanker internal excess flow valve failed to operate. The operator fell to the ground from the tanker ladder and was slightly injured. He, and a plant gasfitter, ran to the office to alert the supervisor.

The vapor cloud drifted to nearby housing and was ignited. The flame flashed back to the tanker, and an intense fire was established.

Attempts to make use of a firewater monitor and hose reel were unsuccessful owing to the intensity of the heat. The monitor local valve was closed and the remote valve open, so it was not possible to actuate the monitor remotely.

At 3:26 P.M., the fire brigade arrived, and a limited flow of water was applied to cool the tanker.

At 3:27 P.M., the tanker relief valve operated, and the escaping gas ignited.

At 3:31 P.M., the tanker ruptured, and a BLEVE fireball erupted, with a diameter of around 100 meters.

One-half of the tanker was projected around 110 meters. Fragments were projected around 300 meters.

Twenty-eight people were injured, and one killed. Five of the burn victims were seriously affected, three of them firemen.

Lessons include:

- the importance of preventive maintenance;
- the importance of testing protective systems, although excess flow valves are not easy to test in service;
- the importance of keeping emergency systems (e.g., firewater monitor systems) in their designed condition;
- the importance of fixed deluge systems on LPG tanker bays (recommended by the coroner);
- the importance of insisting on the highest standards of design, operation, maintenance, supervision, etc., with hazardous materials, of which LPG is one.

(For further information, see Hopkins, 1987.)

13.12 VESSEL RUPTURE AND FIRE AT A GAS PROCESSING PLANT

13.12.1 Introduction

An explosion and fire occurred on 25 September 1998, at the Esso plant at Longford, Australia, which was processing liquid and gaseous hydrocarbons from offshore platforms in Bass Strait. As a result, two people were killed, several were injured, and supply of gas to the city of Melbourne ceased for several weeks, causing massive disruption to industry, businesses, and homes.

In the year leading up to the incident described here, the plant had an excellent occupational safety record, with no lost-time injuries at all.

The postulated sequence of events is fully described in the report of the investigation by Dawson and Brooks (1999). What follows is a very brief summary. It should be noted that there is still speculation about some of the details of the scenario.

13.12.2 Description of the Incident

At Longford, Esso operated three plants in parallel (GP1, GP2, and GP3), all performing broadly similar functions. The incident described here occurred on GP1, the oldest.

The heavier hydrocarbons are separated by gravity and removed before the gas reaches the section of plant shown in Figures 13-10 and 13-11.

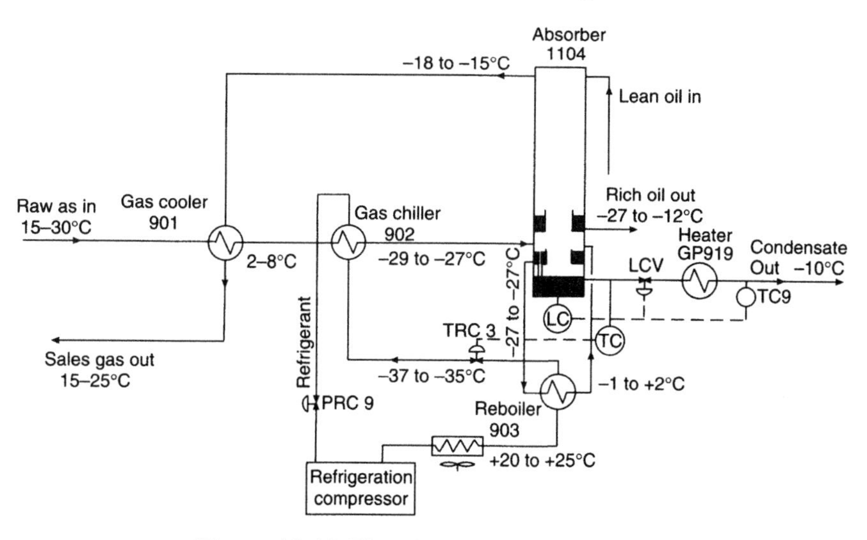

Figure 13-10. Flowsheet of feed to absorber.

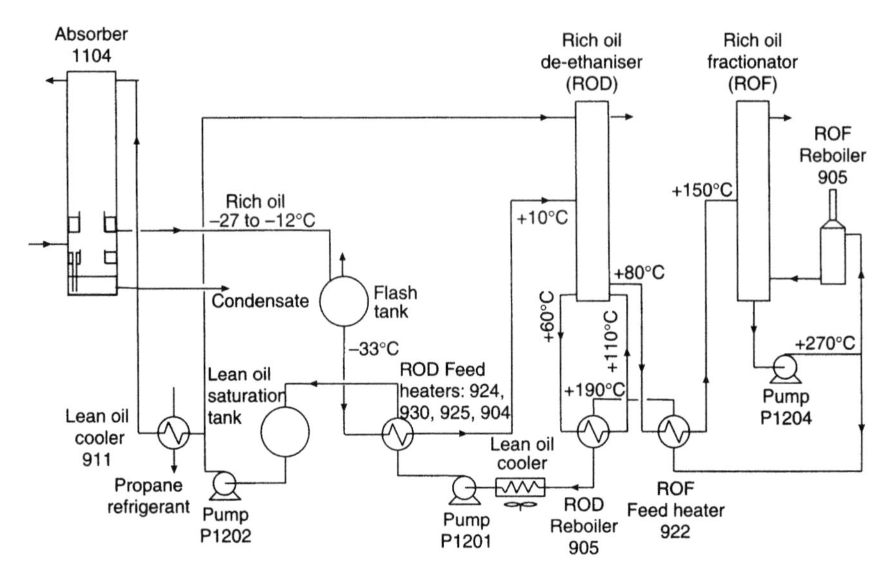

Figure 13-11. Flowsheet of fractionation section.

The mixed gases are chilled and sent to the absorber, where the condensate (mainly LPG) collects in the bottom, the sales gas exits from the top, and other constituents are absorbed by a countercurrent flow of chilled "lean oil." The resulting "rich oil" (i.e., lean oil plus absorbed hydrocarbons) flows via a flash tank through a number of heat exchangers, where it is heated by a counterflow of hot lean oil, to the rich oil demethanizer (ROD).

The rich oil in the ROD reboiler is heated by a counterflow of lean oil and is normally at about 80–100°C. The rich oil then flows to the rich oil fractionator (ROF). The now denuded rich oil, that is, lean oil, is heated by a gas-fired heater to around 270°C and returns to the absorber via the heat exchangers used to heat the rich oil.

Around the time of the incident, the gas from Bass Strait had a much higher content of condensate than normal. The ambient temperatures were low, and there were constant problems with the temperature control valve TRC3. This led to the condensate temperature falling to –20°C rather than its normal level of –10°C. This low temperature, combined with the unusually high flow rate of condensate, overloaded heater GP919 and caused TC9 to override the condensate level controller and throttle the flow of condensate through the heater. The upshot of these and other control problems was that the level of condensate in the absorber rose substantially, allowing condensate to pass out in the rich oil stream.

The excessive quantities of condensate entering the ROD resulted in greatly increased vapor flow up the ROD column and appears to have contributed to flooding and excess liquid flow out of the vapor overheads of the column. There are possibly other contributing factors, but these were not determined with certainty because of damage to the ROD column in the subsequent fire, and uncertainty about some of the actions by operators attempting to restore control of the plant.

The level started to rise in the Lean Oil Saturation Tank. This led its level control system to reduce the inflow of lean oil to the minimum. Pump 1201 tripped, possibly because of low flow.

The ROF feed heat exchanger began to leak.

When Pump 1201 tripped, the level in the Lean Oil Saturation Tank fell, causing Pump 1202 to trip on low level. When the level built up again, the pump was restarted and apparently continued to run in spite of a subsequent low level, possibly because of loose wiring. The pump became overheated and was shut down manually.

There were repeated attempts to restart Pumps 1201 over the next few hours. Then ice was noticed on the pumps, showing the presence of condensate, possibly due to a passing nonreturn valve on one of the 1201 pumps.

Flow continued from the absorber to the ROD for some time, until the gas supply was finally cut off. The temperature of the Rich Oil Flash Tank dropped to around –42°C when the pumps stopped, then fell toward the atmospheric-pressure boiling point of the liquid (–48°C).

This low temperature did not cause failure. It was subsequently found that additional stress was needed to cause failure.

The ROD reboiler 905 now had condensate at a very low temperature on the tube side with low flow, and cold lean oil (possibly with vapor) at around the same temperature on the shell side, with no flow.

To start Pumps 1201 and 1202 it was necessary for Pumps 1204 to be running, so they were started. Hot lean oil entered ROD reboiler 905, and it ruptured violently due to thermal shock to the embrittled steel combined with the repressuring, with a large subsequent fire.

Two people were killed at that time and others injured. The fire was at a location where it damaged pipework and other equipment common not just to Plant 1, but to Plants 2 and 3, causing all plants to be put out of action for several weeks.

13.12.3 Some Features of the Situation Leading to the Incident

- Condensate concentrations in the feed to the plant were unusually high (possibly beyond the design limits of the plant).
- There was acceptance of operating the plant when it was well beyond its design limits, that is, with the condensate level well above its alarm setpoint, without a thorough review of the safety of such a practice.
- The incident would not have occurred if it were recognized that the ROD (and hence the entire Gas Plant 1) must not operate in the absence of flow of lean oil because of the danger of chilling.
- If it had been recognized, efforts to restart pumps GP1201 would have been redoubled, and if not soon successful, GP1 would have been shut down.
- Hazop would have been expected to identify the danger of operating with low or zero flow of lean oil, and hence of chilling ROD.
- Once it was decided to persevere with attempts to restore control to the process without restoring lean oil flow, control difficulties were inevitably going to worsen, and some form of major process deviation, and undefined hazardous situations, became much more unlikely.
- ROD and GP905 were not designed for feasible abnormal conditions of chilling, for example, due to depressuring.

13.12.4 Possible Contributory Factors

- The design of the plant led to ongoing difficulties and a reputation as being very difficult to control.
- There was a failure to recognize the potential for problems (of various kinds) with a plant that is difficult to control, and hence failure to tackle the control problem as such.
- There was acceptance that the condensate level in the absorbers could be permitted to be above the maximum recordable level (loose standards of plant control? Indication of it normally being difficult to control the plant so major deviations were "normal"?)
- Some chart recorders were not operational, possibly delaying recognition of a control problem.
- A previous problem with chilling was not recognized as an example of a warning of a possible serious problem related to difficulty of control, and a threat to the structures of the chilled equipment.

- The absence of management staff on leave, on courses, or at the head office led to difficulty by operators in interpreting the signs, understanding the problem, and hence in recognizing the correct action.
- Following the hydrate incident earlier, there was pressure on Esso to maintain production, leading to reluctance by the operators and the supervisors (particularly in the absence of technically qualified managers) to shut down when the situation became clearly complex and not understood.

13.12.5 Summary of Findings of the Royal Commission of Enquiry

EXXON requires its member companies, including Esso Australia Ltd, to implement its comprehensive Operational Integrity Management System (OIMS). This is part of an EXXON-wide system (Exxon Company International Upstream Guidelines—ECIUG). OIMS sets out 11 elements and nominates an "owner" for each element. The Royal Commission found that there were many shortcomings in the application of OIMS at Longford, in a range of fields such as:

- training, where the main emphasis was on avoiding loss of production and maintaining the specification of the product;
- risk identification, analysis, and management, such as where a planned Hazop study of GP1 was repeatedly deferred year by year;
- management of change, where changes in plant design were not subjected to rigorous safety review;
- quality of supervision, which had allowed sloppy practices to become normal operating procedure;
- review of the implementation and review of the requirements of OIMS, which had not detected the many failures:

  documentation,

  data collection and analysis, particularly failure to use the recorder charts and electronic logs, and

  communications, particularly relating to use of the plant log and verbal communication at shift changeover.

The Commission also commented on the effect of relocation of the plant engineers to Melbourne, and the consequent loss of close informal communication and firsthand observation of the performance of the plant.

For details of these failures, see Royal Commission (1999). For other analyses, see Hopkins (2000) and Kletz (2001).

REFERENCES

Dawson, D. M., and Brooks, B. J., "The Esso Longford Gas Plant Explosion—Report of the Longford Royal Commission." Government Printer for the State of Victoria, June 1999.

Hopkins, A., *Lessons from Longford: The Esso Gas Plant Explosion.* CCH, Australia, 2000.

Hopkins, J. McE.,: "LPG Rail Tank Car Accident Report." Queensland Department of Mines, Gas Operations, 1987.

IChemE, Piper Alpha—*Lessons for Life-Cycle Safety Management.* Institution of Chemical Engineers, Symposium Series No. 122, 1990.

Kalelkar, A. S., "Investigation of Large-Magnitude Releases: Bhopal as a Case Study." Proc. IChemE Conf. Preventing Major Accidents, London, May 1988.

Kletz, T. A., *Lessons from Accidents*, 3rd ed. Butterworth–Heinemann, Oxford, UK, 2001.

Lees, F. P., *Loss Prevention in the Process Industries*, 2nd ed., Appendix 5. Butterworth–Heinemann, Oxford, UK, 1996.

OSHA, "Phillips 66 Company Houston Channel Complex Explosion and Fire." Occupational Safety and Health Administration, Washington, DC, 1990.

Chapter 14

Case Studies and Worked Examples

Notes

1. Most of the case studies are based on real cases in a number of countries, but they have been changed sufficiently to avoid identification of the particular facilities, and have been simplified to clarify the points being illustrated.
2. The worked examples are also simplified to illustrate the methods more clearly.

When you have studied this chapter, you will have a broader understanding of some of the practical applications of the philosophies and methods of risk and reliability management, and a better understanding of some of the applications of the calculation methods covered in earlier chapters.

14.1 CASE STUDY: LPG FACILITY RISK

14.1.1 Introduction

A provincial town had, in the past, been supplied with gas for domestic purposes from a coal gasification plant. Some years ago it had been closed down and replaced with a facility that used LPG. The LPG was delivered to the site by road tankers and stored on site. LPG was piped to an evaporator, and the vapor produced was diluted slightly with air to a level still above the upper flammable limit, but with combustion characteristics similar to the old coal gas. The LPG vapor/air mixture was stored in the

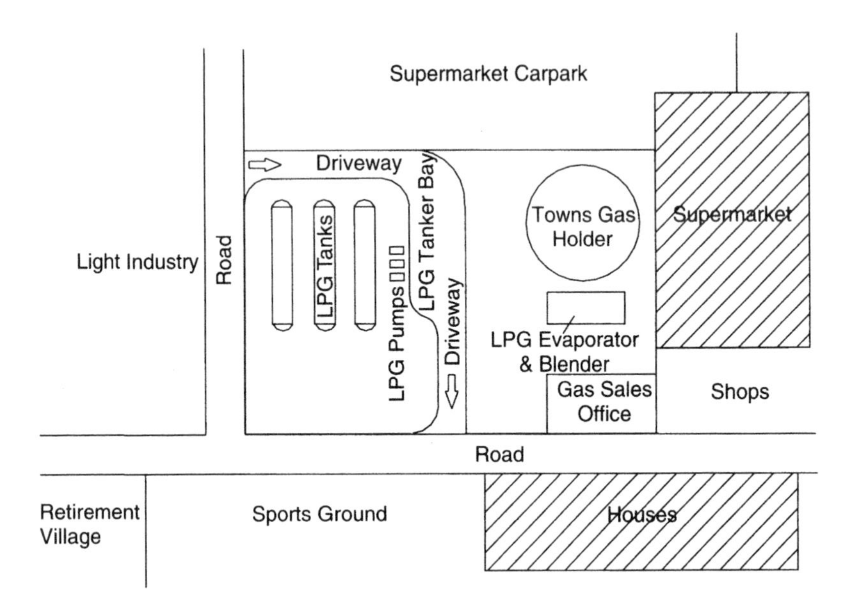

Figure 14-1. Simplified layout of LPG gas facility (not to scale).

town's gas holder, previously used for coal gas, and distributed to the users by means of the original coal-gas network of pipes.

The regulatory authority of the state in which the facility was located introduced legislation requiring all "hazardous" facilities to prepare emergency plans in collaboration with their local councils and emergency services.

The gas company sought the assistance of a consultant to undertake a study with three objectives:

- to inspect the plant and check that it complied with the relevant regulations and national standards for LPG;
- to undertake a quantified risk assessment of the facility and its operation; and
- to advise whether the facility could continue in its existing location, or whether it would need either (a) to be moved to another location, or (b) to be replaced by natural gas by laying a transmission pipeline from the nearest town presently supplied with natural gas. See Figure 14-1.

14.1.2 Description

The features of the facility and its operation relevant to this case study were:

Operation

The plant was staffed 5 days per week during office hours only. The staffing comprised one LPG operator and two sales clerks in the office. The site was locked at all other times.

LPG delivery was by tankers of 18 tonnes, arriving several times per week, typically four times per week in the winter, and twice per week in the summer. The facility was also used for filling small LPG tankers for delivery to farms in the district. Around five to seven such tankers would be filled per week.

Supervision

A supervisor with extensive supervisory experience traveled around the state visiting all the company's facilities, and so spent around 1 day per 3 weeks at this particular facility.

Maintenance

The company had a roving maintenance team that visited regularly every 6 weeks, and promptly on request by either the operator or the supervisor.

Electrical and Instrument Maintenance

The company had a contract with a local electrical contractor to test the alarm and emergency systems, and to undertake any repairs he found necessary.

Auditing

The facility was reportedly audited every 6 months by a team from the head office.

Emergency Services

The town had a volunteer fire brigade, with the members being summoned by a telephone chain and ringing the town's fire bell. They had substantial experience in house fires and wildfire in the surrounding country, but little industrial fire experience and none with LPG. The town had a moderately sized hospital and an ambulance. The police station had a staff of five officers.

14.1.3 The Study

The consultant did a detailed inspection of the facility, checking it thoroughly against the relevant regulations and standards. Apart from some small details, the facility met the requirements.

A conventional quantitative risk analysis was undertaken, using a computer system and typical data for the frequency of the various postulated leaks leading to fire, BLEVE, vapor explosion, etc. The risk contours did not quite meet the requirements of the relevant statutory authority.

In reviewing the systems and procedures used in the facility, the consultant noted a degree of informality, though no major deficiency was noted.

The consultant was concerned about the supervision, maintenance, and security of the facility. The infrequent visits by the supervisor and the maintenance team, and the lack of specialist instrument experience of the local electrical contractor, all seemed out of tune with the requirements of a facility located in the business district of the town, next to a supermarket and its busy parking lot, and near housing and a retirement village.

The consultant asked to visit a broadly similar facility in another town, subject to the same auditing. The one he was shown had reportedly been audited recently, and its firewater deluge had reportedly been tested as part of the audit. But, on inspection, there was good evidence that such a test had not been undertaken for a long time, and there were many other signs of neglect.

The consultant felt that a major incident was entirely credible, regardless of the results of the quantitative risk assessment, because of the extent of usage of the facility by tankers, unloading or loading several times per day, coupled with the very limited technical oversight, maintenance, and auditing.

Further, the volunteer staffing of the local fire brigade, and the inevitable prolonged delay between the onset of a fire and an effective response, were major causes of concern.

Finally, the company was required shortly to embark on emergency planning in partnership with the town council. The scale of the credible incidents (to be evaluated in the planning process) would be certain to become public knowledge, and so cause great anxiety and even outrage among the public. It was likely that the company would be subjected to intense political agitation for it to move.

So the consultant expressed the view that the facility should either be moved or replaced with a supply of natural gas via a transmission pipeline from the nearest source.

This view was initially opposed by the engineering section of the company, which felt that the risk contours could be contracted by a number of minor engineering modifications. But senior management understood that their task was not just to reduce the calculated risk, but to provide a facility that was demonstrably "good practice" and acceptable to the public.

(The Safety Balance in Chapter 8 illustrates their problem well. The inherent hazard—the LPG tanks, tankers, and associated equipment—was substantial and not able to be significantly reduced. The separation between the hazardous features of the facility and the vulnerable surroundings was minimal. Thus the preventive and protective measures would need to be much above the minimum set by the regulations and the national standards if the risk were to be low. The hardware basically met the minimum standards required for a typical facility, but the software—organization, procedures, training, documentation, and culture/climate—would have been difficult to upgrade sufficiently without a major increase in ongoing expenditure.)

14.2 CASE STUDY: SITING OF A HYDROCARBON PROCESSING PLANT

It was proposed that a small hydrocarbon processing plant, producing a range of end products including liquefied hydrocarbon gases, be constructed on a site with a distance of 300 meters between the nearest housing and the boundary of the processing plant. It was recognized that a vapor cloud explosion was an inherent hazard of that type of plant, even though it was intended that the plant be designed, operated, and maintained to the highest standards to keep the likelihood of such an explosion occurring at a vanishingly low level.

It was decided to aim for a design that could not, because of its design features, expose the public in the residential area to a vapor overpressure greater than that which would break windows, but not cause significant structural damage, or personal injury. A blast overpressure of 3 kPa was selected as that maximum pressure.

A calculation was done to determine the largest vapor explosion to meet that criterion. This calculation is set out below, using the methods set out in Section 5.3 of Chapter 5.

Distance:	400 meters
Overpressure:	3 kPa
Scaled distance for 3 kPa:	33 meters/$kg_{TNT}^{0.333}$

Scaled distance = distance/(TNT equivalent mass)$^{0.333}$
that is, TNT equivalent mass = (distance/scaled distance)^3

Therefore

$$\text{TNT equivalent mass} = (400/33)^3 = (12.1)^3 = 1771 \text{ kg of TNT}$$

Assuming the efficiency of the explosion to be 4%, and the heat of combustion of the vapor to be around 10 times that of TNT, then the corresponding mass of vapor can be calculated.

Because

TNT equivalent mass = explosion efficiency × (heat of combustion of vapor/heat of combustion of TNT) × mass of vapor cloud
Mass of vapor cloud = 1771 × (100/4)/10 = 4427 kg of vapor

As a result it was decided that the plant should be designed with limitations on vessel inventories and with automatic leak detection and isolation facilities such that the largest credible open-air vapor explosion would not exceed around 4.5 tonnes.

This simple calculation resulted in a plant that was close to inherently safe in relation to people in the residential area. When a formal quantitative risk assessment was undertaken, initially at the flowsheet stage of design, and later when the plant was complete, the results demonstrated the benefit of the low inventories and the preventive and protective measures adopted.

14.3 CASE STUDY: RISK ANALYSIS OF PROPOSED LPG STORAGE AND PROCESSING PLANT

An existing LPG container-filling plant had a well-established tank farm, which had been designed to high standards with numerous hardware and procedural safeguards, with the design being influenced by a quantitative risk assessment undertaken in parallel, partly for that purpose, and partly to meet regulatory requirements. See Figure 14-2.

The adjoining site held a metal fabrication works, but most of the site near the service roadway was vacant. The metal fabrication works did not wish to expand, so it offered for lease the unoccupied area of the site.

A small organization from a distant city, with extensive experience in transporting and storing LPG, started negotiations to lease the area for two purposes:

- to construct and operate an LPG storage and tanker unloading/loading facility; and
- to construct and operate a small plant to manufacture a specialized liquefied flammable gas, also to be stored in the LPG storage area.

The consultant who had undertaken the original risk assessment for the existing tank farm and LPG container filling operation was engaged to

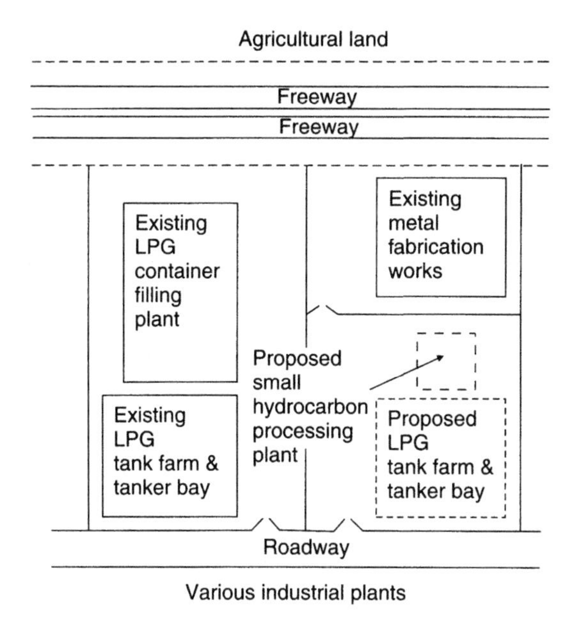

Figure 14-2. Layout of existing and proposed plants (not to scale).

advise on the safety of the design and to undertake the risk assessment required by the statutory authorities.

In the first discussion with the organization's representative proposing the new facility, the consultant commented that, although the assessed risks to the safety of the public from the existing tank farm met the regulatory requirements, it was possible that two tank farms in the same vicinity might result in an excessive cumulative risk to the public.

He then steered the discussion to be able to form a view of the answers to some of the questions listed in Section 3.9 of Chapter 3.

He learned that:

- it was planned that the site be operated without staff, with the tanker drivers performing whatever operations were needed when discharging or filling their tankers;
- the organization had no experience with chemical processing; and
- the small manufacturing plant would be operated by staff from the adjacent LPG filling plant, visiting several times per shift.

With some embarrassment, the consultant expressed the view that, although it was possible that the quantitative risk assessment would generate risk contours that met the regulatory requirements, he would feel obliged to include in his report an assessment of the unquantifiable risks. These would include his view that the organization did not have the experience to operate the manufacturing facility at a low-risk level (in effect, they "did not know what they did not know") and that relying on periodic supervision of the plant by staff from the adjacent LPG filling plant (who had no experience in processing plant operation, either) would also introduce substantial unquantifiable risks. Further, he was uncomfortable with the concept of the proposed LPG tank farm and tanker facility being operated by the tanker drivers without local supervision, and with the organization being based far away with no local technical supervision or support.

The meeting ended, and the consultant expected that the representative would seek a more "flexible" consultant.

Some months later the representative returned, accompanied by someone known to the consultant from a major LPG distribution company. In that second meeting it was explained that the small manufacturing plant would be built, not at the proposed LPG storage, but on a site operated by another organization that operated some similar processes. The plant would be thus operated and supervised by people experienced in processing of hazardous materials. Further, the LPG storage and tanker facility would be jointly owned by the original organization and the LPG major, and would be operated and supervised by the LPG major.

By open, frank discussion of the risks and of the requirements of "good practice" at the conceptual stage of the proposed new plant, the consultant had initiated discussions that led to a substantial reduction in the risks, before any risk calculations had been undertaken.

14.4 CASE STUDY: SITING OF AN LPG TANKER LOADING BAY CONTROL ROOM

An LPG tanker loading bay at the tank farm of a hydrocarbon processing plant was being upgraded, and a new control center for the tank farm and the tanker bay was planned.

There were limitations on the location of the new control room. It had to be reasonably close to the tanker bay for observation of operations there and for easy access to the bay. There were the normal hazards associated with LPG tanker bays, and there was also a major processing plant in the vicinity that presented a risk (minimal but credible) of vapor cloud explosion.

It was decided to construct the control room to be fire- and explosion-resistant. For economic and operational reasons, it was hoped to be able to avoid a fully explosion-proof construction.

A computer-based risk assessment had been undertaken of the site, including the tank farm and the tanker bay. As well as generating contours of the fatality risks to people, the risk assessment generated contours of the frequency of exposure to selected damaging levels of heat radiation and blast overpressure.

A site was selected for the tanker bay that was operationally feasible and that had a very low-calculated frequency of exposure to a high blast overpressure. However, there was a significant (but still low) calculated frequency to a moderate blast overpressure, so it was decided to construct the control room to withstand that level of overpressure.

There was a history, elsewhere in the world (e.g., Flixborough) of control rooms collapsing when exposed to blast through the walls collapsing and the roof falling in. So the design principle adopted required the roof to be supported separately from the walls, and for the walls to be secured against blast both at the bottom and the top. This was achieved in the manner shown in Figure 14-3.

14.5 CASE STUDY: RESITING OF TOXIC GAS PLANT

A long-established chemical factory manufactured a wide range of materials, including a toxic gas. Exposure to low concentrations of this gas causes breathing distress and severe irritation of the eyes.

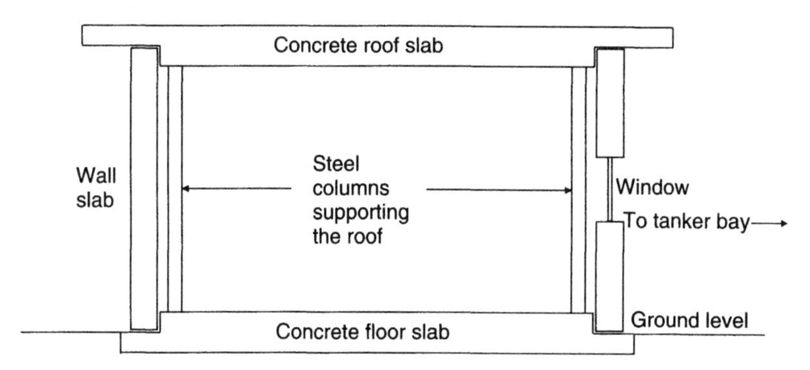

Figure 14-3. Section through control room to show construction principles (not to scale).

As part of improvement of the road network in and around the city near which the factory was located, a freeway was planned that would cross the factory near the toxic gas plant.

The company that operated the factory claimed that it would be unsafe to operate the plant so close to the freeway, and sought sufficient compensation either to relocate the plant or to construct a new one at a safer distance from the freeway. The roads authority requested a risk assessment be undertaken to establish the reality of the claim. As the company had substantial expertise in quantitative risk assessment, they insisted that they undertake the required assessment themselves. The roads authority then appointed an independent consultant (Consultant A) to monitor the data and methods used in the risk assessment, and to provide a report on the validity of the findings of the company's assessment.

The company and Consultant A agreed that the assessed risks showed that the risks would be too high, as they stood at that time, although they disagreed about the extent of the risk, as they disagreed about the frequency data used in the assessment. Consultant A, however, insisted that if the company were to implement a number of plant modifications as recommended by the consultant, the risks would be sufficiently low for the plant to remain where it was. The company disagreed. An impasse developed. See Figure 14-4.

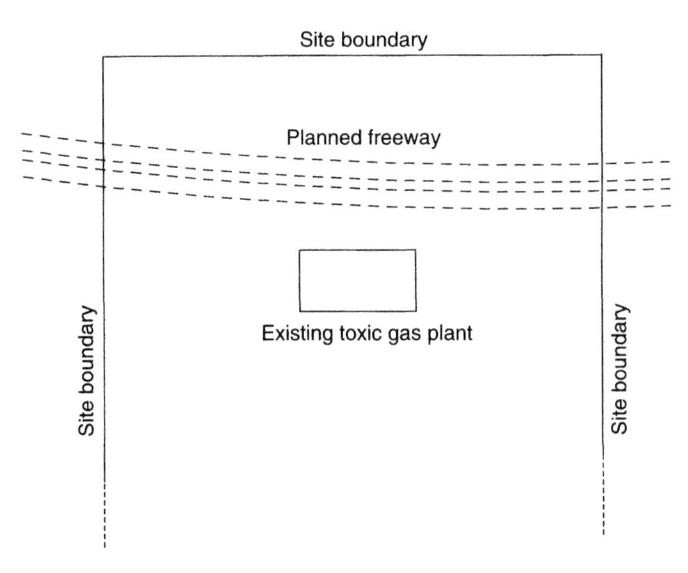

Figure 14-4. Layout of site with toxic gas plant.

Another independent consultant (Consultant B) was engaged by the roads authority, with the agreement of the company, in an attempt to resolve the impasse. He inspected the plant, reviewed both the company's risk assessment and the first consultant's report, and held discussions with the staff of the company and with Consultant A.

He made the following observations:

- the risk assessment method used by the company was reasonable;
- it was not possible to determine with confidence which set of frequency data (company or Consultant A) was preferable, and debate could continue indefinitely;
- quantitative risk assessment assesses only the quantifiable risks, and many of the most likely triggers of incidents are not able to be quantified with any confidence;
- in the event of an escape of the toxic gas, if the wind direction took it over the freeway, the effect would be very serious with numerous high-speed crashes due to drivers being forced to close their eyes;
- in the event of such an event, the company would correctly be held responsible, not the consultant;
- therefore, as the company would bear the responsibility, the company should also have the authority to determine whether they were prepared to operate the plant so close to the freeway. While Consultant A may insist that it would be safe to operate the plant there if it were modified as they recommended, he would not have to bear the consequences in the event of an incident occurring. Therefore Consultant B recommended that the company's view should prevail.

14.6 CASE STUDY: REVIEW OF OIL REFINERY

A consultant was asked to undertake a brief survey of an old oil refinery to identify the main types of improvement that should be considered to bring the refinery as close as practicable to current standards of process safety.

The approach adopted took the form of several discussions with staff responsible for major production units, followed by a brief inspection of those units.

The aim of the discussions was to form an impression of:

- the nature and scale of the processes involved;
- the nature of the hazards presented by those processes;

- the nature of the safeguards in place;
- the types of safety system being used;
- the level of understanding of process safety by the responsible staff; and
- the main features of the plant to be examined during the subsequent inspection.

In effect, the discussions were structured around the six requirements set out by Hawksley, although not in the same sequence.

Then, in the inspection, the following were given particular attention:

- the separation between plants, and sections of plants;
- the physical condition of the plants, including vessels, pipework and fittings, instrumentation, etc.;
- the leak detection and isolation equipment;
- the fire detection and firefighting equipment;
- the manner in which maintenance work was being carried out; and
- the state of alarms in the control room.

The report was structured around Hawksley's principles, discussing where the main strengths and weaknesses appeared to be, and recommending where more detailed study should be undertaken, either by refinery staff or by external resources.

One of the topics recommended for further study was that of installation of remotely operable emergency shutoff valves. Because of the cost of these, it was important that the priority for them be carefully assessed. This was done by assembly of a spreadsheet, with each row representing a pipework system with the potential for a major leak, together with inventory, design, and process information sufficient for an approximate assessment of both the consequences (e.g., potential leak rate, fire intensity, vapor cloud magnitude) and the likelihood of the incident. The spreadsheet was then sorted to identify the pipework systems with greatest fire risk and vapor cloud explosion risk, and the most pressing needs for emergency isolation facilities were identified and ranked. The ranked list was then used as the basis for discussion with refinery staff to assist in reaching agreement about the schedule for installation of the isolation facilities, which, for economic reasons, had to be spread over several years.

14.7 CASE STUDY: REVIEW OF GAS/LIQUID SEPARATION PLANT

A company operated a number of gas and oil wells and a centralized gas/liquids separation plant in an isolated, semidesert area. The natural gas and other products from the plant were piped to distant cities.

The company engaged a major engineering consultancy to undertake:

- a review of the pressure relief and blowdown facilities at the plant;
- a fire safety study; and
- a quantitative risk assessment, including the fatality risks to employees and the risks of inability to supply the market for natural gas.

In addition, although it was not expressly included in the remit of the study, the project leader, as part of his familiarization process, examined the design and condition of the plant and inquired into the "software."

Among the weaknesses identified were:

- a high proportion of the firewater monitors either were located too close to the plant to be operated in the event of a fire, or were too far away and would be unable to cool elevated sections of the plant;
- there were no written operating instructions for most of the plant sections;
- the work permit system was very informally operated;
- there was little testing of protective systems;
- the emergency and firefighting team, although well equipped with vehicles and equipment, was staffed largely with people who had been rejected by operations, and they were not permitted to practice deployment on the plant; and
- the emergency procedures developed by the emergency team had been developed in isolation from the operations staff, who had their own emergency procedures.

The study of the relief and blowdown facilities was conducted by specialist process engineers, assessing the capability of the systems compared with the relevant codes of practice. A number of weaknesses were identified and the appropriate engineering changes specified.

The quantitative risk assessment found that the risks to employees from the quantifiable risks were generally satisfactory and recommended a number of improvements to preventive and protective measures to effect further reduction. The computer system used in the assessment had also been used to calculate the frequency of incidents with various levels of impact (i.e., the frequency of exposure of key sections of the plant to various levels of heat radiation and blast overpressure), with these assessments being used in discussion with production and engineering staff to estimate the time that would be required to restore supply of gas to the market.

The fire safety study, after testing the actual range of firewater monitors and hoses and their flow rates, made a number of recommendations for relocation of monitors, additional fire hydrants, pump capacity, water storage, etc.

The full report was necessarily very lengthy. As an aid to the managers, it was decided to issue the executive summary in the form of a brief, separately bound report of around 30 pages, printed on high-quality paper and illustrated with photographs. This report had a separate chapter for each of the sections of the remit, plus a chapter on software.

Senior managers of the company objected strongly to the chapter on software. They wished to distribute the executive summary widely, to the regulatory authorities, insurers, unions, etc., and did not wish their major managerial deficiencies to be publicized. They argued that software was outside the remit and that the assessment was subjective and unsound, and so should be deleted. They withdrew all their copies of the executive summary.

The consultant, after some consideration, revised the section on software, omitting the details of the identified weaknesses, but setting out the critical importance of a high standard of software, the reliance of the quantitative risk assessment on the existence of a high standard, and recommending that the company undertake urgently a review of the existing state of the software, followed by implementation of any needed improvements.

Then, when the project leader was invited to make an oral presentation of the report to a meeting of the senior managers, he made the following points to them very forcefully:

- a consultant (or indeed anyone) who identifies any weaknesses affecting risks to people is legally and morally required to bring them to the attention of management;
- many such weaknesses had been identified as part of gaining a full understanding of the risks presented by the plant;
- those weaknesses had been brought to the attention of management via the first issue of the executive summary;

- even though it was understood that management had recalled and suppressed that first issue, the consultancy had copies in their files;
- corporate management could not walk away from those weaknesses, and should tackle them urgently.

14.8 CASE STUDY: FAULTY APPROACH TO RISK-BASED INSPECTION

Inspection of pressure vessels at process plants such as oil refineries is a major contributor to production downtime. Of course, it is not the time taken to undertake the inspection that is the problem, but the time taken to shut down the plant, cool it down, and isolate and purge the vessels, and then to reverse the procedure until the plant is operating once more.

An engineer at an oil refinery was given the task of using reliability mathematics to determine the feasibility of extending the period between inspections of pressure vessels from the existing 4 years to around 10 to 15 years. Naturally the engineer was keen to produce the results sought by the refinery manager.

His method was seriously flawed, showing very limited understanding of reliability principles and of good practice, and relying heavily on dubious or wholly unsupported assumptions. For example, he assumed that, as the vessels had been designed with a corrosion allowance for a 25-year life, then it would be reasonable to assume that the probability of failure of each vessel was 1 in 25 per year, without realizing that if this were the case, then in a refinery with numerous pressure vessels there would be many failures per year.

He was required to have his method checked by a risk and reliability consultancy. The consultancy explained the weaknesses and errors in his method and said that because of the absence of essential hard data, any calculated result was only an estimate, and a very dubious one at that.

Some months later, the consultancy received a copy of the final report. Only minor details had been changed, and the report recommended that the inspection period could be increased at once from 4 years to 14 years. Furthermore the report stated in its introduction that the work had been done in conjunction with the consultancy. The consultancy was obliged immediately to write to the refinery to dissociate itself from the report, stating that although a discussion had been held, the method used had not been approved and was not valid.

The root cause of the problem lay with the refinery manager, who had specified the outcome required from an investigation before it was started and put pressure on an engineer to produce that outcome. A second root cause was giving the task to an engineer who lacked the necessary know-how and was not sufficiently experienced to recognize his limitations.

Because of the potential severity of the consequences of vessel failure, rather than relying on very limited and statistically inadequate data, what was needed was a controlled, managed approach. This would entail, for each pressure vessel:

- preparation of a list of the various potential failure modes and causes;
- a corresponding list of the inspection and test methods that could provide early warning of impending failure, together with an estimate of the confidence that could be placed in each method; and
- a statement of the condition that each inspection or test would be expected to reveal at the next periodic inspection.

Then, the interval between tests would be increased slightly (e.g., from 4 to 6 years). The inspection results (e.g., the extent of corrosion) would be compared with the predicted results. Where there was close agreement, it might be decided to extend the inspection interval again. Where there was a significant difference, that would be taken to indicate a degree of inability to predict the condition, and so the inspection interval would be reduced back to the original period. Similarly, if a vessel were nearing the end of its design life, or for some other reason (e.g., the corrosion allowance was nearly all used), the inspection interval would be reduced.

14.9 CASE STUDY: A LEGAL OBSTACLE TO AN AUDITOR

A small company operated an ethanol distillation plant. The feedstock was the by-product of a fermentation process. Because of the nature of the product, the plant was required to be open to inspection by government officials at any time, day or night.

One of the stills exploded one day and slightly injured an operator. The government authority took legal action with the aim of having the operating license withdrawn, as they were concerned that their inspector could be exposed to danger during an inspection.

The court arranged for an independent specialist to undertake a safety inspection. This was to take place in the presence of the legal advisers for both the company and the government authority.

The first question the specialist asked the plant manager was to test the compliance with Hawksley's first principle (see Chapter 10, Section 10.1). He asked: "What sort of hazardous incidents could you, in theory, have at this plant, and how serious could they be?"

The legal adviser to the company immediately instructed his client not to answer. The specialist outlined the six requirements defined by Hawksley as the basis for determining the degree of process safety, and explained the relevance of the question. The legal adviser insisted that his client should not answer.

The specialist explained that he would have to set out in his report that he was prevented from determining whether the plant manager had sufficient understanding of the safety requirements of his plant for any confidence that he could ensure a high level of safety into the future. The legal adviser would not change his advice, so the report to the court carried that observation, along with other observations made when inspecting the plant against the other five requirements defined by Hawksley.

14.10 WORKED EXAMPLE: HEAT RADIATION RISK

A pressurized stock tank containing 100 tonnes of LPG is exposed to fire and ruptures, producing a BLEVE fireball. See Figure 14-5.

What will be:

(a) the diameter of the fireball;

(b) the duration of the fireball;

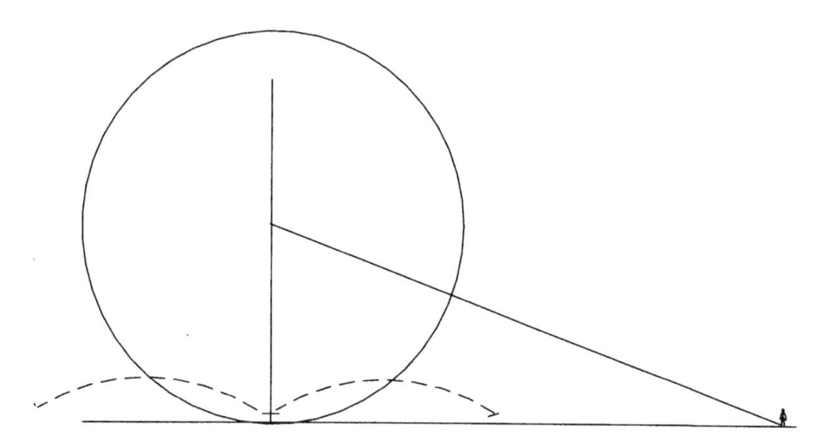

Figure 14-5. Typical configuration of BLEVE for assessment.

(c) the megawatts radiated from the fireball, assuming that the heat of combustion of the LPG is 48,000 kJ/kg and that 35% of the heat of combustion is radiated;

(d) the intensity of the radiant heat (kW/m^2) at ground level at a distance of 350 meters from the stock tank before the BLEVE, for the position of the fireball as shown above with the lower surface of the fireball just touching the ground, with a flame surface emissivity of 250 kW/m^2, and taking account of attenuation of the heat radiation by atmospheric water vapor and carbon dioxide (assuming an average temperature of 20°C and relative humidity of 40%);

(e) the probability of fatality for people exposed to the attenuated heat radiation calculated in (d) for the duration calculated in (b)?

Calculation:

(a) Diameter of fireball:

$$\begin{aligned}\text{Diameter} &= 6.48\,M^{0.325} \text{ meters}\\ &= 6.48 \times 100{,}000^{0.325}\\ &= 273 \text{ meters}\end{aligned}$$

(b) Duration of fireball:

$$\begin{aligned}\text{Duration} &= 0.852\,M^{0.26} \text{ sec}\\ &= 0.852 \times 100{,}000^{0.26}\\ &= 17 \text{ sec}\end{aligned}$$

(c) Megawatts radiated:

$$\begin{aligned}\text{Megawatts radiated} &= (100{,}000 \times 48{,}000 \times 0.35)/(17 \times 1000)\\ &= 98{,}823 \text{ MW}\end{aligned}$$

(d) Heat radiation intensity:

$$\begin{aligned}\text{Distance from the center of the fireball to the observer} &= (300^2 + 175^2)^{0.5}\\ &= 347 \text{ meters}\end{aligned}$$

$$\begin{aligned}\text{View Factor} &= (136.5/347)^2\\ &= 0.155\end{aligned}$$

$$\begin{aligned}\text{Heat radiation intensity (unattenuated)} &= 0.155 \times 250\\ &= 38.7 \text{ kW/m}^2\end{aligned}$$

$$\begin{aligned}\text{Distance from the surface of the fireball to the observer} &= 347 - 150\\ &= 197 \text{ meters}\end{aligned}$$

For a temperature of 20°C and 40% relative humidity, and a distance of 197 meters, the atmospheric transmissivity is around 0.67 (Figure 5-6).

$$\text{Attenuated heat radiation intensity} = 38.7 \times 0.67 = 25.9 \text{ kW/m}^2$$

(e) Risk of fatality:

Probit (Chapter 5.8.3) $Y = -10.7 + 1.99 \ln\{(I^{1.333} \times t)/10{,}000\}$
(where I is W/m^2)

$$Y = -10.7 + 1.99 \ln\{(25{,}900^{1.333} \times 17)/10{,}000\} = 3.56$$

Referring to the probit–probability graph (Figure 5.31), the probability of fatality is about 7%.

14.11 WORKED EXAMPLE: VAPOR CLOUD RISK

Propylene liquid at an ambient temperature of 30°C escapes from a leaking vessel at a rate of 150 kg/sec. The vapor cloud, formed by the leaking liquid, drifts through the congested plant toward a continuously fired furnace, 100 meters distant from the leak source, in a wind of 1 meter per second. The cloud is ignited by the furnace and explodes.

Assume the following data:

- Heat of combustion of propylene vapor: 48,000 kJ/kg
- Atmospheric-pressure boiling temperature of propylene: –46°C
- Specific heat capacity of propylene in the relevant temperature range: 2.3 kJ/kg K
- Latent heat of vaporization of propylene: 420 kJ/kg

(a) What will be the blast overpressure at a distance of 200 meters from the center of the blast?

(b) What is the likely effect of such a blast on conventional brick residential buildings?

(c) What is the estimated probability of fatality to a person in a conventional brick house at that location?

(d) What is the estimated probability of fatality to a person in the open air at that location?

Calculation:

(a) Blast overpressure:

Liquid propylene escape rate = 150 kg/sec
Rate of adiabatic evaporation = $150 \times 2.3 \times (30+46)/420$
= 62.4 kg/sec

This is doubled to allow for evaporation of spray in the air at 30°C. Thus

Vaporization rate = 124.8 kg/sec

The cloud drifts 100 meters to the furnace at 1 meter/sec, so the cloud accumulates for 100 sec.

Cloud mass at explosion = 124.8×100
= 12,480 kg

As the cloud is all in a congested area of the plant, an explosion efficiency of 10% will be assumed. Therefore,

TNT equivalent mass = $12480 \times 0.1 \times 48000/4600$
= 13,022 kg_{TNT}

The distance to the observer is 200 meters, so

Scaled distance = $200/(13{,}022)^{0.333}$
= 200/23.45
= 8.52

From Figure 5-13, blast overpressure is 20 kPag.

(b) Effect of the blast: From Table 5-4, a house would be very severely damaged and may not be repairable. A person in the open air might be injured by missiles projected by the blast. A person in the house would probably be injured, possibly killed.

(c) Probability of fatality in a house: From Figure 5-14, the probability of fatality in a house would be low, but significant, at around 10–20%.

(d) Probability of fatality in the open air: From Figure 5-14, the probability of fatality in the open air would be low, probably less than 5%.

14.12 WORKED EXAMPLE: TOXIC GAS RISK

An isolated plant surrounded by open country has the potential for an accidental release hydrogen sulfide gas from ground level at a rate of 5 m^3 per second. The escape would have a duration of around 5 minutes. The nearest house is 150 meters away to the east over open country. The frequency of that release occurring has been estimated at 0.001 per year.

The meteorological data for the region has been examined, and the following data extracted. When the wind is from the west, that is, blowing toward the house, the probabilities of different wind speeds and atmospheric stability categories are shown in the table below.

Atmospheric Stability	Wind Speed (m/sec)	Probability (%)
A	3	4
C	5	6
E	2	2

What is the fatality risk (due to exposure to the hydrogen sulfide) for a person continuously present at the house, assuming that they cannot take evasive action on detection of the gas?

Calculations

Case A3 (atmospheric stability A; wind speed 3 m/sec): For a distance of 150 meters, and using Figures 5-23 and 5-24,

$\sigma_y = 38$ meters
$\sigma_z = 20$ meters

It is a ground-level release, so the stack factor reduces to 1.0. So,

$$
\begin{aligned}
\text{Concentration of the gas} &= (5 \times 1)/\pi \times 38 \times 20 \times 3 \\
&= 0.000698 \\
&= 698 \text{ ppm}
\end{aligned}
$$

The duration of exposure is estimated to be around 5 minutes.

From Table 5-9, the probit constants for hydrogen sulfide are: $K_1 = -36.2$; $K_2 = 2.366$; $n = 2.5$.

Therefore,

$$\text{Probit } Y = -36.2 + 2.366 \times \ln(698^{2.5} \times 5)$$
$$= 6.3$$

From Figure 5-31, the probability of fatality would be high, assumed here for the example to be around 88%, that is, 0.88.

The frequency of the leak is estimated to be 0.001 per year, and the probability of A3 atmospheric conditions at the time of the leak is 4%, so the frequency of fatality from a leak in A3 conditions is:

$$\text{Fatality frequency} = 0.88 \times .001 \times .04$$
$$= 3.52 \times 10^{-5} \text{ per year}$$
$$= 35.2 \text{ per million per year.}$$

The above calculations are repeated for the other atmospheric cases (C5 and E2), and the results are tabulated below.

	a	b	c	d	e	f	g	e×f×g/ 10,000
Weather case	σ_y (m)	σ_z (m)	*C* (ppm)	Probit	Fatal prob. (%)	Escape freq. (p.a.)	Probability of occurrence of weather case (%)	Fatality risk (p.a.)
A3	36	20	698	6.3	88	0.001	4	0.000035
C5	18	11	2,679	14	100	0.001	6	0.000060
E2	8.5	4.8	13,000	23	100	0.001	2	0.000020

The total fatality risk is thus 95×10^{-6} per year, or 95 per million per year.

14.13 WORKED EXAMPLE: SEPARATION DISTANCE ESTIMATION (A)

One section of a downstream oil and gas processing plant has the potential to generate a vapor cloud explosion of 20 tonnes of propane vapor.

It has been estimated that the frequency of such an explosion is around 1 in 100,000 per year.

The relevant statutory authority has specified that the individual fatality risk at residential areas must not exceed 1 in 1,000,000 per year for a person in a house for 24 hours per day.

What is the required separation distance between that plant section and the nearest residential area?

Calculation

The frequency of the explosion is estimated to be 1 in 100,000 p.a.

The maximum fatality risk at the residential area is 1 in 1,000,000 p.a.

Therefore the maximum fatality risk per explosion is a probability of 0.1, that is, 10%.

From Figure 5-14, the overpressure must not exceed 16 kPag.

From Figure 5-13, the scaled distance must not exceed 10.5.

As scaled distance $\lambda = R/(kg_{TNT})^{0.333}$
then $R = \lambda \times (kg_{TNT})^{0.333}$

The explosion is of 20 tonnes of vapor. The TNT equivalent mass is $(0.04 \times 20{,}000 \times 48{,}000)/4600 = 8347$ kg. Therefore, $R = 10.5 \times (8347)^{0.333} = 10.5 \times 20.2 = 212$ meters.

14.14 WORKED EXAMPLE: SEPARATION DISTANCE ESTIMATION (B)

A quantitative risk assessment is carried out of an LPG storage tank and associated facilities. In that assessment, the possibility of a vapor cloud explosion or a BLEVE is identified. The possible sizes and frequencies of the incidents are estimated as below.

BLEVE:	Mass of LPG involved: 15 tonnes	Frequency: 1×10^{-6} p.a.
Vapor explosion:	Mass of vapor: 25 tonnes	Frequency: 10×10^{-6} p.a.

What separation distance would be required between the LPG facility and the nearest residential area, if the maximum fatality risk in the open air at the residential area is not to exceed 1×10^{-6} p.a.?

The following data will be assumed:

- Heat radiation intensity of the surface of the BLEVE fireball is 250 kW/m^2.
- Heat of combustion of the vapor is 48,000 kJ/kg.

- Efficiency of explosion of vapor cloud is 4% relative to TNT.
- The average atmospheric temperature is 20°C and the relative humidity 50%.

14.4.1 Calculation

A graphical method must be used here, plotting the total fatality risk in the open air against separation distance for several suitable distances (e.g., two or three distances) and interpolating to estimate the required distance.

The following distances will be assessed initially: 100 and 200 meters. From those results, an estimate will be made of a closer value for the separation distance to be assessed.

The calculations for 100 meters are shown in detail below, and the results of all calculations are tabulated.

(a) BLEVE

The radius to the observer is 100 meters. From Figure 5-8, the diameter of the BLEVE would be around 150 meters, so the radius of the BLEVE would be around 75 meters.

The radius from the center of the BLEVE to the observer would be:

Radius $=(100^2+75^2)^{0.5}=125$ meters
View factor $=(75/125)^2=0.36$
Radiation intensity (without attenuation) $=250\times0.36=90$ kW/m^2
Distance from edge of fireball to observer $=125-75=50$ meters

From Figure 5-6, the transmissivity is around 0.75. Therefore the heat radiation intensity would be around $90\times0.75=67.5$ kW/m^2.

From Figure 5-9, the duration of the fireball would be around 11 seconds. Using the probit equation in Section 5.8.3 of Chapter 5,

$$\begin{aligned}\text{Probit } Y &= -10.7+1.99\ln\{(I^{1.333}\times t)/10{,}000\}\\ &= -10.7+1.99\ln\{(67{,}500^{1.333}\times 11)/10{,}000\}\\ &= 5.24\end{aligned}$$

From Figure 5-31, the probability of fatality would be around 60%.

The estimated frequency of the bleve is 1×10^{-6} p.a., so the fatality risk to a person in the open air at the residential boundary would be $1\times10^{-6}\times 0.60\%=0.6$ per million per year (pmpy).

(b) Vapor Explosion

The mass of the vapor would be 25 tonnes. Thus:

TNT equivalent mass $= 25{,}000 \times 0.04 \times 48{,}000/4600 = 10{,}434\ \mathrm{kg_{TNT}}$
Scaled distance $= 100/(10{,}434)^{0.333} = 4.6$ meters/kg$^{0.333}$

From Figure 5-11, the blast overpressure is 55 kPa; from Figure 5-12, the probability of fatality is about 50%. The frequency of the explosion is estimated to be 10×10^{-6} p.a., so the fatality risk would be: $10 \times 10^{-6} \times 0.50 = 5.0$ pmpy

Incident	Ground Level Distance to Observer (meters)	Radius of BLEVE (meters)	Radius to Observer (meters)	View factor	Attenuated Radiation Intensity (kW/m^2)	Fireball Duration (sec)	Fatality Probability (%)	Fatality Risk (pmpy)
BLEVE	100	75	125	0.36	67.5	11	60	0.60
BLEVE	200	75	213	0.124	21.4	11	2	0.02
BLEVE	150	75	168	0.199	36.4	11	8	0.08
BLEVE	125	75	146	0.264	48.9	11	28	0.28

Incident	Distance to Observer (meters)	Scaled Distance (m/kg$^{0.333}$)	Blast Overpressure (kPa)	Fatality Probability (%)	Fatality Risk (pmpy)
Vapor explosion	100	4.6	55	50	5.0
Vapor explosion	200	9.2	18	2	0.2
Vapor explosion	150	6.9	27	6	0.6
Vapor explosion	125	5.7	40	19	1.9

From this, the totals of the risks from both BLEVE and vapor explosion are:

Distance	Fatality Risk (pmpy)
100 meters	5.6
200 meters	0.22
150 meters	0.68
125 meters	2.18

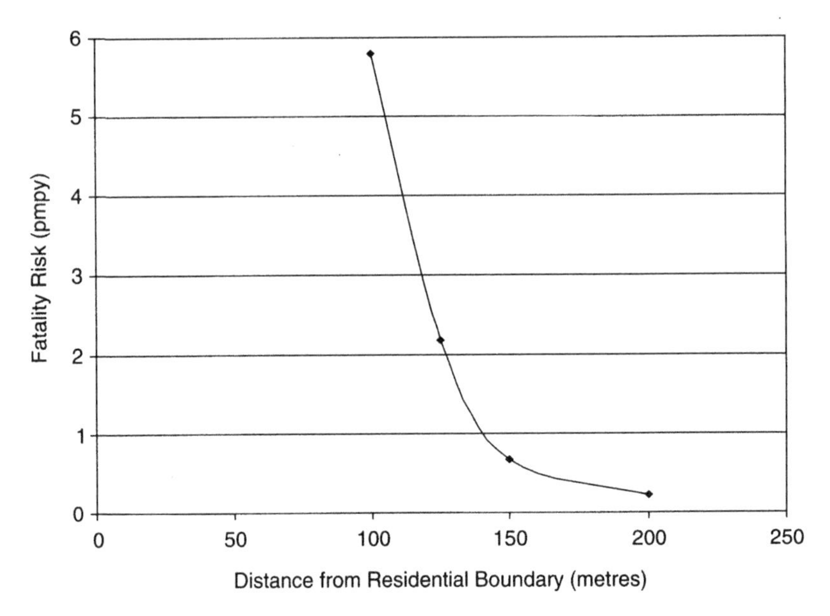

Figure 14-6. Risk versus distance from residential boundary.

These are plotted in Figure 14-6. This graph suggests that the theoretical separation distance would need to be between 100 and 150 meters, perhaps around 140 meters. Although calculations could be performed for additional distances to get a closer estimate of the distance, the uncertainties in the method are such that the apparent extra precision would be illusory. It would be prudent to round the distance up to 150 meters at least.

14.15 WORKED EXAMPLE: FAULT TREE ANALYSIS AND CUTSET ANALYSIS

The above methods of evaluating the reliability of a control and protective system are illustrated by the simple example below. See Figure 14-7. It is important that the particular tank shown in the figure does not overflow.

The level in the tank is controlled by a system comprising a level sensing element LE, which sends an electrical signal to a level controller LC, which sends a pneumatic signal via a solenoid valve SV to the level control valve LCV. The solenoid valve is normally open direct from the level controller through to the level control valve.

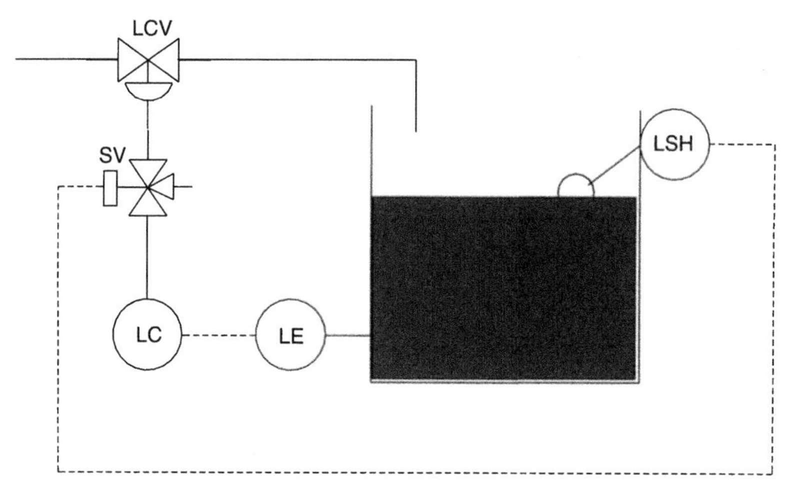

Figure 14-7. Flowsheet of tank system.

If the level rises too high, such that there is a risk of overflowing, the ball float operates the high-level switch LSH, which sends an electrical signal to the solenoid valve SV, which closes off the air signal from the level controller LC and vents the air from the level control valve LCV, which closes under spring pressure, thus stopping flow of the liquid into the tank.

What is the logical structure of failures that will lead to the tank overflowing?

14.15.1 Analysis and Calculation

In constructing the fault tree, the first step is to identify the control system, then the protection system. The control system comprises LE, LC and LCV; the protection system comprises LSH, SV and LCV. The first attempt at the fault tree is shown in Figure 14-8.

Note that LCV appears in both control and protection. This fault tree must be simplified to eliminate double counting. In a simple case such as this, it can be done by inspection, but as an example, Boolean algebra will be used here.

For the purposes of the Boolean algebra, we will use single letters for each element of the fault tree.

Overflow will be represented by the letter T (i.e., the "top event").

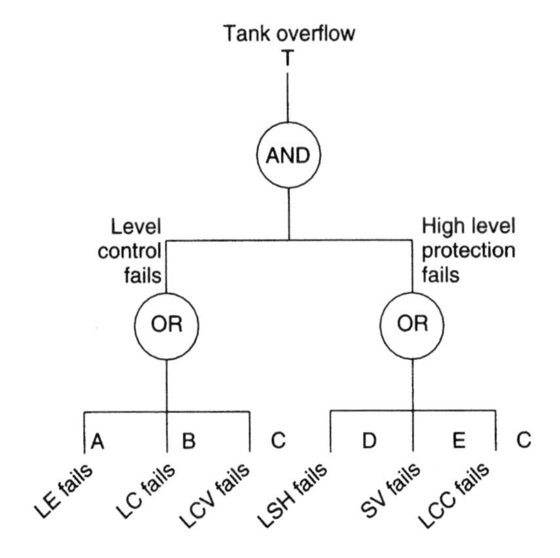

Figure 14-8. Initial fault tree (needing reduction).

LE becomes	A
LC	B
LCV	C
LSH	D
SV	E

The control system fails if A or B or C fail, that is, in Boolean algebra, A+B+C, where "+" means OR.

The protection system fails if D or E or C fail, that is, in Boolean algebra, D+E+C.

The tank overflows if Control fails and Protection fails, that is, Control Fail × Protection Fail, where × means AND.

Thus,

$$
\begin{aligned}
T &= (A+B+C)\cdot(D+E+C) \\
&= (AD+AE+AC+BD+BE+BC+CD+CE+CC)
\end{aligned}
$$

The Boolean identities needed are:

$$
\begin{aligned}
&AA = A \\
&A+A = A \\
&A+AB = A
\end{aligned}
$$

The above expression can be simplified, first by replacing the CC term by C:

$$T=(AD+AE+AC+BD+BE+BC+CD+CE+\overset{\mathbf{C}}{\cancel{CC}})$$

Then all the other terms containing C with other elements can be eliminated, as they are subsets of C (e.g., C+CE=C), that is,

$$T=(AD+AE+\cancel{AC}+BD+BE+\cancel{BC}+\cancel{CD}+\cancel{CE}+C)$$

from which

$$\begin{aligned}T&=AD+AE+BD+BE+C\\&=(A+B)\cdot(D+E)+C\end{aligned}$$

This can be used to construct the reduced fault tree, which has no double counting, as below. See Figure 14-9.

The following elements can be initiating events, requiring protection: LCV, LE, LC. Their failures are called "demands" and must be expressed in frequencies. All other elements are protective elements and must be expressed in probabilities, that is, the probability that they will be in a failed state when a demand occurs—it is not their frequency of failing that is important, it is the probability that they will be in a failed state at the moment when they are needed.

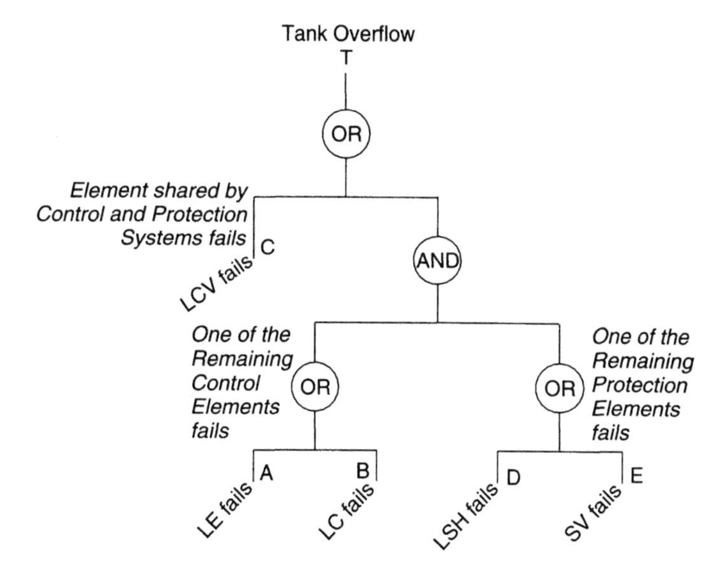

Figure 14-9. Reduced fault tree.

This can be quantified as follows, using illustrative data to clarify an important lesson, discussed later.

LE	0.1 per year
LC	0.1 per year
LCV	0.1 per year
LSH	0.1 per year
SV	0.1 per year
Testing:	4 per year (i.e., every 0.25 years)

The control elements must be expressed in frequency form, as given. The protection elements must be expressed in dimensionless quantities, such as FDT:

$$\text{FDT (LSH)} = 0.5 \times 0.1 \times 0.25 = 0.0125$$
$$\text{FDT (SV)} = 0.5 \times 0.1 \times 0.25 = 0.0125$$

These quantities are used in the reduced fault tree above, as follows. See Figure 14-10.

The lesson demonstrated by the above example is that elements or components that are shared by both control and protective systems usually contribute a major part of the total unreliability of a system.

The general rule is therefore that protective systems should not rely on any components of the control system, unless those components are

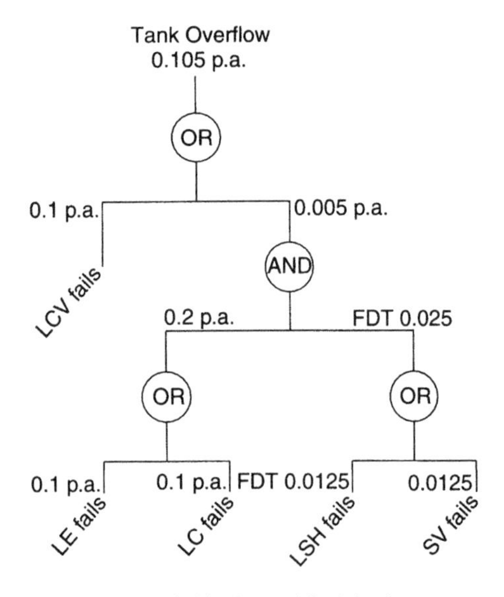

Figure 14-10. Quantified fault tree.

known to be highly reliable, and able to be demonstrated to be in working order.

14.5.2 Cutset Approach

The same problem can be tackled by the cutset approach.

The signal paths are:

LE....LC....LCV (the control signal)
LSH....SV....LCV (the protection path)

First search for any one element that, if it fails, will block all signal paths. Clearly LCV does, so it is one of the cutsets.

Then look for any combination of two elements, which does not contain any single-element cutset, that can block all signal paths. (We exclude any combination that includes a single-element cutset already identified, as the new combination is a subset of the already identified cutset. This is because of the Boolean identity: $A+AB=A$.)

The combinations are:

LE and LSH

LE and SV

LC and LSH

LC and SV

These are also cutsets.

Then look for any combinations of three elements, which do not contain any single-element cutsets or any two-element cutsets, that can block all signal paths. (As there are only two paths in this example, there are no three-element cutsets.)

Thus the complete list of cutsets is (demand elements, i.e., those that must be expressed as frequencies, are in bold type):

1. **LCV**
2. **LE,** LSH
3. **LE,** SV
4. **LC,** LSH
5. **LC,** SV

The numbers in each cutset are multiplied together, and then the sum of the values for each cutset is totaled to give the frequency of total system failure, that is, the value of the "top event."

Only one element in a cutset can be a frequency (in this example, the elements in bold type); all the others must be probabilities (otherwise, we would be multiplying frequencies and getting a result per square year).

The cutsets are quantified using the same data as used in the fault tree method illustrated earlier. As before, elements and numbers in boldface are frequencies; others are FDTs.

LCV: **0.1** = **0.1** p.a.
LE, LSH: **0.1** × 0.0125 = **0.00125** p.a.
LE, SV: **0.1** × 0.0125 = **0.00125** p.a.
LC, LSH: **0.1** × 0.0125 = **0.00125** p.a.
LC, SV: **0.1** × 0.0125 = **0.00125** p.a.
Total **0.105** p.a.

Major advantages of this method are:

- It is often easier to visualize the "signal paths" required for the system to operate safely than to visualize the failure combinations that will lead to system failure, that is, one prepares a list of signal paths instead of a fault tree;
- There are computer programs that will identify the minimal cutsets, which can become complex; and
- there is no need for Boolean algebra, which can become very tedious and error-prone in a complex system.

Appendix A

Conducting a Hazop Study

1. OBJECTIVES

The objective of Hazard and Operability Studies (Hazop) is to facilitate smooth, safe, and prompt commissioning of new plant, without extensive last-minute modifications, followed by trouble-free continuing operation.

The track record of Hazop studies is impressive. Wherever the technique has been applied in accordance with the principles set out below, the results have been:

- smooth, trouble free commissioning and startup;
- greatly reduced (expensive) last-minute modifications;
- well-briefed staff; and
- smooth subsequent operation, (except where Hazop recognized possible problems that were not subsequently resolved).

2. ESSENTIAL FEATURES OF A HAZOP STUDY

Hazop studies can take a variety of different forms, which can lead the casual observer to wonder what it is that makes a Hazop study different from some other form of meeting or review.

The essential features of a Hazop study are as follows:

- It is a systematic, detailed study following a preset agenda.
- It must be conducted by a team comprising members with a variety of backgrounds and responsibilities, representing all the groups with

a responsibility for the operation (e.g., a Hazop of a new project in design would have representatives from design, construction, and ultimate operation).

- It concentrates on exploring the possibility and consequences of deviations from normal or acceptable conditions.
- It is an audit of a "completed" design, that is, the design would be sent for construction in that form if it were not for the Hazop study. (However, it is expected that the Hazop study will result in the design being changed; hence the use of quotation marks around "completed.")

In outline, a study takes the form of a discussion, examining each element of a design or operation in turn, considering a checklist of possible deviations for each element. For each postulated deviation, an attempt is made to envisage ways in which the deviation could occur, and for each such way a judgmental estimate is made both of the severity of the possible consequences and of the likelihood. If the meeting comes to the view that the combination of the severity and the likelihood is sufficient, the deviation is noted as a problem to be resolved. If resolution is likely to require little discussion, it may be tackled in the meeting. Deviations apparently requiring significant effort for resolution are listed for attention outside the meeting.

3. TIMING IN A PROJECT

Normally in a new project, a series of design reviews are held to ensure that the design will be able to perform the duty required. These give rise to series of revisions.

When the design is at a stage ready to be frozen and issued for construction, the Hazop should be scheduled. (This necessitates allowance being made in the project timetable. If the sequence of study of the different sections of the design is planned carefully with the project timetable in mind, the delay to the project need not be severe, and in any case it is usually amply made up by elimination of delays in the commissioning period.)

If it is found, in the Hazop study, that the design is not quite firm, then it is necessary to defer Hazop of that section until it is firm. An attempt to finish off the design during the Hazop will prove to be frustrating, and it undermines a principle of Hazop, that is, to audit a complete design.

4. SELECTING THE TEAM

It is important to assemble a good team. The essential requirements are representatives of all groups involved, such as design, construction, and operation, with the representatives bringing both technical know-how and sufficient organizational seniority to have the agreed actions implemented. There must also be at least one person who knows how to run a Hazop study.

It is not essential to have an independent chairman, but in big projects where the meetings may have a large attendance (e.g., 8 to 10) it can be helpful to have the proceedings led by someone whose prime task is to watch the pace and the dynamics of the meeting. (The role of the study leader or chairman is discussed in more detail later.)

In a study of a design for a small new facility, the team could comprise the following as a bare minimum:

- the design engineer,
- the project engineer responsible for managing the construction, and
- the plant superintendent (designate),

From these, it is important to appoint as leader someone who has experience in the conduct of the studies.

For a larger project, the team might comprise:

- the senior design engineer,
- the design engineer responsible for the section being studied,
- specialists, such as an instrument engineer, or a mechanical engineer,
- the project engineer,
- the plant superintendent designate,
- one or two additional production staff as they are appointed, and
- an independent chairman.

For a study of an existing facility, the team might comprise:

- the plant superintendent (staff member responsible for all plant performance),
- the plant engineer,
- a suitable representative of the technical support departments,
- a supervisor or foreman, and
- possibly an operator or a tradesperson or both.

There is sometimes reluctance to include operators or tradespeople, but if the climate permits, there can be big benefits:

- from their close contact with the operation they can make an important contribution to the understanding of what actually happens;
- they learn more about the way the facility is intended to be operated and why; and
- teamwork is developed.

5. ROOM LAYOUT AND EQUIPMENT NEEDED

One effective way of setting up the room ready for the study is illustrated in Figure A-1.

The features of the layout and equipment are as follows:

- The relevant drawings (P&I diagrams, layouts, detail drawings, etc.) are fastened to a wall.
- The members of the study team sit in a semicircle around the drawings.
- There is no table in the room.

Figure A-1. Arrangement of the room for a Hazop study.

- The study chairman sits at one end of the semicircle, with a bench or shelf at his side, on which is placed the open book of guidewords.
- Next to the wall on the side opposite the chairman is an easel with flipchart paper to be used by team members to illustrate their ideas.
- The study secretary uses a clipboard rather than a table.
- The team members do not have their own copies of the drawing.
- There is only one row of seating.

Briefly, these arrangements have evolved for the following reasons:

- If individual team members have their own drawings, they tend to start private discussions with their neighbors, rather than concentrating on the main discussion.
- An important feature of Hazop studies is the informality of discussion, with members free to get up and go to the drawings or the flipchart board to illustrate a point. Having a table in the room greatly inhibits that freedom of movement.
- With only one drawing in use, it becomes the focus of attention, and changes marked on it are official.
- If some of the team members are in a second row, they become second-class citizens and cannot contribute as effectively to the discussion. Visitors, such as senior managers, can inhibit discussion if they are sitting at the back, so such visitors can be accepted only if they join the semicircle and become team members for the duration of their visit, being expected to contribute with the rest.
- A flipchart board is preferred to a blackboard, as its record is permanent and can be referred to in later studies.

In summary, the equipment needed comprises:

- book of guideword cards,
- bench to stand or prop the book on,
- easel and flipchart board with paper,
- clipboard and record sheets (see later) for the secretary,
- masking tape, etc., for fastening drawings to the wall,
- felt-tipped pens of different colors for use on the flipcharts,
- highlighter pen (yellow or green suggested) for indicating the line under discussion at any time, and
- fine fiber-tipped red pen for marking agreed changes on the drawings.

6. CONDUCT OF THE STUDY

1. At the first meeting of a team, if there is anyone present who has not taken part in a Hazop, the chairman outlines the study procedure. This normally takes around 5–10 minutes. This may cover the following points:
   - objectives of Hazop;
   - essential features of Hazop;
   - the fact that Hazop focuses on identifying abnormal circumstances that could upset normal operation;
   - that fact that, because of this focusing on abnormalities, and the team approach, it is normal in Hazop for even the best designs to be found to have potential for improvement, and it is no reflection on anyone if faults are found; and
   - a brief outline of the steps in a study.
2. The chairman then asks someone with a good understanding of the design to outline the broad purpose of the section of plant covered by the drawing under study, and its normal mode of operation. This should be kept to just an outline, as the details will be covered later in the discussion. (Allow 5 to 10 minutes.) Following that, questions are invited where clarification of the purpose or mode of operation is needed, but questions about detail are deferred until later.
3. The detailed study of the first section of pipeline (or the first step in a batch operation) then starts. The chairman marks the selected section with the highlighter pen, using a dotted line. He then asks someone to describe the line and its associated equipment including the vessels at each end, its purpose, its normal operating condition, and its normal method of operation. There is then a short period of general discussion, limited to around 10 minutes as there is a tendency for the discussion to become a random questioning of design features that will be more systematically covered later.
4. The chairman then uncovers the first guideword, such as "Flow—high." He asks three questions:
   - "Is too high a flow conceivable?"
   - "What are the possible consequences of too high a flow?"
   - "How likely is it?"

If the group, in discussion, forms the view that the combination of the severity of the consequences of an event related to too high a flow (e.g., a leak) with the likelihood of its occurrence is unacceptable, then the event is defined as a problem needing resolution.

Resolution of an identified problem can be undertaken in the meeting if the expectation is that it will be complete in around 5 minutes, but if it is apparent that more time will be needed, or if someone outside the meeting needs to be consulted, or files consulted, or a calculation done, then resolution outside the meeting should be arranged.

The secretary should record:

- resolved problems with their solution, and
- unresolved problems, and the person nominated to arrange for resolution outside the meeting,

Where appropriate, the solution is marked up on the drawing on the wall using the red felt-tipped pen.

Generally no record should be kept of the discussion where no problem is found, as this inhibits the free flow and creativity of the meeting.

When no further problems are identified for the first guideword, the chairman turns to the next guideword. (There is no reason why the card cannot be turned back if someone later thinks of an avenue to be explored.)

When all the first group of guidewords have been used for the first section of pipeline, or the first step in a batch operation, then the chairman marks in that section of pipeline with the highlighter pen, using a continuous line as a sign that that section is complete.

Suggested guidewords for this stage follow. This selection is designed primarily for study of continuous process plants or operations.

FLOW:	leak, high, low, reverse, two-phase
LEVEL:	high, low
TEMPERATURE:	high, low
PRESSURE:	high, low, vacuum
REACTION RATE:	fast, slow
QUALITY:	concentration, impurities, cross-contamination, side reactions, inspection and testing
PHYSICAL DAMAGE:	impact, dropping, vibration
CONTROL:	response speed, sensor and display locations, interlocks
PROTECTION:	response speed, independence, testing

Other selections of guidewords suitable for batch operations, and for materials handling applications, are as follows.

For materials handling operations:

LOAD:	overload, underload
LOCATION:	wrong horizontal/vertical location
DIRECTION:	to one side, upward, downward, reverse
SPEED:	too fast, too slow

For sequential batch operations:

TIMING:	start too early, too late; stop too early, too late; duration; sequence

5. The next section of pipeline is then selected and marked with a dotted line using the highlighter pen, and the above process is repeated.
6. When all pipelines or all stages of a batch operation have been covered, then the chairman moves to the second group of guidewords, which are used to guide an overview of the whole drawing.

These overview guidewords are as follows:

MATERIALS OF CONSTRUCTION:	suitability for abnormal process conditions, corrosion, erosion
SERVICES NEEDED:	air, nitrogen, water, steam, power, etc.
COMMISSIONING:	authorities, training, supervision, compliance checking
STARTUP:	sequence, problems
SHUTDOWN:	isolation, purging
BREAKDOWN:	loss of services, "fail safe" response, emergency procedures
ELECTRICAL SAFETY:	area classification, electrostatic discharge, grounding
FIRE AND EXPLOSION:	prevention, detection, protection, control; automatically actuated isolation valves
TOXICITY:	acute, long-term; adequacy of ventilation
ENVIR'L CONTROL:	effluent: gaseous, liquid, solid; noise; monitoring
ACCESS:	for operation, maintenance, means of escape
TESTING:	raw materials, products, equipment, alarms and trips
SAFETY EQUIPMENT:	personal equipment, safety showers
OUTPUT:	sources of unreliability, bottlenecks
EFFICIENCY:	potential for loss of material or performance

Although this second group of guidewords appears formidable, discussion with them rarely takes more than around 30 minutes, as nearly all of the issues will have been raised earlier in the line-by-line discussions.

7. When the overview is complete, the chairman signs the drawing as complete and arranges for issue of the record sheets and for follow-up of the outstanding actions.

7. SELECTION OF GUIDEWORDS

The guidewords in the lists above have been found appropriate to continuous plant in numerous projects and have also been used for batch processes and materials handling operations, but they have not been as extensively tested in those latter applications. It is the responsibility of the team leader to ensure that all the appropriate guidewords are used for any particular study.

There is nothing sacred about any particular set of guidewords. There are many variations in use. But they all have a common factor: they prompt discussion about all significant types of deviation from all the required "qualities" such as level, temperature, flow, and sequence. So, when planning a Hazop study for an unusual process or operation, the leader should (preferably in discussion with others) identify the important qualities and modify the guidewords as necessary to ensure that all significant deviations will be discussed.

It is better to have too many guidewords than too few. If a particular guideword is inapplicable in a particular case, it can be passed over with no loss of time.

8. DURATION OF STUDY

Half a day is a good duration, that is, around 3.5 hours including a coffee break in the middle. It is possible to have two such sessions in the one day, but that leaves the participants no time to attend to the actions arising from the studies or to their other work.

In a large project, however, it is perfectly satisfactory for different teams to be operating concurrently; a "morning" team and an "afternoon" team, preferably with the same chairman.

9. STUDY RECORDS OR MINUTES

For a small study, such as for a small modification of an existing plant, the minute sheet may be of the form shown in Figure A-2. However, where several meetings will be needed, and many changes are thus to be expected, follow-up is aided by using a separate sheet for each identified problem as shown in Figure A-3.

The secretary, whose main task is to record the details of the identified problems and the nature of the solution agreed, or the nature of the

HAZOP STUDY RECORD SHEET NUMBER:..................

PLANT NAME:... DATE:......./......../.......

PLANT SECTION:.................................... PRESENT:......................................

No	Description of Problem	Solution or Action	Respons.

Figure A-2. Hazop study record for small study.

investigation to be undertaken outside the meeting, needs to be very familiar with the project and competent to interpret the thrust of the discussion in deciding the wording to be used in the minutes. (Any attempt to use an administrative assistant leads to loss of time while someone dictates the wording and the rest of the people at the meeting have their say about how they think it should be worded!) Unfortunately, this leads to the problem that the secretary tends to be inhibited from full participation in the discussion because of his or her preoccupation with the minutes. The best way around this, which has been found so far, is for the discussion to move on while the secretary makes note of the problem, and he asks for a pause only if he becomes aware that he is missing something to which he can make a special contribution.

After each meeting, the secretary sends a photocopy of each minute sheet to those named for action on it (either to have a problem resolved, or to implement a solution agreed at the meeting), and keeps the master sheet in a folder. As each action is completed, the lower half of the minute sheet is filled in by the person responsible, and a photocopy of the completed sheet sent to the secretary, who files it in the master folder, removing the uncompleted original sheet. Then the status of the actions can be easily seen at any time by flipping through the master folder and noting which of the original sheets have not yet been replaced by one with the lower half completed.

HAZOP STUDY RECORD SHEET NUMBER:...........

PLANT NAME:....................................... DATE:......../......./.......

PLANT SECTION:................................... PRESENT:..

IDENTIFIED PROBLEM PERSON RESPONSIBLE FOR

ACTION:...........

INTERIM ACTION DEFINED AT MEETING

DEFINED FINAL ACTION

SIGNED:........................

DATE:....../......./........

Figure A-3. Hazop study record for individual identified problem.

10. ROLE OF THE CHAIRMAN

Whether the study chairman is one of the project team, or someone from another area, the role is the same: to ensure that the technical result of the Hazop is sound, without inefficient use of people's time.

A sound technical result is one in which all the significant hazards and operational problems have been identified, and a proper balance found between eliminating the problems and managing them.

It is sometimes said that Hazop results in overdesigned plants. It is the responsibility of those present to ensure that this does not happen, and it is the responsibility of those selecting the participants that they choose people whom they can rely on to find a proper balance between eliminating and managing problems.

The chairman must constantly aim to have the team achieve these objectives. In doing so, the points he or she should pay attention to may be summarized as:

- "group dynamics,"
- technical standard, and
- pace.

Apart from the usual methods of discussion leadership, leadership of Hazop studies has some special points worth noting. In the field of group dynamics, the main points are:

- watching that those who may have a contribution to make on a particular topic have the opportunity to do so;
- ensuring that debates or arguments are resolved on rational grounds rather than on seniority or force of personality; and
- keeping track of all the points raised during discussion of any particular guideword, to ensure that those which are slightly peripheral to the main thrust of the discussion are not forgotten but picked up and considered before moving on to the next guideword.

Maintaining a good technical standard is not difficult if the right people have been selected for the study team. However, even the best team may need an occasional prod. The main points are:

- ensuring that important topics, or critical sections of plant or operations, are fully discussed;
- maintaining an independent judgment about the technical standard of the solutions agreed at the meeting, and about the feasibility and durability of managerial action defined to cope with residual risks where it is decided that these cannot reasonably be further reduced; and
- aiming to have problems tackled systematically, seeking to:

  reduce or eliminate the inherent hazard or problem,

  improve containment or control, so that the likelihood of a problem arising is reduced,

  improve protective systems and response, to improve the chance of stopping an initiated problem early ("nipping it in the bud"),

  limit the damage by providing separation, or by strengthening the buildings or structures potentially exposed, and make sure that a mature judgment is reached about which hazards should be reduced or eliminated by expenditure on equipment, and which should be recognized and managed. It is sometimes helpful to remind the team that it is their responsibility to ensure that no one outside the meeting can fairly make the accusation that the results of the study are unbalanced.

A common problem of Hazop studies is that they slow down in the interests of not missing anything. The result is that they take so long that no-one can spare the time for future studies, or else those chosen are too junior and inexperienced for the studies to be effective. It is important that the chairman keep the study moving on at a good pace. It is better to find 90% of the problems (i.e., be confident of finding all the major problems) and continue doing studies in future, than to do one study to perfection and then stop.

The chairman must be constantly trying to get the group to move on, while being alert to any issue that still needs exploration.

Pitfalls (or "Traps for the Unwary")

Once learned, the techniques of Hazop are so simple to follow that those involved sometimes forget important principles. Some of the difficulties that have been encountered by various teams and companies include the following:

- A team may decide, in the interests of speed, to dispense with the book of guidewords and to scan over the drawing using their "understanding of Hazop principles." In fact, they are demonstrating their ignorance of Hazop principles, as they are not complying with the requirements for:

  systematic study, and

  detailed study.

- A senior engineer once said, on first having the Hazop techniques explained to him: "That is a very useful procedure. We have a plant at the detailed design stage right now. I'll take the drawings home this evening and Hazop it myself." He was overlooking the requirement for the study to be undertaken by a team comprising members with a variety of backgrounds and experience.
- A team may record a list of changes identified in a Hazop study session, only to have the list reviewed by a more senior person outside the meeting who then deletes or modifies a number of the changes. Such actions will rapidly reduce a Hazop study to impotence. Anyone who

feels a need to exercise managerial authority over the actions decided on at a Hazop study must either

join the team, so that he can base any decisions on the insights gained in the study, or

change the membership of the team to include people he is confident will ensure that all decisions are appropriate.

- A team may decide to concentrate on the sections of the plant which handle the most hazardous materials, and not bother with noncritical sections such as the water treatment plant. Although any faults in the water treatment plant may not have the potential to cause major hazards, nevertheless it has been common experience that study of such sections reveals numerous design faults that would otherwise be found during the commissioning period and corrected then with much greater expense and delay.

  (One reason for this is that the "straightforward" sections of the design are often given to the more junior staff and are not checked as thoroughly because of their simplicity.)

- Similarly, it may be decided not to study those sections of plant which are supplied as "packaged units," such as standby generators or small boilers. Again, experience shows that there are often major weaknesses in the way in which these sections of plant have been designed to fit into, and interact with, the whole.
- A design may be submitted for Hazop before some details have been finalized in the belief that "we can decide that during the Hazop study." Such resolution of options can only be undertaken by suspending the normal Hazop procedures, which severely interrupts the flow of the discussion. What develops is an increasingly unstructured discussion, which becomes a design working meeting, rather than a systematic audit of a completed design. The drawing should be sent back for further work and completion before being taken further in Hazop.

EXAMPLES OF APPLICATION OF HAZOP

1. A plant for making a common plastic from a hazardous liquefied flammable gas was submitted to a Hazop study. This resulted in more than 1300 actions being identified, mostly minor, but some more important. Most of them related to avoidance of operational problems, with around 30% being related to safety and environment protection. An assessment of the benefits of the study showed that the cost of carrying out the study

(staff time, cost of changing the design, cost of additional equipment, etc.) would be recouped twice before the plant started up by avoiding the need to make essential changes during construction, and would return around 30% per year subsequently because of the improved performance of the plant in ways which it would be uneconomic to attempt if the changes had to be made to physical plant rather than on the drawing board.

2. Automatic operation of a large fuel depot, with limited manning, was studied to check the proposed manner of operation and the controls and alarms to be incorporated in the computer-based control, protection, and security interlock system. Numerous potential problems were identified, leading to a much more reliable, secure, and operable system.

Appendix B

Failure Mode and Effect Analysis (FMEA) and Failure Mode and Effect Criticality Analysis (FMECA)

1. INTRODUCTION

Failure Mode and Effects Analysis (FMEA) is a form of nonquantitative analysis that aims to identify the nature of failures that can occur in a system, machine, or piece of equipment by examining the subsystems or components in turn, considering for each the full range of possible failure types and the effect on the system of each type of failure. Failure Mode and Effects Criticality Analysis (FMECA) is an extension of FMEA that assigns a rating to both the severity of the possible effects and their likelihood, enabling the risks to be ranked.

FMEA and FMECA are most applicable when only one type of impact is being considered, such as production loss, or safety, or environmental damage, not a combination of them. Where a combination of types of impact is to be considered, it is preferable to use rapid ranking that is structured and designed for computer sorting, and is thus more flexible.

However, both FMECA and rapid ranking use the same basic approach:

- subdivision of the equipment or operation into "bite-sized" subsections,
- for each subsection, systematically examining the potential for each of a checklist of possible failures or mishaps,
- for each identified failure or mishap, estimating the potential severity of the consequences and the likelihood, and
- determining a risk score for each identified failure or mishap.

2. FAILURE MODE AND EFFECTS ANALYSIS (FMEA)

FMEA is often undertaken by one person alone, but to meet the requirements of effective risk identification it is very desirable that it be undertaken by a small team (e.g., three people) with a variety of backgrounds (such as design, production, maintenance).

Typical applications of FMEA include:

- identification of specific scenarios when undertaking a risk assessment of equipment or machinery;
- identification of specific scenarios when undertaking a risk assessment of a modification to equipment or machinery; and
- identification of the specific scenarios when studying a single risky activity.

The steps in FMEA are:

1. Define the scope of the study, by defining the limits of the machine, machine section, system, or subsystem to be studied. This is often specified most clearly by listing the main features which are included, and the main features (if any) which are explicitly to be excluded.
2. Decide the level of analysis. This can be difficult to decide, as there is always the possibility that going into extra depth of detail may uncover a further problem that needs to be tackled. However, if the dangers of "analysis paralysis" (i.e., spending so much time on analysis that the marginal cost greatly exceeds the marginal benefit and other productive efforts grind to a halt) are borne in mind, a reasonable balance can be found. The level of analysis is determined by the selection of the elements for study. A detailed study of a machine could, in the extreme, consider each individual physical component in turn as an element for separate study, whereas in a broad study the main subsystems may be regarded as the elements.

3. For the types of element selected, identify and list the variety of failure modes possible. For individual mechanical components these could include:
   - mechanical breakage,
   - excessive wear,
   - corrosion, and
   - deformation (elongation, compression, bending).

   For electrical components the failure modes could also include:
   - open circuit,
   - short circuit,
   - increased resistance,
   - reduced resistance, and
   - insulation breakdown.

   Instrument failures could include:
   - reading too high,
   - reading too low,
   - seizing/not moving, and
   - responding too slowly.

   On the other hand, if the analysis is limited to consideration of subsystems, rather than components, the failure modes would be those for such subsystems. They could include:
   - premature operation,
   - failure to operate when needed,
   - intermittent operation,
   - failure to cease operation when needed,
   - loss of output or failure during operation, and
   - unsatisfactory output.
4. For each element (whether component or subsystem) to be studied, consider each of the listed possible failure modes, and identify the effect on the machine or system as a whole and the relative importance of those effects. These effects could include:
   - injury to people,
   - damage to the environment,

- damage to equipment,
- loss of production,
- reduced quality of production, and
- increased cost of operation.

5. For each failure mode for each element studied, identify:
   - the means of preventing the failure by design, operating and maintenance practices, and management;
   - the means of detecting the failure and responding effectively to it; and
   - means (if any) of limiting the impact of the failure, particularly by design changes.
6. On completion of the analysis, review the options for reduction of the likelihood or effects of the failures, and document the recommendations for action.

3. FAILURE MODE AND EFFECTS CRITICALITY ANALYSIS (FMECA)

Failure Mode and Effects Criticality Analysis is an adaptation of FMEA to enable a semiquantitative examination of the risks arising from the potential failures. In essence, it entails assessment of the severity of the impact from each potential failure scenario (i.e., combination of failure type and failure cause) and of the frequency of occurrence. The risks associated with each failure scenario is then determined by some means, such as:

- multiplication (if the consequences and likelihood have been expressed in numbers representing their actual magnitude);
- a risk matrix (if the consequences and likelihood have been expressed in words—e.g., low, medium, high—or defined on some nominal scale such as low = 1, high = 5).

A simple nonquantitative form of risk matrix is illustrated in Figure B-1. Other more quantitative forms are shown elsewhere in this text, and in Appendix D of AS/NZS 4360-1995, "Risk Management."

Typical applications for FMECA include:

- identification and ranking of the specific scenarios when studying a single risky activity;

Severity / Frequency	LOW	MEDIUM	HIGH
HIGH	Medium Risk	Medium-High Risk	HIGH RISK
MEDIUM	Low-Medium Risk	Medium Risk	Medium-High Risk
LOW	Low risk	Low-Medium Risk	Medium Risk

Figure B-1. Example of risk matrix for use with FMECA.

- identification and ranking of specific scenarios when undertaking a risk assessment of equipment or machinery; and
- identification and ranking of specific scenarios when undertaking a risk assessment of a modification to equipment or machinery.

If the severity of the consequences and the frequency of occurrence can be estimated (in broad groups with scales that are orders of magnitude (e.g., 1, 10, 100, 1000 for the severity, and 10, 1, 0.1, 0.01 per year for the frequency), then the risk matrix can be calibrated accordingly and the risk magnitude calculated in numerical terms as the product of the two scales. Cells in the matrix which lie on a diagonal can be seen to have equal risk magnitudes (but this only applies if the severity and frequency scales are calibrated using the same steps between them).

It should be noted that both FMEA and FMECA rely on the judgment of those undertaking them, as do most simple methods of risk assessment.

Appendix C

Application of Rapid Ranking to Improve the Reliability of Supply to the Market

BACKGROUND

Plant XYZ is the sole source of supply of natural gas for domestic and industrial use to several cities and provincial centers.

During recent months, the design of the plant has been carefully reviewed in relation to its potential impact on the safety of people, and a program of work is in progress implementing the recommendations of those reviews. Concurrently, a program of implementation of improved or new safety-related procedures is being undertaken.

In view of the reliance of the population centers on gas supplied by Plant XYZ, it is critical that the plant continue to operate without interruption to that supply. Recently duplicate units were constructed and commissioned for several critical functions, thus providing improved security of supply in the event of breakdown of any single unit.

It is recognized that some parts of the plant are not provided with backup. Further, there are other sections of the plant where a single serious incident (e.g., fire) could conceivably damage critical equipment such that production is interrupted.

Thus it has been recognized that the vulnerability of the plant to single incidents needs to be examined.

Such an examination, to be rigorous, would require a large amount of highly detailed investigation, much of which would, in the event, be found to have been of plant sections where the potential for interruption of production was small. It was therefore decided to undertake a "screening" review, to identify those units or sections of plant that presented the main risk of such interruption of production, and to define the options for each such units or sections, the options possibly including a more detailed and narrowly focused study.

Objectives

The objectives of the assessment are:

1. to identify which units at Plant XYZ appear to present most risk of causing interruption to full supply of gas to the market;
2. to identify the main options for eliminating or reducing or managing those risks; and
3. to provide adequate documented support for those findings.

Scope

The assessment is to be completed within around 2 months. Thus it will not be possible to investigate in detail. This means that it will not be possible to:

- prepare a comprehensive list of the possible scenarios by which continuity of supply could be lost;
- undertake a statistically based quantitative assessment of either the extent or the probability of any interruption to supply; or
- determine the absolute value of the risks of such interruptions.

The assessment is thus to develop a relative ranking, based on subjective, but experienced, judgment.

Limitations

There is an inevitable trade-off between the level of detail of the assessment and the level of confidence that all the significant risks have been identified. This is a short-listing investigation, with the inherent possibility that some significant scenarios may be overlooked.

METHOD

Introduction

The method is based on identification of physical features of the plant and the related systems and procedures that introduce greater or lesser risk of interruption to production of natural gas for the market.

This in turn entails identification of features which:

- present a potential for an incident (e.g., fire, explosion, mechanical breakdown), or
- are vulnerable to impact from such an incident (e.g., damage).

By examination of the types of incident and the related vulnerabilities, estimates can be made of the severity of the consequences of incidents, that is, the extent of any interruption to supply to the market (magnitude and duration).

For those postulated incident/vulnerability combinations that are judged to have significant impact on production, the features of the plant and the related systems and procedures which introduce greater or lesser likelihood of occurrence of the impact are considered, resulting in a judgment about the likelihood.

A matrix can then be constructed for the various levels of impact and the various levels of likelihood of that impact, and the plant units will be categorized according to those combinations.

For those plant units which are judged to have combinations of consequence and likelihood that constitute the most significant risks, consideration will then be given to the options for action aimed at eliminating or reducing those risks. Where plant units are judged to have potential incidents with serious consequences, but with low likelihood, consideration will be given to means of managing those risks so as to keep the likelihood low, and to detect any increase in the likelihood.

The method is designed such that it can be undertaken in either of two ways:

(a) nonquantitatively, using a simple rating system of HIGH/MEDIUM/LOW for both consequence and likelihood assessments, and deriving the risk by use of a risk matrix; OR

(b) semiquantitatively, by subjectively assigning "scores" to both consequence and likelihood, the scores being derived as $\log_{10}$ of the estimated consequence (equivalent full days loss of supply) and likelihood (frequency per year).

Consequence Estimation

Approach

The Hazard/Vulnerability Worksheet (Table C-1) is used to assist in deciding the consequence rating or consequence score. It requires those undertaking the study to consider each class of equipment (e.g., structure, vessels, pipework—see the Vulnerability Checklist in Table C-2) in the plant section being studied, together with the sources of inherent hazard (e.g., extreme process conditions, inventories of hazardous materials—see the Hazard Sources Checklist in Table C-3) in that plant section.

Use of the Vulnerability/Hazard Sources Worksheet

By examining each plant unit for hazard sources, and looking for and considering the vulnerabilities which may be affected by those hazard sources, the worksheet is completed, placing a sequential reference number (starting at 1 for each unit) in the cell for each combination which warrants further consideration. These combinations are referred to from here on as "potential incidents."

As an aid to memory, a separate note is made about each numbered potential incident, for example, in some form such as the Potential Incident Log in Table C-4.

Qualitative Consequence Rating

If it has been decided to use a qualitative rating, a judgment is made whether the possible consequences of each potential incident would be Minor/Medium/Serious. It would be helpful early in the estimation process to define "benchmark" incidents of each of those three categories.

Semiquantitative Consequence Score

If it has been decided to use a semiquantitative assessment, an estimate is made of the probable duration and extent of the loss of supply to the market. The magnitude of the impact on the ability to supply the market depends on

Table C-1
Hazard/Vulnerability Worksheet

VULNERABILITIES / HAZARD SOURCES	Structures	Vessels	Pipework	Cabling	Instruments	Motors	Pumps	Machinery	Impact from Elsewhere	Service: Power	Service: Water	Service: instruments	Service: Computer	Service: Relief/Blowdown	Service: Compr. Air	Service: Drainage	Other	Comments
Demanding Process Conditions																		
Flow																		
Pressure																		
Temperature																		
Corrosion																		
Other																		
Process Material Energy: large inventory or availability of:																		
Flammable liquid																		
Flashing flammable liquid																		
Flammable gas																		
Compressed non-flam gas																		
High-pressure water																		
Other																		
Mechanical Energy/Momentum																		
Linear motion (e.g., vehicles)																		
Rotational motion																		
Vibration																		
Seals																		
Other																		
Effects from Nearby Units																		
(Unit)																		
Ignition Sources																		
Flame -continuous																		
-intermittent																		
-other																		
Electrical -continuous																		
-intermittent																		
-static electricity																		
Hot surfaces																		
Other																		
Other																		

Table C-2
Vulnerability Checklist

Structures
Vessels
Pipework including joints
Cables
Instruments
Motors
Pumps
Machinery
Impact on nearby/other plant units
Sources of essential services:
• Power
• Water
• Instrument signaling
• Compressed air
• Relief and blowdown
• Drainage
• Computing

Table C-3
Hazard Sources Checklist

Demanding Process Conditions:
• Flow
• Pressure
• Temperature
• Corrosion
• Other
Process Material Energy
Large inventory or availability of:
• Flammable liquid (e.g., oil)
• Flashing flammable liquid (e.g., propane)
• Flammable gas
• Compressed nonflammable gas
• High-pressure water
• Other
Mechanical Energy or Momentum
• Linear motion (e.g., vehicle collision)
• Rotational energy (e.g., turbines)
• Vibration
• Seals (e.g., on pumps)
• Other
Ignition Sources
• Flame
• Continuous (e.g., fired heaters)
• Intermittent (e.g., welding)
• Other
• Electrical
• Continuous (e.g., switch room)

Table C-3
Continued

- Intermittent (e.g., power tools)
- Static electricity
- Other
- Hot surfaces
- Other

Impacts from Nearby/Other Plant Units
- Fire
- Explosion

Table C-4
Potential Incident Log (Accompanies Hazard/Vulnerability Worksheet)

PLANT UNIT____________
DATE ____/___/___

Description of Potential Incident	**Consequence Rating/Score**
1.	
2.	
3.	
4.	
5.	
6.	
7.	
8.	

the extent of the reduction of supply rate (e.g., full cessation of supply, partial, negligible) and the duration of that reduction (e.g., 1 day, 1 week, 1 month).

A simple logarithmic scoring system based on the duration of the loss of supply, measured in 100% lost days, can be developed to match those periods of day, week, and month, for example, Table 3-6 in Chapter 3.

When, for each plant unit, the identified potential incidents have been rated or scored according to the effect and duration of the impact on production, those ratings/scores are examined for any evident anomalies (i.e., ratings/scores that appear too high or too low in relation to others), and the appropriate adjustments are made.

A cutoff level may then be defined at a level where the magnitude of the Consequence Rating/Score is sufficiently low not to warrant further attention at this stage, such as where the impact on production is insufficient to have any significant effect on the consumers in the market. Those above this cutoff level are then assessed for their likelihood, so as to enable an indicative assessment of the relative level of risk that they present.

Table C-5
Likelihood Factors Checklist

Relative Complexity of Hardware:	• Pipework, jointing, etc.
	• Small bore pipework, branches, etc.
	• Control instrumentation
	• Mechanical arrangement
	• Structural requirements or arrangement
	• Other
Relative Complexity of Systems and Procedures:	• Operational requirements or procedures
	• Maintenance requirements or procedures
	• Emergency response requirements or procedures
	• Other
Relative Difficulty of:	• Understanding the process and its control
	• Controlling the plant within design parameters (e.g., frequency of alarms and trips)
	• Understanding hazards/scenarios
	• Monitoring plant condition
	• Monitoring operational practices
	• Monitoring maintenance practices and standards
	• Monitoring emergency preparedness
Frequency of:	• Process upsets
	• Maintenance activities

Likelihood Estimation

Introduction

The likelihood of incidents is affected by many factors, including those shown in the Likelihood Factors Checklist in Table C-5. Additional items can be added to cover plant-specific hazards and situations.

Use of the Likelihood Factors

The likelihood factors are set out across the top of the Potential Incident Likelihood and Risk Estimation Worksheet (Table C-6).

The plant is examined in relation to the Likelihood Factors, and a mark is placed in each cell where one of the likelihood factors applies to a particular potential incident.

Qualitative Likelihood Rating

If it has been decided to undertake a qualitative rating, then a judgment is made, by consideration of the likelihood factors for each potential incident, whether its likelihood is

Table C-6
Potential Incident Likelihood and Risk Estimation Worksheet

From the Hazard/Vulnerability Worksheet and the Potential Incident Log.			Likelihood estimation: either as a Likelihood *Rating* or a Logarithmic *Score*																													
	LIKELIHOOD FACTORS	A. CONSEQUENCE RATING/SCORE	Complexity of Hardware	Pipework, incl. joins	Small bore pipework	Control instrumentation	Mechanical arrangement	Structural requirements	Other	Relative Procedural Complexity	Operational requirements	Maintenance requirements	Emergency response requirements	Other	Relative difficulty of . . .	understanding the process	controlling the process	understanding hazards & scenarios	monitoring plant condition	monitoring operational practices	monitoring maintenance practices	monitoring emergency preparedness	Other	Frequency of . . .	process upsets	maintenance activities	other human involvement	other	B. LIKELIHOOD RATING . . . OR	C. INITIATION FREQUENCY SCORE	D. MITIGATION FAILURE PROB. SCORE	E. RISK RATING/SCORE
Selected Potential Incident Details																																
NO HAZARD SOURCE	VULNERABILITY																															

(A) likely,

(B) reasonably possible, or

(C) highly unlikely.

That rating is inserted in column B of the Potential Incident Likelihood and Risk Worksheet (Table C-6).

Semiquantitative Likelihood Score

If it has been decided to use a semiquantitative approach, by consideration of the likelihood factors for each potential incident a judgment is made of the initiation frequency, and the appropriate Initiation Frequency Score (e.g., Table 3-6) is inserted in column C of the Worksheet.

(Because it is not possible to estimate the likelihood in quantitative terms with confidence without extensive mathematical and behavioral analysis, it is necessary to use a simple scoring system which can be used by consensus among experienced production and engineering staff.)

Similarly, an estimate is made of the probability of failure of the measures available for mitigation of the incident, for example, Table 3-7.

The frequency of the estimated consequences occurring is the initiation frequency multiplied by the probability of the mitigation measures failing to prevent the incident from developing to that extent. Where a logarithmic scoring system is used, the initiation frequency score plus the mitigation failure score is the logarithm of the frequency.

Risk Level

Introduction

Risk is the product of severity of consequence multiplied by the frequency or likelihood.

Risk can be estimated either purely qualitatively, or semiquantitatively, as below.

Qualitative Risk Estimation

Table C-7 may be used to estimate the level of risk.

(Note that there are potential problems and anomalies with any qualitative "Risk Matrix" such as that in Table C-7. If the intervals between various levels of consequence and of likelihood are not the same, the risk levels

Table C-7
Qualitative Risk Matrix

	Consequence (from column A)		
Likelihood (from column B)	**Minor**	**Medium**	**Serious**
A: Likely	Medium Risk (3)	High Risk (4)	Very High Risk (5)
B: Reasonably Possible	Low Risk (2)	Medium Risk (3)	High Risk (4)
C: Highly Unlikely	Very Low Risk (1)	Low Risk (2)	Medium Risk (3)

Key:

Highest Priority for Risk Reduction

Needs ongoing management to keep the likelihood low

on the diagonals are not equal, potentially leading to distortion of the ranking of the risks.)

The risk levels may be summarized as lying on a scale from 1 to 5. But note that these numbers may not represent the relative magnitudes of the various risks at all accurately. For example, there may be a greater difference between (say) risk levels 2 and 3 than between (say) risk levels 3 and 4. And it is always possible, if the differences between the three consequence levels and the three likelihood levels are not the same, that (say) risk level 4 may in some cases be actually less than risk level 3.

Semiquantitative Estimation

To derive the semiquantitative risk, add the scores for Consequence, Initiation Frequency, and Mitigation Failure Probability, that is, columns A+C+D=E. (Note that adding the logarithmic scores is equivalent to multiplying the real values.)

Because the Initiation Frequency Score has been adjusted upward by adding 5 to the $\log_{10}$ of the Initiation Frequency, interpretation of the meaning of the Risk Score needs to take account of this adjustment.

A Risk Score of (say) 5 in column E is equivalent to the following risks of loss:

- 1 day of supply per year; or
- 10 days of supply with a probability of 10% per year (or once per 10 years on average); or

- 100 days of supply with a probability of 1% per year (or once per 100 years on average), etc.

Similarly, a Risk Score of 4 in column E is similar to the above losses but with a likelihood 10 times lower in each case.

The identified risks should be ranked in two lists:

- in descending order of risk score; and
- in descending order of consequence score.

The risks near the top of List A are those with highest priority for risk reduction.

The principal options for risk reduction are:

- Hardware: Eliminate, modify, duplicate, relocate
- Systems and procedures: modify and improve:
- Organization, staffing, communications, training: Modify and improve
- Emergency capability: Modify and improve
- Culture/climate: Improve
- Other

The risks near the top of List B (where they are not on List A) are those with highest priority for careful ongoing management to ensure that their likelihood remains low. This necessitates establishment of a "quality" operation (design, operation, maintenance, etc., that are "fit for purpose"), setting up a risk-specific monitoring program, and periodic auditing.

OPTIONS FOR ACTION

Review of the Need for Action

The Potential Incident Likelihood and Risk Assessment Worksheet for each plant unit summarizes, in its last column, the pattern of risks the plant unit presents to continuity of production.

It may be helpful to tabulate these for all plant units as shown in Table C-8.

The findings assembled for each plant unit are then examined in turn, so as to identify the practicable options for handling the risks presented by that unit. These options would be briefly outlined for each plant unit.

Table C-8
Suggested Form of Site–Wide Risk Summary of Number of Sources of Risk per Plant Unit

Risk Level / Plant Unit	Very Low	Low	Medium	High	Very High
Unit A					
Unit B					
Unit C					
Unit...					
Etc.					

OUTCOME

The end result of the above type of assessment is:

1. agreement about which of the plant units present most risk of incidents causing loss of production of a scale that could have a serious impact on the market;
2. a list of the options for reduction of the risks, awaiting decision and approval of expenditure on detailed evaluation of the options or more detailed examination of specific risks;
3. a list of the risks which most need careful ongoing management, to keep their likelihood low, and suggested initiatives to provide that level of management; and
4. assembled documentation prepared in the course of the assessment, which enables the bases and limitations of the assessment to be penetrated and understood.

From the above documentation, a summary document can be easily prepared setting out the key findings and options.

Such a summary document may comprise the following:

One page per plant unit, summarizing:

- the principal hazards of the plant unit, and the related vulnerabilities;
- the principal factors influencing the likelihood;
- the estimated possible impact on the market, and the likelihood category; and
- the principal options for action.

A summary of the site-wide risks by plant unit, such as Table C-8.

Appendix D

An Illustrative Audit Checklist for Process Safety and Reliability [Structured around Hawksley's Six Principles][1]

1. INTRODUCTION

This checklist can also be used as an aide-memoire when considering what should be included in routine monitoring of plant safety and reliability. The checklist can be edited to suit the particular type of plant or operation being audited, and expanded into the form of a worksheet with sufficient space for comments.

(Note: The checklist is designed to illustrate broad principles. For any specific plant the audit checklist should be tailored, perhaps using this checklist as a starting point, to match the plant-specific hazards and potential sources of unreliability.)

If the plant being audited is unfamiliar to the auditor, or if the auditor wishes to test the plant manager's understanding of the hazards, it is helpful to

[1] This also draws on unpublished work by D. Drewitt.

discuss both (a) the factors introducing the potential for an incident and (b) the possible severity of such an incident with the plant manager (i.e., Sections 2 and 3) before continuing with the later topics. This helps the auditor recognize the sections of the plant, and the software, that most need attention.

2. SUMMARY OF INHERENT HAZARDS

Discussing these with the plant management achieves two principal goals:

- it helps the auditor to identifying those plant sections and features that introduce a likelihood of an incident; and
- the understanding displayed of these by plant management reveals how well the 1st Hawksley requirement is met)

2.1 Raw Materials, Intermediates, and Products (with Inventories).

- Flammables:
- Explosive:
- Unstable:
- Toxics:
- Environmentally Active:

2.2 Process Types

2.3 Ignition Sources

2.4 Range of Process Temperatures

2.5 Range of Process Pressures

2.6 Corrosion and Erosion Hazards

2.7 High-Speed Machinery

2.8 Complex/Sensitive Equipment

2.9 Domino Potential

2.10 Other

3. MAGNITUDE OF POTENTIAL ACCIDENT, LOSS, ENVIRONMENTAL DAMAGE

This identifies the degree of impact or disruption possible from each credible type of incident. The understanding displayed of these also tests the first Hawksley requirement and helps to clarify the sections of the plant needing the most detailed audit.

3.1 Fire
3.2 BLEVE
3.3 Vapor Explosion
3.4 Flash Fire
3.5 Dust Explosion
3.6 Toxic Gas Escape
3.7 Toxic Smoke from Fire
3.8 Environmental Damage from Contaminated Firefighting Water Runoff
3.9 Environmentally Damaging Spill
3.10 Breakdown of Critical Machinery/Equipment
3.11 Domino Incident
3.12 Other

4. SOFTWARE, PART A

Discussion of this section tests the understanding of hazards, incident potential, safeguards, etc., and suggests what software needs personal inspection.

4.1 Training and Testing in Principles of Process Safety
4.2 Training and Testing in Specific Hazards of the Plant
4.3 Training and Testing in How Incidents Could Arise
4.4 Training and Testing in Safeguards
4.5 Participation in Plant Safety Inspections
4.6 Participation in Investigations of Process Accidents and Incidents

5. FACILITIES AND EQUIPMENT

5.1 General Design and Condition
- 5.1.1 Design and Structure:
- 5.1.2 Materials:
- 5.1.3 Hard Paving:
- 5.1.4 Passive Fireproofing:
- 5.1.5 Blast Proofing:
- 5.1.6 Fire Break Walls:
- 5.1.7 Blast relief:

5.2 Separation

5.2.1 Spacing between Blocks:

5.2.2 Spacing within Blocks:

5.2.3 Location of Pumps Handling Flammable Liquids:

5.2.4 Control Room Location:

5.2.5 In-Plant Dikes:

5.2.6 Bunding of Storages:

5.3 Ventilation

5.3.1 Location of Machinery Handling Flammables:

5.3.2 Congestion of Plant:

5.3.3 Control Room Ventilation, and Prevention of Ingress of Flammable or Toxic Vapor:

5.4 Instrumentation

5.4.1 Instrumentation for Control:

5.4.2 Alarms and Trips:

5.4.3 Apparent Condition:

5.4.4 Calibration and Testing:

5.5 Process Lines, Valves, and Fittings

5.5.1 Robustness:

5.5.2 Valving:

5.5.3 Emergency Isolation Valves (EIVs) and Excess Flow Valves (EFVs):

5.5.4 Apparent Condition:

5.5.5 Use of Blank Flanges and Spades:

5.5.6 Management of Leaks:

5.5.7 Potential for Cross-Contamination:

5.5.8 Facilities for Relief and Blowdown:

5.5.9 Provision for Purging and Washout:

5.6 Heating and Cooling Systems Available to Plants on the Site

5.6.1 Reserve Capacity:

5.6.2 Historical Reliability:

5.6.3 Potentially Serious impact of Loss of Cooling:

5.7 Electric Power Supply and Distribution on the Site

5.7.1 Historical Reliability of Supply:

5.7.2 Potential Severity of Impact of Blackout:

5.7.3 Design (e.g., Radial Where Ring Reticulation Required):

5.7.4 Apparent Condition (Especially Condition of Flameproof Equipment):

5.7.5 Classification of Areas:

5.8 Other Services

5.8.1 Compressed Air:

5.8.2 Instrument Air:

5.8.3 Nitrogen:

5.8.4 Emergency Power:

5.9 Means of Disposal from the Site of Flammable and Toxic Waste

5.9.1 Disposal Routes and Destinations: Flammable Waste:

5.9.2 Disposal Routes and Destinations: Toxic Waste:

6. SOFTWARE, PART B

Systems and procedures, operational knowledge, documentation, etc.

6.1 General Systems and Procedures.

6.1.1 Safety Policy and Environmental Policy:

6.1.2 Documented Arrangements for Implementing the Policy:

6.1.3 Security:

6.1.4 Records of Hazardous Incidents (Safety, Environmental):

6.1.5 Accident Investigation:

6.1.6 Safety/Environmental Auditing:

6.1.7 Work Program for or Progress with Implementing Findings of Previous Audits:

6.2 Operational Systems and Procedures

6.2.1 Documented Standard Operating Instructions:

6.2.2 Unusual Occurrence Reporting:

6.2.3 Permit to Work Procedures:

6.2.4 Housekeeping:

6.3 Maintenance Systems and Procedures

6.3.1 Documented Standard Maintenance Procedures:

6.3.2 Modification Control:

6.3.3 Condition Monitoring:

6.3.4 Protective System Testing:

6.3.5 Shutdown Planning:

7. SOFTWARE, PART C

Organization, staffing, job training, communications.

7.1 Operations

7.1.1 Operational Management Selection:

7.1.2 Operational Management Development, Promotion:

7.1.3 Operator Selection, Promotion:

7.1.4 Operator Training, Testing:

7.1.5 Operational/Production Organization and Working Relationships:

7.2 Engineering and Maintenance

7.2.1 Engineer Selection:

7.2.2 Engineer Professional Development, Promotion:

7.2.3 Plant-Specific Trades Training:

7.2.4 Engineering/Maintenance Organization and Working Relationships:

7.3 Other

7.3.1 Specialist Technical Resources Available:

7.3.2 Regular Meetings including Formal and Informal Review of Process Safety:

8. EMERGENCY RESPONSE CAPABILITY

Fire, toxic gas escape, environmental, etc.

8.1 Works Fire/Emergency Team

8.1.1 Staffing:

8.1.2 Equipment:

8.1.3 Media Stocks:

8.1.4 Training and Practices:

8.2 External Fire Team and Emergency Services

8.2.1 Liaison:

8.2.2 Joint Exercises:

8.2.3 Response Time:

8.2.4 Equipment/Staffing:

8.3 Water Supplies for Firefighting

8.3.1 Source(s):

8.3.2 Pumping:

8.3.3 Reticulation System:

8.3.4 Testing:

8.3.5 Firewater Drainage; Collection Areas:

8.4 First Aid Fire Extinguishers: Summary

8.4.1 Types:

8.4.2 Numbers:

8.4.3 Locations:

8.4.4 Inspection:

8.5 Fixed Active Fire Protection: Summary

8.5.1 Hydrants/Monitors:

8.5.2 Sprinklers:

8.5.3 Fire Hoses, Hose Reels

8.6 Alarm Systems on the Site

8.6.1 Times When the Site or Buildings Are Unmanned

8.6.2 Location/Marking of Alarm Points:

8.6.3 Flammable Gas/Fire Detector Points:

8.6.4 Annunciation:

8.6.5 Testing:

8.6.6 Link to External Brigade and Emergency Services:

8.7 Environmental Emergency Handling

8.7.1 Catchment Facilities for Drainage and Contaminated Fire-fighting Water:

8.7.2 Spill (etc) Detection and Response Arrangements:

8.8 Emergency Plans and Training

8.8.1 First Aid Training:

8.8.2 Extinguisher/Breathing Apparatus Training:

8.8.3 Emergency Plans:

8.8.3.1 Fire

8.8.3.2 Environmental

8.8.3.3 Other

8.8.3.4 Emergency Practices:

9. SAFETY PROMOTION AND ATTITUDE BUILDING

9.1 Means of Actively Promoting the Safety Policy

9.2 Commitment by Management Visible at All Levels

9.3 Apparent Effectiveness of the Promotion (e.g., Evidence of Widespread Commitment):

Appendix E

Brief Outline of Process Safety Legislation in the United States

1. INTRODUCTION

As in other countries, safety legislation in the United States has a long history, with legislation relevant to process safety possibly starting with the Rivers and Harbors Act of 1899 and the Explosives Transportation Act around the start of the 1900s. There are now numerous agencies, acts, and regulations related to safety, both federal and state administered.

From the viewpoint of process plant risks and process safety generally, the most notable now are the National Environmental Policy Act of 1969 under which the Environment Protection Agency (EPA) was established, and the Occupational Safety and Health Act of 1970, under which the Occupational Safety and Health Administration (OSHA) was established. Both of these two organizations are involved in process safety, the EPA starting from pollution prevention and control, and OSHA from workplace safety, but a degree of overlap now exists (which is clearly preferable to leaving a gap).

The following is only an outline to introduce the approach. The full documents should be consulted for a complete and detailed understanding.

2. OSHA

The administration of the Occupational Safety and Health Act entails two major parts: development of standards; and enforcement. Process safety management is specifically addressed in OSHA Department of Labor Section 110.119 Process safety management of highly hazardous chemicals (29 CFR 1910.119). This applies to any process that involves more than specified threshold quantities of chemicals listed in an appendix.

It requires the responsible organization to have in place a defined range of the elements of effective process safety management. For example, these include:

- employee consultation and involvement;
- preparation of written information about the process relevant to its safety, including details of the hazards and the technology used;
- a process hazard analysis, with a documented priority order for undertaking the studies based on a rationale that takes account of factors related to the severity and the likelihood of potential incidents;
- written operating procedures;
- training;
- dealing with contractors;
- pre-startup safety review;
- mechanical integrity;
- hot work permit system;
- management of change;
- incident investigation;
- emergency planning and response; and
- compliance audits.

Appendices include a list of highly hazardous chemicals with their threshold quantities; a nonmandatory guide and recommendations for implementing and maintaining a good standard of process safety management; and a list of sources of information and references.

The responsible organization is required to comply within defined time limits, and to demonstrate compliance on request by the Department of Labor, but there is no routine requirement for submission of reports for approval.

3. EPA

Under the Clean Air Act, the EPA administers Part 68, the Chemical Accident Prevention Provisions (40 CFR 68.10). This has some similarity to the OSHA requirements above, but also some marked differences.

It applies to any fixed installation having more than a threshold quantity of a regulated substance in a process. The list of regulated substances is more extensive than that of 29 CFR 1910.119, although the threshold quantities are mostly similar where materials are on both lists.

Three levels of program are provided for.

- Program 1 is for processes that have had no accidental release (of defined type) in the past 5 years; are at a distance from a "public receptor" greater than the range of a defined "worst case" scenario; and have emergency procedures coordinated with the local emergency planning and response organizations. In broad terms, they are processes with minimal potential for offsite impact which have a good history of operation and with which the local emergency planning and response organizations are familiar.
- Program 2 is for processes that do not meet the requirements for Program 1 or the specification of Program 3 (below).
- Program 3 is for processes that do not meet the requirements of Program 1, but are covered either by 29 CFR 1910.119 or are of specified industry types (e.g., oil refineries, petrochemical and chemical manufacture, pulp mills, etc.)

The organization is required to submit a Risk Management Plan (RMP) that includes all processes that are covered by the Regulation. Basic contents required in each RMP include:

- an offsite consequence analysis;
- a 5-year accident history;
- information about the emergency response program;

In addition to the basic requirements, the content of the RMP depends on which Program the particular process is subject to.

For those processes eligible for Program 1, the particular contents are:

- documentation of the worst case scenario and demonstration that the nearest public receptor is beyond its range;
- documentation of the 5-year accident history;
- assurance that the response actions have been coordinated with the local emergency planning and response agencies; and

- a certification that the worst-case scenario, the accident history, and the coordination of the response actions meet the requirements, and that no further actions are needed to prevent offsite impacts from accidental releases.

For those processes subject to Program 2, in addition to meeting the basic requirements, additional requirements include the following:

- development and implementation of a management system of risk management program elements with a nominated responsible person;
- conducting an offsite consequence analysis for worst-case and alternative-case scenarios, including a definition of the effects on the surrounding population and environment;
- assembling safety-related information and conducting a hazard review;
- implementing a defined range of "good practice" in operation and maintenance;
- undertaking compliance audits and incident investigation; and
- implementing a prevention program.

For those processes subject to Program 3, in addition to meeting the basic requirements, additional requirements are similar in concept to those of Program 2, but more rigorous. For example, a Process Hazard Analysis is required instead of a hazard review and the "good practice" requirements are more detailed and tighter.

Comment: In none of the above is the probability or likelihood of incident specifically addressed. However, it is implicit, in a nonquantitative manner, in the requirement for worst-case and alternative-case scenarios to be addressed. Worst-case scenarios (normally) have extremely low probabilities, because they rely on the simultaneous existence of a number of unfavorable conditions or occurrences, or because their potential has led to very tight preventive measures being adopted. The alternative-case scenarios, although intended also to be of low probability, are those arising from the type of mishap that occurs from time to time in industry in spite of preventive efforts. Examples include hose coupling failures, leaking pipe joints, and overfilling of vessels.

Index

Printed and bound by CPI Group (UK) Ltd, Croydon, CR0 4YY
08/05/2025
01864816-0004